Laser Photoionization Spectroscopy

Laser Photoionization Spectroscopy

VLADILEN S. LETOKHOV

Institute of Spectroscopy
USSR Academy of Sciences
Troitzk, USSR

1987

ACADEMIC PRESS, INC.
Harcourt Brace Jovanovich, Publishers

Orlando San Diego New York Austin
Boston London Sydney Tokyo Toronto

ACADEMIC PRESS, INC.
Orlando, Florida 32887

United Kingdom Edition published by
ACADEMIC PRESS INC. (LONDON) LTD.
24–28 Oval Road, London NW1 7DX

Library of Congress Cataloging in Publication Data

Letokhov, V. S.
Laser photoionization spectroscopy.

Bibliography: p.
Includes index.
1. Laser spectroscopy. 2. Photoinoization.
I. Title.
QD96.L3L47 1986 621.36'1 85-28596
ISBN 0-12-444320-6 (alk. paper)

PRINTED IN THE UNITED STATES OF AMERICA

87 88 89 9 8 7 6 5 4 3 2 1

Contents

Preface

For more than 25 years one of the most remarkable inventions of the twentieth century—the laser—has continued to develop rapidly and find ever increasing application. Among the first application fields for lasers were spectroscopy and substance analysis on its base. Lasers have made it possible not only to improve the characteristics of already known spectroscopic methods but also to devise absolutely new techniques having no analogs in traditional spectroscopy. One such new method is the laser photoionization spectroscopy discussed in this book. Laser photoionization spectroscopy is based on the resonance ionization of atoms or molecules as a result of successive absorption of several laser photons.

The most distinctive feature of the photoionization technique is its exceptionally high sensitivity (the method is capable of handling even single atoms or molecules), which can be combined with high spectral or temporal resolution and high selectivity to detect the desired atomic or molecular species. All this makes the method considered here a very effective tool that holds much promise for both research and practical purposes. The development of this technique during the past ten years has been associated with the progress in the field of tunable dye lasers, and it may be expected that as various types of these lasers are developed further to become simpler and more reliable use of the method will spread further to laboratories of varying research.

The present book is intended for a wide readership—physicists, chemists, analysts, geochemists, etc., who need the most novel laser spectroscopy techniques to solve nontrivial problems. It is written so as to be intelligible to undergraduate and postgraduate students and also to specialists in other fields of research remote from laser spectroscopy. It is exactly for this reason that an attempt has been made in this book to balance theoretical and experimental information.

The book is based on research work carried out at both the author's laboratory at the Institute of Spectroscopy of the USSR Academy of Sciences during the period 1970–1985 and many laboratories in a number of countries.

It consists essentially of two parts. The first part (Chapters 1–6) deals with the fundamentals of the technique for atoms and molecules, while the second part (Chapters 7–12) considers its applications and modifications. The first two chapters are introductory and outline the history of the method (Chapter 1) and some elements of the theory of resonance interaction between laser light and two- and multilevel quantum systems, which are essential to the understanding of the fundamentals of the technique (Chapter 2). The next two chapters are devoted to the resonance photoionization of atoms through their excited states (Chapter 3) and photoionization detection of atoms with an ultimate sensitivity (up to single atoms) based on this method (Chapter 4). Chapters 5 and 6 consider the extension of the laser resonance photoionization technique to molecules, including the detection of molecules by this method in combination with mass spectrometry.

The second half of the book, devoted to the applications and modifications of the technique, begins with Chapter 7, which discusses the acquisition of various spectroscopic data on atoms and molecules by means of this method, including experiments with a high spectral or temporal resolution. Chapter 8 describes the analytical application of the technique, i.e., ultrasensitive detection of traces of atoms in various objects under a variety of conditions. Discussed in Chapter 9 are the ways to extend the method to cover the case of detecting traces of very rare isotopes. Attention here is concentrated on the problem of attaining selectivity much in excess of the capabilities of all previously known methods of spectroscopy. Chapter 10 is devoted to the experimental uses of the photoionization technique in molecular, chemical, and statistical physics. The last two chapters deal with the modifications of photoionization spectroscopy, Chapter 11 considering the combination of resonance photoionization with the measurement of the energy spectrum of photoelectrons and Chapter 12, the possibilities of localizing the point of exit of photoelectrons or photoions, which potentially opens up a way to develop a photoionization microscope possessing a high (up to a few angstroms) spatial resolution. Thus, the reader will find in the book information on not only the history of the technique and its present status but also on its prospects that he or she will, I hope, help to realize.

In preparing the manuscript of the book, the first of its kind in the world literature, I used widely the possibility of discussing and preparing its individual chapters with my students and colleagues at the Institute of Spectroscopy. In this connection, I would like to express my gratitude to Dr. V. S. Antonov for his assistance in preparing Chapters 5 and 6, Dr. G. I. Bekov (Chapter 8), Dr. A. A. Makarov (Chapters 2 and 9), Dr. V. I. Mishin (Chapters 3 and 4), and V. N. Fedoseev (Section 3.4 and the Appendix). My thanks are due especially to them all and also to other participants in numerous experiments carried out at the Institute of Spectroscopy including those performed in cooperation with other research institutes.

I am also grateful to the authors and publishers of the articles, journals, and books cited by me for their kind permission to use their results in this book. Finally, I wish to thank my present publishers for their help in publishing the book.

1. Introduction

The task of detecting atoms and molecules is well-known in the field of experimental atomic and molecular physics. The theoretical limit of detection sensitivity is a single particle (atom or molecule), and such a sensitivity is quite sufficient, for even a single particle contains all the information there is about its structure. However, in actual physical experiments that could be performed up to now, spectra could be measured only if the sample under investigation contained a rather large number of particles (around 10^{10}–10^{20}, depending on the method used and the object of study). Therefore, the problem of enhancing the sensitivity of spectroscopic methods remained pressing for the entire 100-year period of development of optical spectroscopy.

At the same time, experiments involving the detection of single nuclei have long been known in nuclear physics. Suffice it to recall the scintillation counter with which Sir Ernest Rutherford observed individual alpha particles. Later on, the Geiger counter, the Wilson cloud chamber, the spark track chamber, and other detectors were invented that made it possible to detect isolated nuclei and elementary particles, reveal their structure, and measure their energy.

What is the reason for such a great difference between the detection methods used in atomic–molecular physics and in nuclear physics? This difference is, obviously, due to the fact that experimentalists in nuclear physics must deal with, first, charged particles and, second, particles of high kinetic energy. Such particles, as a rule, interact fairly effectively with a macrodetector. In optical physics, we are most commonly concerned with neutral atoms and molecules in the ground or low-lying quantum states a few tenths of an electron volt in energy. The interaction of such particles with a macrodetector causes it to suffer but a very weak perturbation which can barely be perceived against the background of fluctuation noise.

1.1. Can Single Atoms Be Detected by Spectral Means?

The question naturally arises whether it is possible to reach the detection sensitivity adequate to experiment with individual neutral particles, and if it is, how it can be done. Experience gained in the field of experimental nuclear physics enables us to draw two conclusions immediately.

First, it is necessary to try and excite the *internal degrees of freedom* of such a particle, seeing that the translational degrees of freedom of a neutral particle are extremely difficult to influence. Second, we should try to *convert the neutral particle to a charged one*, for its kinetic energy could then be increased with ease. In this case, the whole store of nuclear physics techniques could be applied to detecting the charged particle thus formed.

The first way—excitation of internal degrees of freedom has long been used successfully in experiments with atomic and molecular beams such as the famous Lamb–Retherford experiment (1947) in which the Lamb shift in hydrogen was discovered. In this experiment (Fig. 1.1), a beam of atomic hydrogen became excited upon collisions with electrons so that some hydrogen atoms got into the $2S_{1/2}$ metastable state. For such metastable atoms, one-quantum ground-state transitions are forbidden, and so the excited hydrogen atoms with an energy of around 6 eV could reach the detector. The detector in this experiment was a surface ionization type which caused ionization of

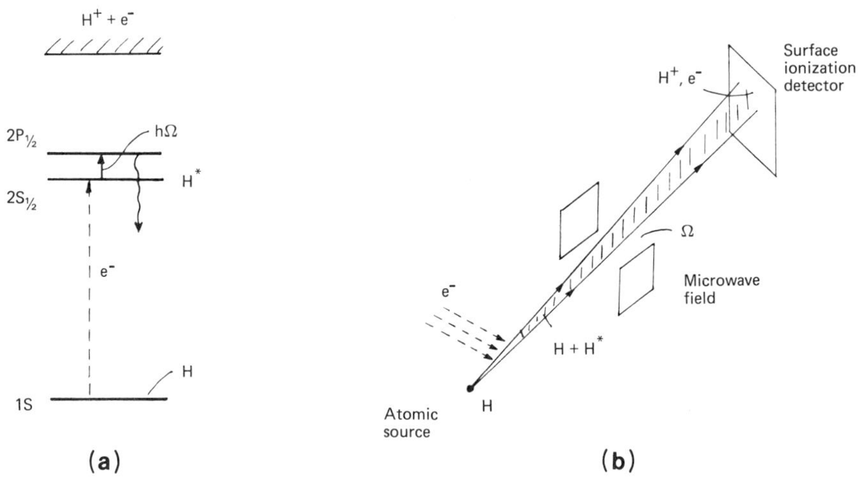

FIG. 1.1. Ionization detection of metastable H atoms in the Lamb–Retherford experiment with an atomic beam. (a) Diagram of quantum transitions and (b) geometry of the atomic beam experiment.

the excited hydrogen atoms interacting with it. The flux of metastable hydrogen atoms, hence the ion current, could be varied by changing the population of the metastable state with the aid of a microwave field whose frequency Ω corresponded to the $2S_{1/2} \rightarrow 2P_{1/2}$ transition. The experiment made use of the high sensitivity of ionization detection of excited atoms to register a very weak absorption line in the microwave range. Of course, such a situation is unique insofar as the species subject to detection are long-lived metastable atoms having substantial internal energy, and so, unfortunately, this exceptionally sensitive technique cannot be used for regular measurements with other atoms.

It is only natural to try to use excitation of atoms with a view to detecting excited particles by the ionization technique. From the standpoint of excitation selectivity, optical radiation is most suitable for this purpose. The probability that an atom will be excited by light to transit from level 1 to 2 with the transition energy being E_{21} and degeneracies of levels g_i is defined by the simple relation

$$W_{12} = \frac{g_2}{g_1 + g_2} \exp\left(\frac{-E_{21}}{kT_{\mathrm{rad}}}\right) \ll 1, \tag{1.1}$$

where T_{rad} is the optical radiation temperature at the transition frequency $\omega_{21} = E_{21}/\hbar$. The probability of excitation of atomic states a few electron volts in energy is very low because the radiation temperature provided by common light sources is such that $kT_{\mathrm{rad}} \ll E_{21}$. In essence, condition (1.1) describes the equality of the radiation and atomic temperatures. To attain a high probability of atomic excitation, the following condition must be satisfied:

$$kT_{\mathrm{rad}} \gg E_{21}, \tag{1.2}$$

i.e., use must be made of a light source providing a very high radiation temperature. The laser is precisely this type of light source.

In principle, to register the absorption of light by an atom, use can be made of one of the following possibilities (Fig. 1.2): (a) measurement of the attenuation of the exciting laser beam intensity (the absorption method); (b) measurement of the intensity of photons reradiated and scattered in different directions and at different frequencies (the fluorescence method); (c) registration of the change in the direction of motion of the excited atom (the resonant deflection method); and (d) direct registration of the excited atom, e.g., by way of its ionization. Let us touch upon each of these possibilities. The reader will find more detailed information on the various techniques of laser detection of individual atoms in numerous reviews (Letokhov, 1977a, 1978, 1980; Fairbank and She, 1979; Hurst *et al.*, 1979a; Balykin *et al.*, 1980b; Alkemade, 1981).

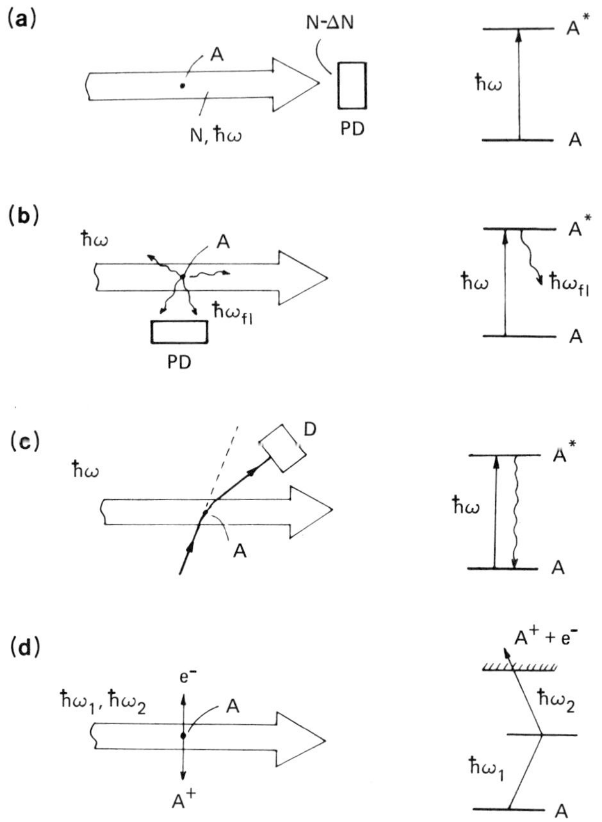

FIG. 1.2. Laser spectral methods of detection of atoms (molecules): (a) absorption method based on light transmission measurement; (b) fluorescent method; (c) resonant deflection method; and (d) stepwise resonant photoionization method.

We can try to detect an atom by registering the attenuation of laser radiation intensity due to a photon being *absorbed* by the atom. A single atom can absorb one photon during a minimum τ_{exc} for which the probability that the atom will be induced to move into an excited state is close to unity. In other words, the quantity τ_{exc} is determined by the inverse of the rate of stimulated transitions between the two states of the absorption transition, i.e.,

$$\tau_{exc} = (\sigma_{12} I)^{-1}, \tag{1.3}$$

where σ_{12} is the stimulated transition cross section and I the photon flux intensity (in photons/cm$^2\cdot$s). On the other hand, the total number of photons that can reach the photodetector during the time τ_{exc} is $n_{\text{ph}} = IS\tau_{\text{exc}}$, where S is the cross-sectional area of the laser beam in the atomic excitation region. In the case of an ideal photodetector, the only irreducible noise source is the photon fluctuation noise, owing to which the photodetector current fluctuates on the average by the amount (for more detail see the book by Kingston, 1978)

$$\delta n_{\text{fl}} \simeq \sqrt{n_{\text{ph}}}\,. \tag{1.4}$$

Clearly a single atom cannot be registered by this method because the mean value of the photon number fluctuations on the detector is much greater than the decrease of the photon flux intensity due to the absorption of a photon, i.e., $\delta n_{\text{fl}} \gg 1$. It is exactly for this reason that the minimum number of atoms N_{at} that can be detected by the absorption method is, in an ideal case, determined by the need for the photodetector signal to exceed the irreducible laser beam fluctuation noise:

$$N_{\text{at}} \geqslant \delta n_{\text{fl}} \simeq \sqrt{n_{\text{ph}}} = \sqrt{IS\tau_{\text{exc}}} = (S/\sigma_{12})^{1/2} \gg 1, \tag{1.5}$$

and so the registration of individual atoms is impossible.

Another possible method is the *fluorescence* technique whereby excited atoms are detected by their spontaneous emission following their excitation (Fig. 1.2b). In this case, the laser beam whose quantum fluctuations limit the sensitivity of the absorption method does not directly strike the photodetector, and so one can expect an increase in the detection sensitivity. This is in fact true. If the observation time T_{obs} is long enough, i.e., $T_{\text{obs}} \gg \tau_{\text{exc}}$, a single atom can be made to absorb and then reemit photons many times. In the limit of sufficiently fast excitation when $\tau_{\text{exc}} \lesssim A_{21}^{-1}$, the factor $(TA_{21}) \gg 1$, and so the reradiation events can be repeated. Even if one considers the actual efficiency η_{coll} of collecting the photons reradiated into a solid angle of 4π sr and the finite quantum efficiency of the photodetector, $\eta_{\text{det}} > 1$, the number of photoelectrons $K_{\text{ph.el.}}$ produced by a single atom during the observation time can be made perceptible:

$$K_{\text{ph.el.}} = \eta_{\text{coll}}\eta_{\text{det}}(TA_{21}) \gg 1. \tag{1.6}$$

In this case, individual atoms can be assuredly detected, for the multiphotoelectron signal materially exceeds the noise level inherent in photodetectors of uniphotoelectron nature. Such experiments on the laser fluorescence detection of single atoms were carried out in 1977 in the United States and USSR (Balykin *et al.*, 1977; Greenlees *et al.*, 1977). The multiphotoelectron fluorescence spectroscopy technique has important advantages over the common fluorescence spectroscopy methods as far as the problem of selective detection of isolated atoms is concerned (Balykin *et al.*, 1980a). Essential to the success of such experiments is the use of a laser providing for so high an excitation rate as is comparable to A_{21} and of an atom capable of multiple, cyclic excitation and radiative relaxation. By no means can such a cyclic excitation–relaxation scheme be realized for all multiple-level atoms; and, moreover, this is almost impossible to do for molecules because of their rich vibrational–rotational electronic level structure. Besides, the time of detecting an atom by this technique is much longer than the limiting decay time exactly because of the need for multiple atomic excitation (typical values of T_{obs} are around 10^{-3}–10^{-4} s). Therefore, the fluorescence technique is not universal and does not solve the problem of detecting single atoms and molecules.

The detection of an atom can also be based on the change in the trajectory of its motion consequent upon its interaction with a laser light field (Fig. 1.2c). In this case, there are at least two possibilities. First, the laser field can change the quantum state of the atom and thus influence the trajectory of its motion in externally applied fields. The "magnetic detection" of atoms in beams was successfully realized (Duong and Vialle, 1974). The capabilities of the method were experimentally demonstrated in measuring the hyperfine splitting and isotope shifts of radioactive isotopes (Huber *et al.*, 1978; Liberman *et al.*, 1979). Second, in the case of multiple photon reradiation, the atom can change the direction of its motion as a result of the recoil effect. This phenomenon was successfully demonstrated in experiments on the isotope-selective *photodeflection* of Ba atoms (Bernhardt *et al.*, 1974, 1976). If used in conjunction with the laser cooling of atoms, this technique may also prove useful for the detection of individual rare atoms (Balykin *et al.*, 1984). However, the methods of laser or magnetic deflection of atoms impose rather stringent requirements on the atomic transition used (cyclicity of interaction or the presence of a magnetic moment) and are, therefore, not universal.

There remains one more way—to detect excited atoms directly. A most simple method for detecting excited atoms against the background of unexcited ones is their *photoionization* by a radiation of frequency ω_i satisfying the condition (Letokhov, 1969, 1970)

$$E_i - \hbar\omega_{12} < \hbar\omega_i < E_i, \tag{1.7}$$

where E_i is the atomic ionization potential (Fig. 1.2d). To fully convert each atom in resonance with a multiple-frequency laser radiation, it is necessary that the intensities (or fluences) of the laser pulses ensure absorption saturation at each excitation stage. The great difference between the resonant excitation cross section σ_{12} and the ionization cross section σ_{2i} require that the laser pulse intensities be different (Letokhov, 1969, 1970):

$$I_1 \gtrsim (2\sigma_{12}\tau_{12})^{-1}; \qquad I_2 \gtrsim I_1(\sigma_{12}/\sigma_{2i}). \tag{1.8}$$

This technique makes use of the ability of multiple-frequency laser radiation to effect through stimulated quantum transitions a rapid resonant excitation of atoms or molecules into high-lying states from which they are subsequently ionized by laser radiation or some other effective means (electric field, collisions, etc.). The atomic or molecular ions thus produced are then registered with 100% probability by the known methods. The method is fairly universal, for it does not require any special scheme of energy levels or their type and decay rates either; it is, in principle, applicable to any atom or molecule. The unique characteristics of the technique employing stimulated quantum transitions of an atom or a molecule make it exceptionally valuable and promising for optical spectroscopy. It was precisely this fact that gave rise to the writing of the present monograph.

1.2. History of the Technique

Resonant photoionization spectroscopy came into being in the early 1970s in the course of the development of new methods for exerting selective influence on atoms and molecules by means of laser radiation. Two-step resonant atomic photoionization was proposed to be used for the purposes of separating isotopic atoms and detecting trace elements (Letokhov, 1969, 1970).

The first experiments on the two-step resonant photoionization of Rb atoms were performed at the Institute of Spectroscopy of the USSR Academy of Sciences as far back as 1971 (Ambartzumian *et al.*, 1971; Letokhov and Ambartzumian, 1971). A schematic of such an experiment is presented in Fig. 1.3. The radiation of a tunable dye laser with a wavelength $\lambda_1 = 7950$ Å (excited some Rb atoms into the $5p^2P_{1/2}$ state. The photoionization of the excited atoms was effected by means of the second-harmonic output pulses from the ruby laser used to pump the dye. The energy of the second-harmonic quantum was high enough to bring about the photoionization of the excited Rb atoms ($E_i - \hbar\omega_1 = 2.62$ eV) but insufficient for photoionization of the ground-state ones ($E_i = 4.18$ eV), exactly as required by condition (1.7). The laser pulse parameters enabled us to attain saturation of the resonant transition and about 0.1% probability for the ionization of the excited atoms during a single pulse, so that the total two-step photoionization yield amounted to about 10^{-3}. In the works cited earlier, the condition was formulated for complete conversion of a neutral atom into a photoion, namely, that absorption saturation must be ensured at each excitation stage.

In later years, several two-step photoionization experiments were performed with Na atoms in a beam (Duong *et al.*, 1973), U atoms in a beam

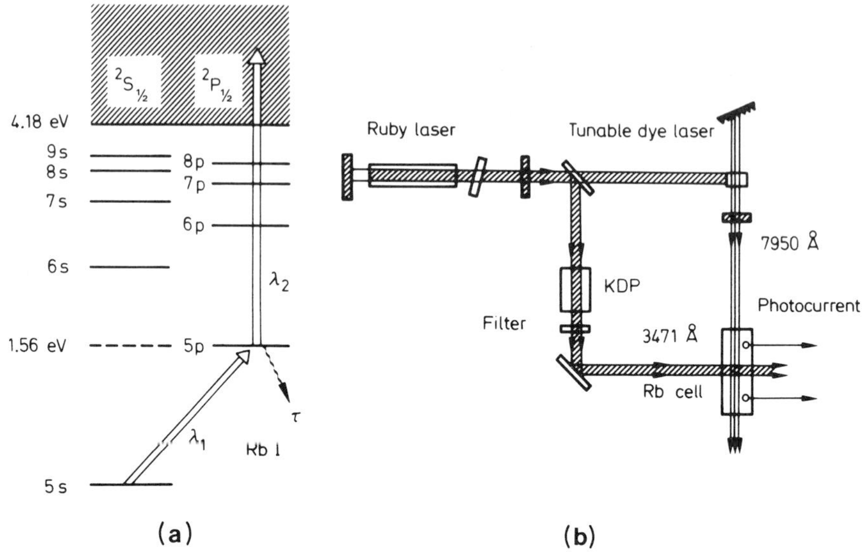

FIG. 1.3. First experiment on stepwise resonant photoionization of Rb atoms. (a) Energy levels of Rb and relevant transitions and (b) experimental setup. [From Ambartzumian *et al.* (1971).]

(Tuccio *et al.*, 1974), and Ca atoms in a beam (Brinkman *et al.*, 1974), excited by the radiation of CW dye lasers. The ionization yield in these experiments was very low because of the impossibility of saturating the transition into continuum. To ease the task of meeting the condition of absorption saturation at the ionization stage, it was suggested that atoms should first be excited in several steps into high-lying discrete levels in the vicinity of the ionization threshold and then ionized by means of an electric field (Ivanov and Letokhov, 1975). Corresponding experiments on the two-step excitation of Rydberg states in Na atoms, followed by their ionization by an electric field, were carried out in 1975 (Ambartzumian *et al.*, 1975a; Ducas *et al.*, 1975). This armed the experimentalists with a simple method for detecting single atoms.

At the same time, it was suggested that a selective two-step photoionization of molecules in a mass spectrometer beam should be used in order to enhance the sensitivity and selectivity of molecular mass spectrometry (Ambartzumian and Letokhov, 1972). Still earlier, the two-step photoionization of UF_6 molecules was proposed as a method for separating uranium isotopes (Robieux and Auclair, 1963). It should be noted that in those years the attention of investigators was centered primarily, with a few exceptions, on the use of selective ionization precisely for isotope separation. I remember one occasion when Doctor B. Snavely spoke at the Gordon Conference on Stable Isotopes in Pacific Grove, California, in 1974. When he demonstrated a ^{235}U-photoion yield spectrum with a distinct hyperfine structure and remarked on some possible uses of multistep photoionization for similar spectroscopic purposes, the audience was frankly incredulous and there was even laughter in the hall. Today, we know that the resonant multistep ionization technique is the most sensitive method for studying the isotope and hyperfine structures of atoms with short-lived nuclei available in very small quantities.

However, already at that time it was clear that resonant multistep photoionization opens up possibilities of developing a laser detector for individual atoms and molecules (Moore, 1973; Letokhov, 1975a). In particular, in my lectures and papers read in 1974–1976 and published a year or two later, I considered various fields of application for the laser photoionization method of detecting solitary atoms and molecules (Letokhov, 1976a,b; 1977a,b). Considered as one of the important uses of the photoionization detection of individual atoms were the tasks of detecting excited atoms, transuranic elements, and so on (Letokhov, 1976a). Special emphasis was placed on the principal possibility of creating a laser optical–mass spectrometer, based on multistep molecular photoionization, for detecting trace concentrations of complex molecules. The idea of the proposed device is illustrated in Fig. 1.4.

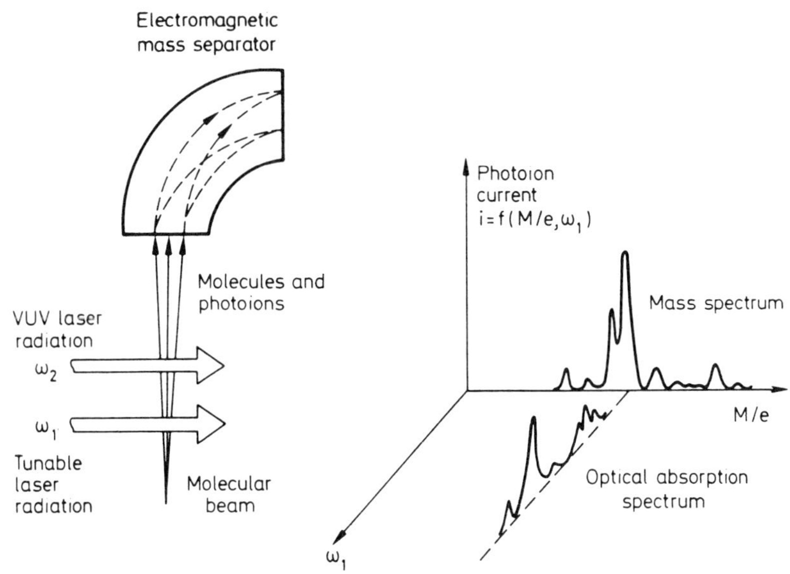

FIG. 1.4. Idea of the two-dimensional optical–mass spectrometer based on stepwise resonant photoionization of molecules. (From Letokhov, 1976a,b.)

The first step in the realization of this idea was the successful experiment on the two-step photoionization of the H_2CO molecule (Letokhov, 1976a; Andreyev *et al.*, 1977). Finally, the selective photoionization of certain molecular bonds in large-sized molecules under the effect of picosecond laser pulses was proposed for use in the visualization of biomolecules adsorbed on the needle surface of an ion projector (Letokhov, 1975b,c).

In the same years, many experiments on the photoionization detection of excited atoms were performed. Worthy of notice are the experiment on detecting metastable He (2^1S and 2^3S) atoms by the method of laser photoionization in the vicinity of the ionization limit (Stebbings *et al.*, 1973), the experiment on the two-photon excitation and subsequent photoionization of high-lying states in Cs (Popescu *et al.*, 1974), the experiment on the two-step photoionization of metastable He (2^1S) atoms in absorption saturation conditions (Hurst *et al.*, 1975), the experiment on the two-photon excitation and subsequent collisional ionization of high-lying states in Ca (Esherick *et al.*, 1976), experiments on the three-step excitation of Rydberg states in U and their subsequent ionization by an electric field or an IR laser pulse (Solarz *et al.*, 1976), and experiments aimed at studying the efficiency of the two-step photoionization of Rb atoms in conditions of absorption saturation at both excitation stages (Ambartzumian *et al.*, 1976a).

Naturally enough, the numerous works listed above led to the realization in 1977 of the first successful experiments on detecting individual atoms by the two-step photoionization method. In the work done at Oak Ridge, USA (Hurst *et al.*, 1977a), Cs atoms contained in a buffer gas were excited and ionized by same laser pulse. The electron–ion pairs formed in the process were registered by means of a proportional counter. In such an experiment, the residence time of the detected atoms in the irradiation region is sufficiently long, since it is determined by the time it takes the atoms to diffuse in the buffer gas (Fig. 1.5a). Such an experiment, however, is characterized by a rather low spectral resolution. In the work done at the Institute of Spectroscopy of the USSR Academy of Sciences (Bekov *et al.*, 1978a), Na atoms

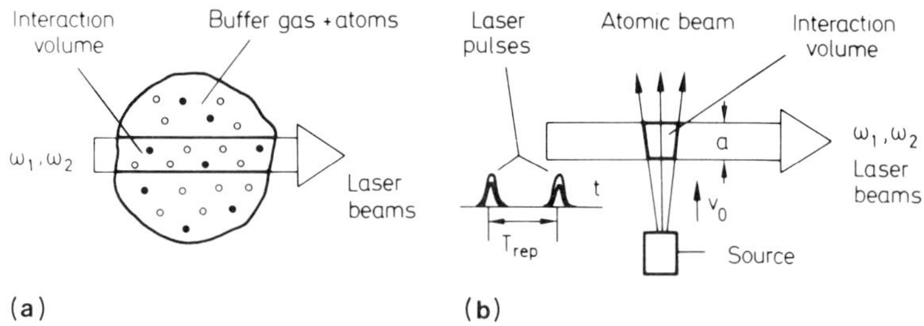

FIG. 1.5. Two types of experiments on laser resonant ionization detection of single atoms. (a) Low-resolution buffer-gas experiment and (b) high-resolution atomic beam experiment.

were subject to detection in a beam, the atoms suffering excitation of their Rydberg states, followed by ionization with the aid of an electric field pulse, as it had previously been suggested (Ivanov and Letokhov, 1975). The detection of atoms in a beam (Fig. 1.5b) makes it possible to avoid the collisional broadening inherent in the proportional counter and thus attain a high spectral resolution. The atomic beam method also allows for its use in conjunction with a mass spectrometer to select photoions. A large number of original works concerning the photoionization detection of atoms were published in the wake of the above-cited ones, and the method itself was analyzed in a series of review articles (Hurst *et al.*, 1979a; Balykin *et al.*, 1980b).

At the same time, the methods of multistep and multiphoton resonant photoionization of molecules, especially in combination with the mass spectrometry of the photoions produced in the process (Antonov *et al.*, 1977a, 1978; Zandee and Bernstein, 1979; Rockwood *et al.*, 1979), underwent successful development. Following this road, they succeeded in realizing an experiment for detecting isolated molecules in a mass-spectrometer chamber by the photoionization method (Antonov *et al.*, 1981a). At present, the laser resonant photoionization of molecules is a rapidly developing area of scientific exploration. If only a few pertinent works were performed by 1978, their number today has reached several hundreds, and they have become the subject matter of a series of reviews (Johnson, 1980; Antonov and Letokhov, 1981; Schlag and Neusser, 1983; Antonov *et al.*, 1984).

1.3. Characteristics of the Photoionization Method

Photoionization spectroscopy, based on the multistep resonant excitation and subsequent ionization of atoms and molecules, is a unique method wherewith all the ultimate characteristics can be realized. In this respect, the photoionization method is unrivaled. Table 1.1 summarizes all the main characteristics of the method and their limiting values.

As noted earlier, the *sensitivity* of the method is high enough to permit detection of single atoms or molecules. The method makes it possible to detect, during a single laser pulse, an atom or a molecule of a certain type in the same initial quantum state. As far as atoms are concerned, photoionization detection with so high a sensitivity is an ordinary thing. There are no principal difficulties here; the only thing to do is to ensure absorption saturation at each excitation stage. Under such conditions, a 50–100% efficiency can be attained in converting a resonant atom to an ion. The high sensitivity of the photoionization method enables us to clearly observe fluctuations in the number of atoms in the observation region, i.e., to observe directly the corpuscular, atomistic structure of matter.

TABLE 1.1. Ultimate Characteristics of Laser Resonant Photoionization Spectroscopy

Characteristics	Theoretical limit
Sensitivity	Single atom or molecule
Spectral resolution	$\Gamma_{\text{hom}} = 10^6 - 10^{13}\ \text{s}^{-1}$
Time resolution	$1/\Gamma_{\text{hom}} = 10^{-6} - 10^{-13}\ \text{s}$
Selectivity	$S = S_1 S_2 \cdots S_n = 10^{15} - 10^{20}$
Versatility	Any atom or molecule

The *spectral resolution* of the photoionization method is governed by the multistep excitation linewidth of atoms or molecules, which in turn depends on the individual linewidths at each excitation stage (or, to be more exact, is determined by the least linewidth in each step of the laser linewidth). The existing store of laser spectroscopy methods makes it possible to eliminate the Doppler broadening, so that the spectral resolution of the method will be defined by the homogeneous linewidth Γ_{hom} (see Shimoda, 1976; Letokhov and Chebotayev, 1977). It is an important point that the ultimate sensitivity (of the order of a single atom or molecule) inherent in the photoionization spectroscopy technique can readily be combined with the exceptionally high spectral resolution of the Doppler-free nonlinear laser spectroscopy technique (Letokhov and Chebotayev, 1977, p. 141). That this is possible was demonstrated in the well-known experiments on the Doppler-free two-photon photoionization spectroscopy of positronium (Chu and Mills, 1982).

The *time resolution* of the photoionization method also reaches maximum possible values. This is due to the fact that the registration of a particle occurs in the course of a single irradiation event whose duration in an optimal case is shorter than the lifetime of each of the excited intermediate states. For atoms, this time usually comes to a few tens of nanoseconds, whereas for molecules, especially those contained in solutions, it is reduced to a few picoseconds. Therefore, where excitation and ionization are effected by means of laser pulses of appropriate duration, a time resolution of the order of the relaxation time of the intermediate excited states can be attained. If the homogeneous broadening of the absorption line exceeds the broadening due to the population relaxation of the excited states, a higher time resolution, determined by the inverse of the homogeneous linewidth Γ_{hom}^{-1}, can, in principle, be obtained. With this technique, a subpicosecond resolution was attained in experiments with the molecules (Greene and Farrow, 1983). It should be noted that the photoionization technique used in conjunction with picosecond pulses can help investigate the dynamics of excited states in molecules under collisionless conditions even at appreciable gas pressures.

An important feature of the multistep excitation and ionization method is its potentially exceptionally *high selectivity*. With ordinary spectroscopy, we are accustomed to the selectivity being limited because of the overlapping of the wings of close atomic or molecular spectral lines. For this reason, it is in principle impossible to optically detect, for example, a radiocarbon atom (^{14}C) against the background of ordinary carbon atoms (^{12}C) in so far as their concentration ratio $^{14}C/^{12}C$ in nature amounts to 10^{-12}–10^{-16}. Despite the well-defined isotope shift of their spectral lines (which is especially noticeable in the vibrational–rotational spectrum of CO molecules), the infinitesimal overlapping of the spectral line wings limits the minimum concentration of ^{14}C that can be detected against the background of ^{12}C by ordinary spectral means. It is different with the multistep excitation method, wherewith the possibility exists of multiple isotope-selective excitations at each stage (Letokhov and Mishin, 1979). Such a multiple selection of the excited particle bears a resemblance to the multiple spectrum monochromatization in double and triple monochromators, but, as distinct from these, it is effected at different lines $\lambda_1, \lambda_2, \lambda_3$ of the absorption spectrum of the particle.

The multiplication of excitation selectivities, for example, in a three-step excitation of an isotopic atom, i.e.,

$$S = S_1 S_2 S_3, \tag{1.9}$$

where S_1 is the excitation selectivity at the ith stage, enables a total selectivity $S \gtrsim 10^{17}$–10^{20} to be attained. This means that a single radioactive atom can be detected in 10^{-3}–1 g of matter. Since this detection method does not rely on the radioactive decay of a nucleus, it can be applied with equal success to long-lived radioisotopes with a decay time of 1–10^6 years, which in principle cannot be detected in the preceding concentrations and amounts of matter by any nuclear physics technique. Thus, the photoionization spectroscopy method has approached in sensitivity the techniques used to register particles in nuclear physics, but substantially surpasses them in potential detection selectivity.

Finally, the photoionization method is *universal*, for it is in principle applicable to any atom or molecule. Also, the method is insensitive to the type of decay the excited levels suffer. It is more easily applicable to atoms, especially a complex atom having spectral lines in the region well-covered by tunable dye lasers. Some experimental difficulties arise in detecting hydrogen, halogen, and noble gas atoms whose excitation lines lie in the poorly mastered vacuum ultraviolet region. However, the two- and three-photon excitation techniques help solve this problem in principle, although the efficiency and selectivity of ionization are in this case appreciably reduced.

Certain difficulties arise also in the photoionization detection of some molecules because of both the lack of suitable excitation lines in the visible and UV regions and the competition between the photoionization of the molecules and their photofragmentation into neutral products. Nevertheless, an experiment was successfully performed wherein a few naphthalene molecules contained in the irradiation volume were detected by a photoionization–mass spectrometry method (Antonov *et al.*, 1981). A combination of selective photoionization and mass-spectrometry) methods seems to be especially versatile as far as the detection of trace amounts of complex molecules is concerned. Such a combination method for detecting molecules may lead to the development of a laser "smell detector", for its sensitivity and selectivity can be the same as those of the olfactory organs of animals and man. Academician P. L. Kapitza believed that "to catch up with the dog's sense of smell" is one of the problems physics will have to tackle in the future (Kapitza, 1966). It is quite probable that photoionization optical–mass spectrometry will help solve this problem.

An interesting feature of photoionization spectroscopy, which distinguishes it from the other laser spectroscopy methods, is the possibility of *collecting* the detected particles, atomic ions in particular. This means the possibility of repeated experiments being conducted with the same particles. This seems to be impossible at first glance because of the destruction of the particles upon their photoionization. However, this is not so for atomic photoions, which are easy to neutralize and store for a long period of time by implanting them into the surface layer of a target-collector (Letokhov *et al.*, 1977, p. 183). In this way, the main disadvantage of the photoionization method, namely, the destruction of the particles being detected, can be eliminated, at least for atoms. For molecules, however, the method remains destructive because of their inevitable dissociation and fragmentation upon photoionization.

The unique combination of laser photoionization spectroscopy parameters in a single method is perhaps due to the fact that the method relies solely on the use of stimulated transitions in atoms or molecules, which occur under the effect of intense laser pulses. In other words, stimulated quantum transitions whose rate can easily be controlled by varying the laser radiation intensity enable the experimentalist to control in a large measure the energy deposition into a given atomic or molecular species.

This is the main distinctive feature of this method as opposed to other methods using uncontrollable spontaneous and collisional processes. By way of example, compare the photoionization method with the laser optogalvanic technique (Green *et al.*, 1976). On the face of it, these methods seem quite similar. In actual fact, as will be apparent from Fig. 1.6, this is not so at all. In the case of resonant photoionization spectroscopy (Fig. 1.6a), laser radiation ensures excitation of an atom to such a high energy level as makes its ionization inevitable. Even if collisions are necessary for this to occur, as is the case with Rydberg atoms, the highly excited atom has not the slightest

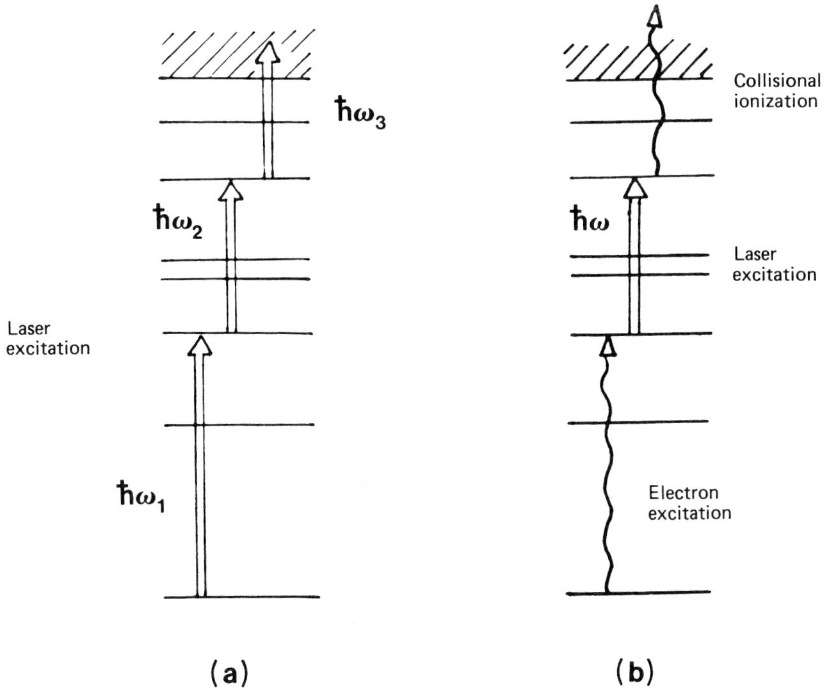

FIG. 1.6. Comparison of quantum transition schemes for (a) resonant photoionization spectroscopy and (b) optogalvanic spectroscopy.

chance to avoid ionization. It is for this reason that the entire atomic ionization process with this method is characterized by a high selectivity yield. As to the optogalvanic spectroscopy technique (Fig. 1.6b), laser radiation here only slightly shifts the ionization equilibrium in a flame or electric discharge, where all sorts of charged particles are produced as a result of a great variety of processes.

1.4. Photoionization Spectroscopy Versions

In resonance photoionization of a neutral particle (atom or molecule), the absorption by the particle of a few laser photons gives rise to an easily detectable pair of charged particles—a photoion and a photoelectron. In most experiments, the number of charged particles, i.e., the photoionization efficiency, as a function of the laser wavelength at some one or, in principle, several resonance excitation step is subject to measurement. This simplest

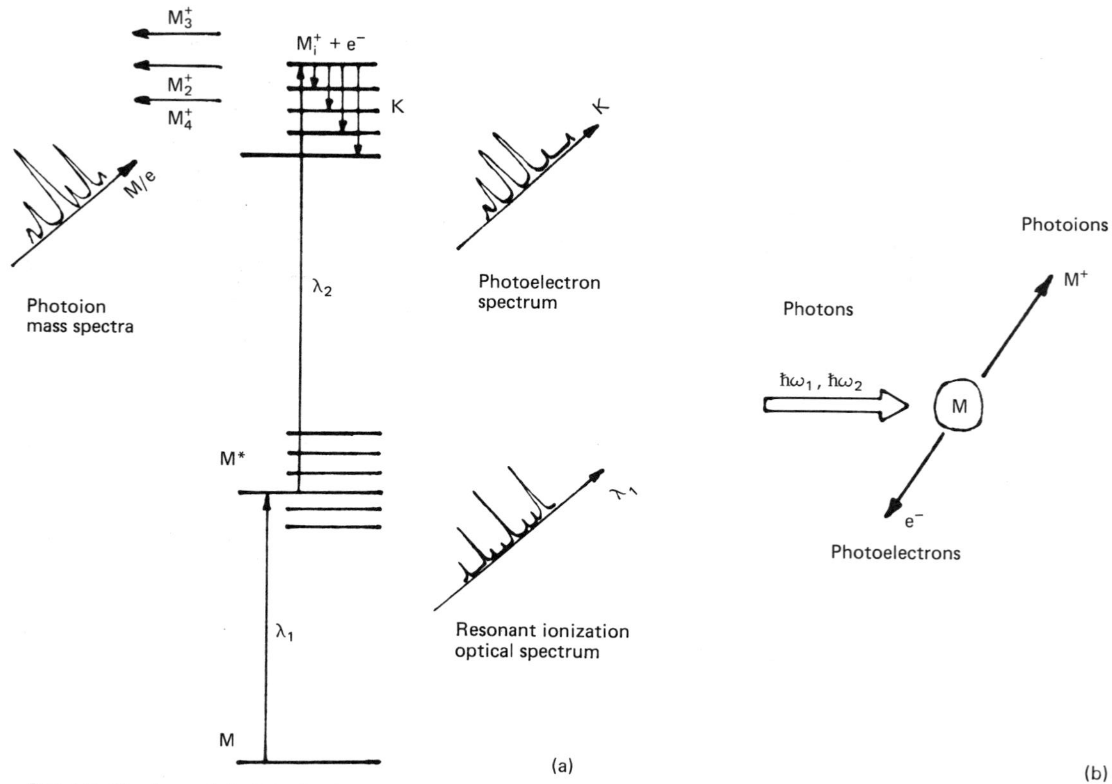

FIG. 1.7. Versions of photoionization spectroscopy wherein not only is the dependence of the multiphoton ionization efficiency on the laser wavelength subject to measurement, but also the mass spectrum of photoions and energy spectrum of photoelectrons. (a) Energy level diagram and (b) collision of a neutral particle with laser photons.

version of photoionization spectroscopy is referred to as *resonance photoionization spectroscopy* (Fig. 1.7). The technique provides fairly accurate information (see Section 1.3) on the structure of the quantum transitions of particles and is discussed in a number of chapters of this book (Chapters 3, 4, 7, 8, and 10).

An excited molecule can be ionized by several pathways, charged fragments of various masses being formed in the process. A mixture of charged particles of various masses is also formed in the course of photoionization of a mixture of isotope atoms or molecules or else of molecules with close absorption lines. To identify more accurately the charged particles formed, we can additionally analyze the mass spectrum, i.e., change over to *photoionization mass spectrometry* (Ambartzumian and Letokhov, 1972), as illustrated in Figs. 1.4 and 1.7. This spectroscopic technique is being widely used in experiments with molecules and mixtures of isotopic atoms. Chapters 5, 6, 9, and 12 of this book are devoted to a detailed description of these techniques.

Finally, the photoelectron formed as a result of the decay of the ionic state of a particle acquires some excess energy, equal to the difference between the initial energy of the particle and the ion energy. This energy excess is transferred to electron translation, i.e., it is converted to kinetic energy. If it populates different energy levels, there occurs a whole spectrum of energies of the photoelectrons produced in the course of ionization. Measurement of the photoelectron energy spectrum provides additional information about the particle being ionized and lies at the root of one or more photoionization spectroscopy versions—*photoelectron spectroscopy* of excited states with resonance ionization (Fig. 1.7). This photoionization technique is discussed in Chapter 11.

Thus, photoionization spectroscopy allows the combination of optical information channel with other very effective channels, those of information on the mass spectrum of the photoions produced and on the kinetic energy of the photoelectrons associated with them. In other words, all the participants in the collision of a neutral particle with a beam of laser photons (Fig. 1.7b) become objects of measurements. Such as approach is characteristic of modern techniques applied in nuclear physics and particle physics but is quite new to optical spectroscopy.

2. Resonant Excitation of High-Lying States

Photoionization spectroscopy is based on the laser resonant excitation of particles into high-lying quantum states that are easy to detect by ionization. What is more, to attain maximum detection sensitivity (on the order of single particles) it is necessary that the probability of such excitation should be close to 100%. Therefore, to gain an understanding of the characteristics of photoionization spectroscopy, we should consider first of all the resonant excitation of particles upon their interaction with laser radiation in conditions where stimulated quantum transitions between energy levels are dominant, i.e., in conditions of absorption saturation. Of course, consideration should be given not only to the simple case of excitation of a two-level system by one-quantum transitions, but also to the excitation of a multiple-level quantum systems under the effect of multiple-frequency laser radiation and multiple-photon excitation of high-lying states. This provides a basis for the qualitative analysis of most experimental situations. Given below is the absolute minimum of information on resonant excitation that is necessary for subsequent discussion. For more detailed analysis, the reader is referred to special monographs and reviews: e.g., two-level resonance including coherent interaction effects (Sargent *et al.*, 1974; Allen and Eberly, 1975), two- and three-level resonance, two-photon excitation (Shimoda, 1976; Letokhov and Chebotayev, 1977), and multilevel systems and their multiphoton excitation (Feld and Letokhov, 1980).

2.1. Resonant Excitation of Two-Level Systems

Consider first a simple two-level particle interacting with a resonant light field in the absence of damping and relaxation (Fig. 2.1a). This model makes it possible to reveal the specific features of the coherent resonant interaction of a quantum system with a light field.

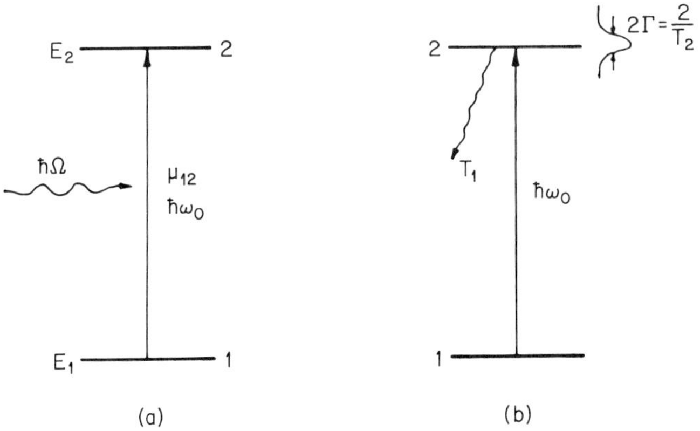

FIG. 2.1. Resonant excitation of (a) an ideal, relaxation-free two-level system and (b) a two-level system with a population relaxation time T_1 and a phase relaxation time T_2.

2.1.1. Ideal Two-Level System Free from Relaxation

2.1.1.1. Schrödinger Equation. The state of a quantum system with the energy levels E_1 and E_2 in an external laser-light field is fully described by the wave function Ψ obeying the time-dependent Schrödinger equation,

$$i\hbar\dot{\Psi} = \hat{H}\Psi, \tag{2.1}$$

where $\hat{H}$ is the system's Hamiltonian, which includes the unperturbed Hamiltonian of the system H_0 and the operator of interaction between the system and the light field, $\hat{V}$:

$$\hat{H} = \hat{H}_0 + \hat{V}. \tag{2.2}$$

In the case of electric dipole interaction between the light field and the particle, the interaction operator $\hat{V}$ has the form

$$\hat{V}(t) = -\hat{\boldsymbol{\mu}}\mathbf{E}(t), \tag{2.3}$$

where $\hat{\boldsymbol{\mu}}$ is the electric dipole moment operator of the particle and $\mathbf{E}$ the electric field strength of the light wave.

According to the superposition principle, the wave function Ψ of the two-level system in the field may be written in the general form as

$$\Psi = a_1(t)\Psi_1 e^{-i(E_1/\hbar)t} + a_2(t)\Psi_2 e^{-i(E_2/\hbar)t}, \tag{2.4}$$

where Ψ_1 and Ψ_2 are the wave functions of the stationary (time-independent) energy states of the system, with the factors oscillating at frequencies $E_k/\hbar$ being written down explicitly and $a_1(t)$ and $a_2(t)$ the probability amplitudes of the system being in the state 1 and 2, respectively. The temporal evolution of the probability amplitudes obeys the following set of equations obtained by substituting Eq. (2.4) into the initial Schrödinger equation (2.1) (Dirac, 1930, 1958):

$$
\begin{aligned}
i\hbar\dot{a}_1 &= V_{21}a_2e^{-i\omega_0 t},\\
i\hbar\dot{a}_2 &= V_{12}a_1e^{+i\omega_0 t},
\end{aligned}
\tag{2.5}
$$

where $\omega_0 = (E_2 - E_1)/\hbar$ is the resonant transition frequency and V_{12} the matrix element of the field-system interaction operator defined by the expression

$$
V_{12}(t) = \int \Psi_1^* V \Psi_2 \, dq, \qquad V_{12} = V_{21}^*. \tag{2.6}
$$

Here, q stands for the coordinates of the quantum system and the asterisk denotes a complex conjugate.

2.1.1.2. Probability Amplitude Evolution. Rabi Frequency. For most laser spectroscopy involving stimulated transitions of the quantum system between its energy levels, the laser-light wave can be represented in the form of a linearly polarized coherent wave with the frequency ω and amplitude $\mathscr{E}$:

$$
\begin{aligned}
E(t) &= \mathbf{e}\mathscr{E}\cos\Omega t\\
&= \tfrac{1}{2}\mathbf{e}\mathscr{E}(e^{i\Omega t} + e^{-i\Omega t}).
\end{aligned}
\tag{2.7}
$$

In this case, the interaction energy $V_{12}(t)$ defined by Eq. (2.6) has the following simple form:

$$
V_{12} = -\mu_{12}\mathscr{E}\cos\Omega t, \qquad V_{21}^* = V_{12}, \tag{2.8}
$$

where $\mu_{12} = \mathbf{e}\boldsymbol{\mu}_{12}$ is the projection of the transition dipole moment matrix element onto the light-wave polarization vector. In cases where $\boldsymbol{\mu}_{12}$ is isotropically oriented relative to $\mathbf{e}$, its value will be taken as the average over all possible orientations:

$$
\mu^2 = \langle(\mathbf{e}\boldsymbol{\mu}_{12})^2\rangle = \tfrac{1}{3}|\boldsymbol{\mu}_{12}|^2. \tag{2.9}
$$

With resonant interaction, the light-field frequency Ω is close to the transition frequency ω_0, so that the frequency difference $\Delta = \Omega - \omega_0$ is much less than the field frequency:

$$
|\Delta| = |\Omega - \omega_0| \ll \omega_0, \Omega. \tag{2.10}
$$

Also, the energy of interaction of the light field with the two-level system is usually much less than the quantum transition energy:

$$
\mu\mathscr{E} \ll \hbar\omega_0. \tag{2.11}
$$

This is only natural, because otherwise the interaction of the field with the quantum system will no longer be resonant. In this case, all the time-dependent terms can be greatly simplified by treating them only in reference to fast motions of the $e^{i\omega_0 t}$ type. With this approximation, known as the rotating wave approximation (Feynman *et al.*, 1957), the probability amplitudes a_1 and a_2 oscillate only at the slow frequency Δ. Under these conditions (which generally hold true), instead of Eq. (2.5) we have for the probability amplitudes the following set of rather simple equations:

$$\begin{aligned} i\hbar\dot{a}_1 &= -\tfrac{1}{2}\mu\mathscr{E}e^{i\Delta t}a_2, \\ i\hbar\dot{a}_2 &= -\tfrac{1}{2}\mu\mathscr{E}e^{-i\Delta t}a_1, \end{aligned} \tag{2.12}$$

which are easy to solve.

Let the particle at first be definitely at the lower level,

$$|a_1(t=0)|^2 = 1, \qquad |a_2(t=0)|^2 = 0, \tag{2.13}$$

and the light field with the constant amplitude,

$$E(t) = \begin{cases} \mathscr{E}\cos\Omega t, & t \geqslant 0, \\ 0, & t < 0, \end{cases} \tag{2.14}$$

be switched on at the moment $t = 0$. The solution of the set of Eqs. (2.12) can then easily be found:

$$\begin{aligned} a_1(t) &= \{\cos[(\gamma/2)t] - i(\Delta/\gamma)\sin[(\gamma/2)t]\}e^{i(\Delta/2)t}, \\ a_2(t) &= i(\mu\mathscr{E}/\hbar\gamma)\sin[(\gamma/2)t]e^{-i(\Delta/2)t}, \end{aligned} \tag{2.15}$$

where γ is the Rabi frequency, defined by the expression (Rabi, 1937)

$$\gamma^2 = \Delta^2 + (\mu\mathscr{E}/\hbar)^2 = \Delta^2 + \gamma_0^2, \tag{2.16}$$

Here, $\gamma_0 = \mu\mathscr{E}/\hbar$ is the Rabi oscillation frequency in the case of exact resonance ($\Delta = 0$).

2.1.1.3. Probability of Excitation. The squared probability amplitude $|a_2(t)|^2$, i.e., the probability that the two-level particle will be in the upper state 2 at the moment t, is

$$|a_2(t)|^2 = (\mu\mathscr{E}/\hbar\gamma)^2\sin^2(\gamma/2)t = [1 - (\Delta/\gamma)^2]\sin^2(\gamma/2)t. \tag{2.17}$$

So, the probability of the particle being excited to the upper level, W_{12}, is given by

$$W_{12}(t) = \sin^2[\tfrac{1}{2}(\Delta^2 + \gamma^2)^{1/2}t]L(\Delta/\gamma), \tag{2.18}$$

where $L(x)$ is the Lorentzian line form factor normalized to unity at maximum:

$$L(x) = 1/(1 + x^2). \tag{2.19}$$

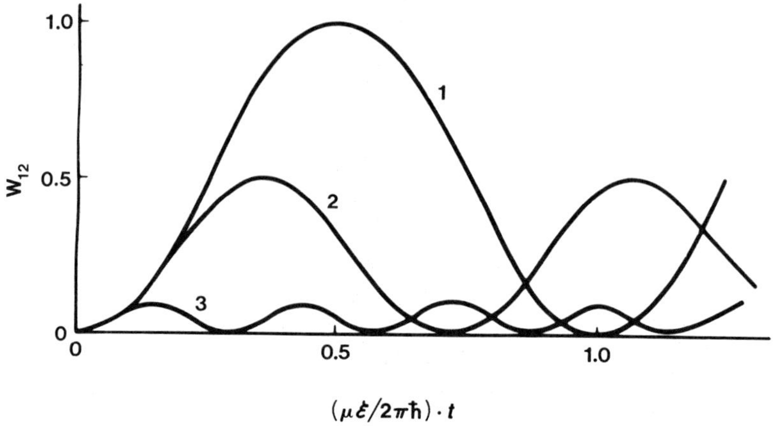

FIG. 2.2. Coherent oscillations of the excited-level population in an ideal, relaxation-free two-level system in a resonant laser-light field switched on at the moment $t = 0$ at various values of the detuning Δ of the field frequency Ω from the exact resonance frequency ω_0: (1) $\Delta = \Omega - \omega_0 = 0$; (2) $\Delta = \mu\mathcal{E}/\hbar$; and (3) $\Delta = 3\mu\mathcal{E}/\hbar$.

In the case of exact resonance ($\Delta \ll \gamma_0$),

$$W_{12}(t) = \sin^2(\gamma t/2) = \tfrac{1}{2}(1 - \cos\gamma_0 t), \tag{2.20}$$

while in the case of highly off-resonance conditions ($\gamma \gg \Delta$), the following approximate expression for W_{12} holds true:

$$W_{12}(t) = (\gamma/\Delta)^2 \sin^2(\Delta/2)t = (\gamma/2\Delta)^2(1 - \cos\Delta t). \tag{2.21}$$

In the case of exact resonance (see Fig. 2.2), complete excitation of particles to the upper level can be achieved during the time $t_{\text{inv}} = \pi/\gamma_0 = \pi\hbar/\mu\mathcal{E}$. However, this is, in principle, impossible to do where the field frequency Ω is detuned from resonance. To properly consider the case of detuning from resonance ($\Delta \neq 0$), we should first of all overstep the bounds of the highly idealized model of an isolated two-level particle and an ideal coherent light field.

2.1.2. Two-Level System Subject to Population and Phase Relaxation

2.1.2.1. Population Relaxation and Phase Relaxation. A real two-level particle is inevitably subject to the delay of its excited state, at least owing to spontaneous emission (Fig. 2.1b). Giving consideration for the excited state decay at once makes the absorption spectral line $1 \to 2$ have a finite width even in the absence of the external field (Weiskopf, 1933). In actual fact, the interaction of the particle with the surroundings, which can be considered as a thermal bath, can additionally shorten the lifetime of the excited state. In the general case, level population relaxation to the equilibrium state is called *longitudinal relaxation* and is characterized by the longitudinal relaxation time T_1 (Bloch, 1946).

The population relaxation, the interaction of the particle with its surroundings, causes "interruption" of the particle's wave-function phase in each of the steady states, which is not necessarily accompanied by the decay of the particle to the lower state. The time of such a "phase relaxation" is denoted by T_2, and the relaxation itself is frequently referred to as the *transverse relaxation* (Bloch, 1946). The phase relaxation has no effect on the level population relaxation, but widens the spectral line of the $1 \to 2$ transition. The homogeneous half-width Γ_{hom} of the Lorentzian (2.19) is related to the time T_2 by the simple equation

$$\Gamma_{\text{hom}} = 1/T_2. \tag{2.22}$$

2.1.2.2. Density Matrix Equations. It is convenient to describe the excitation of a quantum system in the presence of population and phase relaxations by considering the evolution not of the probability amplitudes $a_i(t)$ themselves but of their combinations $a_i^*(t)a_j(t)$ averaged over the ensemble. This is a standard procedure in quantum mechanics, according to which the quantity $\langle a_i^*(t)a_j(t)\rangle$ is the element $\rho_{ji}(t)$ of the state density matrix (see, for example, Sargent *et al.*, 1974):

$$\rho_{ji}(t) = \langle a_i^*(t)a_j(t)\rangle. \tag{2.23}$$

The above-introduced population probability of the state i is described by the corresponding diagonal element of this matrix:

$$n_i(t) = \langle a_i^*(t)a_i(t)\rangle = \rho_{ii}(t). \tag{2.24}$$

The population evolution of the levels 1 and 2 in an external resonant laser-light field is described by equations that follow from the Schrödinger equation after the appropriate averaging in accordance with Eq. (2.23) and include terms accounting for the level population relaxation:

$$\begin{aligned} dn_1/dt &= (i/\hbar)\mathbf{E}(\boldsymbol{\mu}_{12}\rho_{21} - \boldsymbol{\mu}_{21}\rho_{12}) - (1/T_1)(n_1 - n_1^0), \\ dn_2/dt &= -(i/\hbar)\mathbf{E}(\boldsymbol{\mu}_{12}\rho_{21} - \boldsymbol{\mu}_{21}\rho_{12}) - (1/T_1)(n_2 - n_2^0), \end{aligned} \tag{2.25}$$

where n_i^0 is the population probability of the level i in the absence of the laser-light field. The change of the level population depends on the off-diagonal elements $\rho_{12}(t)$ and $\rho_{21}(t)$ of the density matrix, which describe the polarizability of the two-level particle, i.e., the high-frequency dipole moment $\mathbf{P}(t)$ induced in the particle by the laser-light field, i.e.,

$$\mathbf{P}(t) = \boldsymbol{\mu}_{12}\rho_{21} + \boldsymbol{\mu}_{21}\rho_{12}. \tag{2.26}$$

The evolution of the off-diagonal matrix elements is described by equations that also follow from the Schrödinger equation after the appropriate averaging and introduction of the phase relaxation time T_2, i.e., the time of the particle polarization decay:

$$\begin{aligned} d\rho_{12}/dt &= i\omega_0\rho_{12} + (i/\hbar)\mathbf{E}\boldsymbol{\mu}_{12}(n_2 - n_1) - (1/T_2)\rho_{12}, \\ d\rho_{21}/dt &= -i\omega_0\rho_{21} - (i/\hbar)\mathbf{E}\boldsymbol{\mu}_{21}(n_2 - n_1) - (1/T_2)\rho_{21}. \end{aligned} \tag{2.27}$$

2.1.2.3. Polarization and Population Equations. The set of Eqs. (2.25) and (2.27), which looks rather cumbersome, can be reduced to a more descriptive form by using the rotating wave approximation and introducing instead of the four elements $\rho_{ij}(t)$ the following physically observable quantities. We introduce the difference in population probability between the levels 2 and 1 in the form

$$\mathcal{N}(t) = n_2(t) - n_1(t), \tag{2.28}$$

which, by virtue of the condition $n_1 + n_2 = 1$, is related to the level populations by

$$n_1 = \tfrac{1}{2}(1 - \mathcal{N}); \qquad n_2 = \tfrac{1}{2}(1 + \mathcal{N}). \tag{2.29}$$

Let us next represent the polarization oscillating with the frequency Ω of the light field in Eq. (2.7) in the form

$$\mathbf{P}(t) = \mathbf{e}\mathscr{P}(t)\cos[\Omega t + \varphi(t)]. \tag{2.30}$$

Then, instead of the set of Eqs. (2.25) and (2.27), the following simple set of equations can be obtained with the rotating wave approximation:

$$\begin{aligned} \dot{\mathcal{N}} + (1/T_1)(\mathcal{N} - \mathcal{N}_0) &= -(1/\hbar)\mathscr{P}\mathscr{E}\sin\varphi, \\ \dot{\mathscr{P}} + (1/T_2)\mathscr{P} &= (\mu^2/\hbar)\mathcal{N}\mathscr{E}\sin\varphi, \\ (\Omega - \omega_0)\mathscr{P} &= (\mu^2/\hbar)\mathcal{N}\mathscr{E}\cos\varphi. \end{aligned} \tag{2.31}$$

We consider now the response of the two-level system to the switching on of the light field of the form of Eq. (2.14) for the simple case of exact resonance:

$$\Omega = \omega_0. \tag{2.32}$$

In this case, the change of the level population difference $\mathcal{N}$ is described by the simple equation with constant coefficients

$$\ddot{\mathcal{N}}(t) + (1/T_2 + 1/T_1)\dot{\mathcal{N}}(t) + [\gamma_0^2 + 1/(T_1T_2)]\mathcal{N}(t) = \mathcal{N}_0[1/T_1T_2)]. \tag{2.33}$$

The behavior of the two-level system depends materially on the relation between two parameters: the Rabi frequency $\gamma_0 = \mu\mathscr{E}/\hbar$ and the parameter $(1/T_2 - 1/T_1)$. In most cases, phase relaxation takes its course much faster than level population relaxation ($T_2 \ll T_1$), and therefore, to simplify expressions, the quantities $(1/T_2 \pm 1/T_1)$ will hereinafter be considered equal to $1/T_2 = \Gamma_{\text{hom}}$.

2.1.2.4. Coherent and Incoherent Interactions. In the case of a sufficiently strong light field, where

$$2\gamma_0 = 2(\mu\mathscr{E}/\hbar) \gg 1/T_2, \tag{2.34}$$

the change of the level population is described by the approximate expression

$$\mathscr{N} = \mathscr{N}_0 \exp(-t/2T_2) \cos \gamma_0 t. \tag{2.35}$$

In this case, the populations of states 1 and 2 oscillate with the Rabi frequency, as is shown in Fig. 2.2. The only difference is that the oscillations decay exponentially during the phase relaxation time T_2. If condition (2.34) is satisfied, the interaction of the two-level system with the laser-light field is said to be *coherent*.

Let the light-field amplitude now fail to satisfy condition (2.34):

$$2\gamma_0 \ll 1/T_2. \tag{2.36}$$

In this case, the evolution of the level population difference $\mathscr{N}$ is described by the expression

$$\mathscr{N}/\mathscr{N}_0 \simeq 1/(1 + G) + G/(1 + G) \exp[-(t/T_1)(1 + G)], \tag{2.37}$$

where G is a dimensionless saturation parameter defined as

$$G = \gamma_0^2 T_1 T_2 = (\mu\mathscr{E}/\hbar)^2[1/(T_1 T_2)] = \sigma(\omega_0)I/T_1, \tag{2.38}$$

with $I = (c/8\pi)(\mathscr{E}^2/\hbar\Omega)$ being the radiation intensity in photons per square centimeter seconds and $\sigma(\Omega)$ the cross section of the stimulated transition between levels 1 and 2 at the light-field frequency Ω equal to ω_0:

$$\sigma(\Omega) = (4\pi\omega_0\mu^2/c\hbar\Gamma_{\text{hom}})L[(\Omega - \omega_0)/\Gamma_{\text{hom}}]. \tag{2.39}$$

In this case, the populations of the two levels tend to become equal aperiodically. If condition (2.36) is satisfied, the interaction of the two-level quantum system with the laser-light field is said to be *incoherent*. The qualitative distinction between the evolutions of the population of, e.g., the upper level n_2 in the case of coherent and incoherent interaction is illustrated in Fig. 2.3.

2.1.2.5. Rate Equation Approximation. Saturation Power and Energy. With the incoherent interaction approximation, the set of Eqs. (2.31) is greatly simplified. In this case, the polarization $\mathscr{P}$ of the two-level system depends in a steady-state manner of the light-field amplitude $\mathscr{E}$:

$$\mathscr{P} = (\mu^2/\hbar\Gamma_{\text{hom}})\mathscr{N}\mathscr{E} \sin\varphi, \tag{2.40}$$

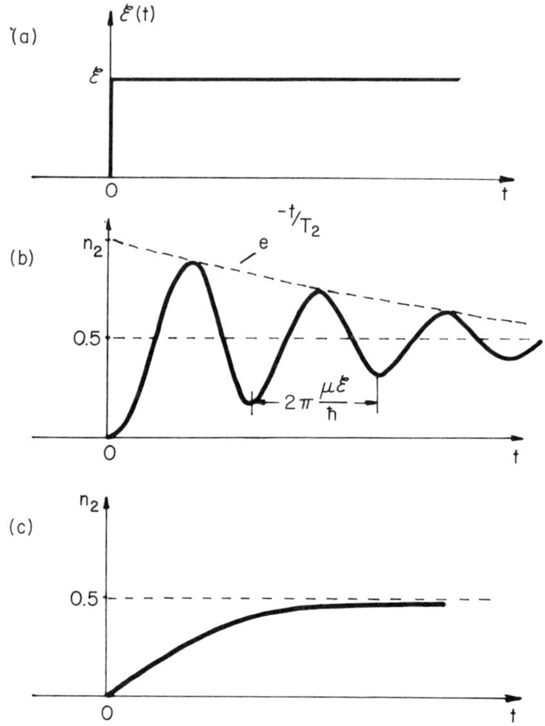

FIG. 2.3. Response of a two-level system to (a) a pulsed resonant laser-light field $\mathscr{E}(t)$, (b) coherent interaction at $2(\mu\mathscr{E}/\hbar) \gg 1/T_2$, and (c) incoherent interaction at $2(\mu\mathscr{E}/\hbar) \ll 1/T_2$ (n_2 is the relative population of the excited level and $\mathscr{E}$ the light-field amplitude).

and so the interaction of the system with the light field can be described by the simple *rate* equation following from Eqs. (2.31) and (2.40):

$$d\mathscr{N}/dt + (1/T_1)(\mathscr{N} - \mathscr{N}_0) = -2\sigma(\Omega)I\mathscr{N}, \tag{2.41}$$

where $\sigma(\Omega)$ is the radiative transition cross section at the light-field frequency Ω, and I is the radiation intensity (in photons/cm^2 s). The population of each level, n_1 and n_2, can be found by solving Eq. (2.41) with due regard for the relation $n_1 + n_2 = 1$.

For instance, under steady-state conditions when the laser-pulse duration τ_p is much longer than the population relaxation time T_1, the level population difference is defined by the very simple relation

$$\mathscr{N}(t) = \mathscr{N}_0/(1 + G) = \mathscr{N}_0(1 + I(t)/I_{sat})^{-1}, \tag{2.42}$$

where I_{sat} is the absorption *saturation intensity*, given by

$$I_{sat} = 1/2\sigma(\Omega)T_1, \qquad \tau_p \gg T_1. \tag{2.43}$$

When the saturation is strong enough ($I(t) \gg I_{sat}$), there evidently occurs equalization of populations of the levels interacting with the radiation. This is natural, for there occurs establishment of equilibrium between the two-level system and the light field having a very high effective radiation temperature ($kT_{rad} \gg \hbar\omega_0$).

In the opposite case of excitation by a short light pulse ($\tau_p \ll T_1$), we can omit the longitudinal relaxation term in Eq. (2.41) and obtain a simple relation for the population difference change:

$$\mathcal{N}(t) = \mathcal{N}_0 \exp\{-[1/2\sigma(\Omega)]\} \int_{-\infty}^{t} I(t')\,dt' = \mathcal{N}_0 e^{-\Phi/\Phi_{sat}}, \tag{2.44}$$

where Φ is the laser pulse fluence defined as

$$\Phi(t) = -\int_{-\infty}^{t} I(t')\,dt', \tag{2.45}$$

and Φ_{sat} is the absorption saturation fluence given by

$$\Phi_{sat} = 1/2\sigma(\Omega), \qquad \tau_p \ll T_1. \tag{2.46}$$

Again, when the saturation is sufficiently strong ($\Phi \gg \Phi_{sat}$), there occurs equalization of populations of the two levels.

2.1.2.6. Power Broadening. Of great importance for spectroscopy is the excitation linewidth in conditions of saturation, which cannot be avoided if the probability of exciting the atom is to be maximized. Solving the set of Eqs. (2.31) for the steady-state case, we can obtain the dependence of the level population difference on the detuning $\Delta = \Omega - \omega_0$,

$$\mathcal{N}(\Omega)/\mathcal{N}_0 = 1 - \Gamma_{hom}^2 G/[(\Omega - \omega_0)^2 + \Gamma_{hom}^2(1 + G)]. \tag{2.47}$$

According to Eq. (2.32), the probability of excitation to the upper level is

$$n_2(\Omega) = \tfrac{1}{2}\{\Gamma_{hom}^2 G/[(\Omega - \omega_0)^2 + \Gamma_{hom}^2(1 + G)]\}, \tag{2.48}$$

where for the sake of definiteness the particle is assumed to be initially at the lower level ($\mathcal{N}_0 = -1$). Figure 2.4 shows the frequency dependence of the upper level excitation probability at various degrees of saturation. As can be seen, the excitation spectrum *power broadening effect* is manifest in the graph.

As follows from Eq. (2.48), the effective half-width of the excitation spectrum is given by (Karplus and Schwinger, 1948)

$$\Delta\omega_{eff} = \Gamma_{hom}\sqrt{1 + G} = (1/T_2)[1 + (\mu^2 T_1 T_2/\hbar^2)\mathcal{E}^2]^{1/2}. \tag{2.49}$$

The broadening of spectral lines due to excitation by a strong light field should be taken into account when interpreting the results yielded by the various spectroscopic techniques based on the use of stimulated quantum transitions.

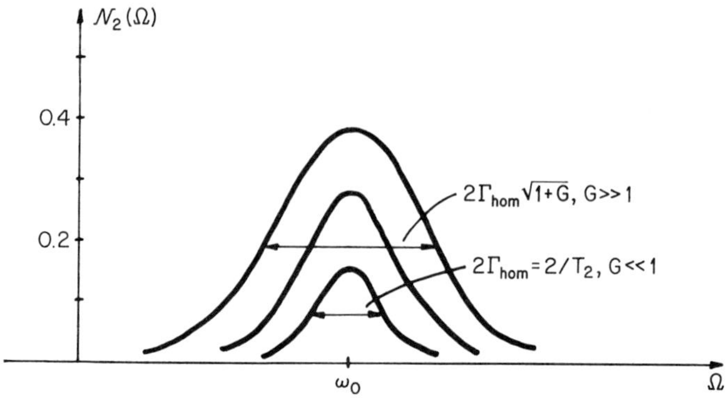

FIG. 2.4. Excitation resonance broadening of a two-level system with increasing laser radiation intensity (Ω is the field frequency, ω_0 the central transition frequency, $\Gamma = 1/T_2$ the homogeneous spectral-line half-width, and G the saturation parameter).

2.1.2.7. Field Splitting of Levels. The broadening of the excitation spectrum in a strong light field is amenable to a simple physical explanation, especially so if we consider the case of a strong field where a particle oscillates between two levels. For the sake of simplicity, we consider the case of exact resonance ($\Delta = 0$). Then, according to Eq. (2.15), the probability amplitudes $a_k(t)$ oscillate as expressed by

$$a_1(t) = \cos(\gamma_0/2)t; \qquad a_2(t) = i\gamma_0 \sin(\gamma_0/2)t, \tag{2.50}$$

while the squared probability amplitudes $|a_k(t)|^2$ describing the level populations oscillate with the frequency γ_0.

In accordance with Eq. (2.4), the wave function of the two-level system may be represented, considering Eq. (2.50), in the form

$$\Psi = \frac{1}{2}\sum_{k=1,2} \exp[-i(E_k/\hbar + \gamma_0/2)t] + \exp[-i(E_k/\hbar - \gamma/2)t]. \tag{2.51}$$

It is at once evident from the preceding expression that the particle's oscillation of frequency γ_0 between the two energy levels E_k results in the *field splitting* of each level into two sublevels $E_k \pm (\gamma_0/2)\hbar$ spaced a distance $\hbar\gamma_0$ apart. Such split energy levels, referred to as quasi-energy levels, are of a common origin associated with the properties of the periodic wave function of the two-level system in an external periodic field of constant amplitude (Zel'dovich, 1973).

The field splitting of energy levels (see Fig. 2.5) can also be observed in transitions from level 1 or 2 to some third level. As a result, the spectral line

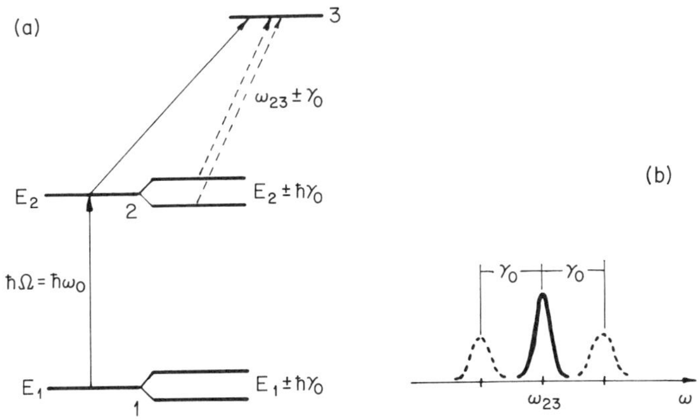

FIG. 2.5. Level splitting of a two-level system in a strong resonant light field, which can be observed on the transition to some third level.

on the $2 \rightarrow 3$ transition gets split into two lines spaced a frequency interval apart (Fig. 2.5b). This phenomenon was first observed in microwave spectroscopy (Autler and Townes, 1950). In the optical region of the spectrum and in the presence of phase relaxation, such splitting can apparently be observed only if condition (2.34) is satisfied, i.e., when the response of the two-level system to the laser-light field is truly oscillatory.

The field splitting of energy levels in conditions of a strong saturation must be taken into account in multistep excitation where the saturation of one quantum transition may affect the excitation where the saturation of quantum transition (see Section 2.3).

To conclude this simple discussion of the resonant excitation of a two-level particle under saturation conditions, let us make a few obvious remarks.

2.1.2.8. Pulsed Excitation. All the above relationships and conclusions also hold true in cases where a laser pulse of a gradually rising amplitude is used instead of the step pulse described by Eq. (2.14), provided that $\Omega = \omega_0$. The only thing to do is to make the following substitution in Eqs. (2.33)–(2.37):

$$\mathscr{E}t = \int_{-\infty}^{t} \mathscr{E}(t')\,dt'. \tag{2.52}$$

2.1.2.9. Level Degeneracy. Usually, the quantum levels of a real atom or molecule are degenerate. In this case, the description of the interaction of a field with a two-level system becomes very complicated. In particular, there is no longer any simple and graphic picture of the particle's Rabi-frequency oscillations between the two levels. There are instead the particle's oscillations between individual sublevels with frequencies of their own, which combine to smooth out the oscillations of the net level populations.

However, in the case of incoherent interaction, the picture looks simpler, for there are no level population oscillations at all. In this case, to describe the evolution of level populations, use can be made as before of rate equations such as Eq. (2.41), but quantities n_1 and n_2 must be understood to be the total populations of each level. The level population difference given by Eq. (2.28) then must be modified as follows:

$$\mathcal{N}(t) = n_2(t) - (g_2/g_1)n_1(t). \tag{2.53}$$

In the case of a strong saturations ($G \gg 1$), there occurs equalization not of the total level populations n_k, but of sublevel populations, that is,

$$n_1/g_1 \simeq n_2/g_2. \tag{2.54}$$

To excite as many particles as possible, use should preferably be made of that quantum transition for which the condition $g_2 \gg g_1$ is satisfied. In this case, the condition $n_2 \gg n_1$ can be realized. For instance, for the allowed $\mathrm{S} \to \mathrm{P}$ dipole transition ($J_1 = 0, J_2 = 1$), $g_1 = 2J_1 + 1 = 1$ and $g_2 = 2J_2 + 1 = 3$, and so 75% of the particles can be excited, provided that saturation is strong enough.

2.1.2.10. Nonmonochromaticity of Exciting Radiation. With resonant excitation, account should be taken of the nonmonochromaticity of the actual laser radiation. If the laser radiation spectrum width $\Delta\omega_{\mathrm{L}}$ is less than Γ_{hom}, use then can be made of all the above formulas without any alterations. In the opposite case where

$$\Delta\omega_{\mathrm{L}} \gg \Gamma_{\mathrm{hom}} = 1/T_2, \tag{2.55}$$

the nonmonochromaticity of laser radiation affects excitations in two ways. It is clear that only part of the spectrum participates in excitation. The principal role is played by that part of the spectrum $I(\Omega)$ that falls within the limits of the saturated absorption linewidth [see Eq. (2.49)]. Besides, the nonmonochromaticity of multimode laser radiation with random field phases in different modes limits the coherence of radiation only within the time interval $\tau_{\mathrm{coh}} \simeq 1/\Delta\omega_{\mathrm{L}}$. The incoherence of radiation causes "phase interruptions" in the interaction between the light field and the two-level system. If the frequency of such "interruptions" exceeds the Rabi frequency, i.e.,

$$\Delta\omega_{\mathrm{L}} \simeq 1/\tau_{\mathrm{coh}} \gg \gamma_0, \tag{2.56}$$

the interaction becomes incoherent even if condition (2.34) is satisfied. In this case, the evolution of level populations is described by simple rate equations similar to Eq. (2.41).

The interaction of a two-level system subject to relaxation with a nonmonochromatic radiation can be coherent only if the following conditions are satisfied:

$$\gamma_0 = \mu\mathscr{E}/\hbar \gg 1/T_2 \qquad \text{and} \qquad \Delta\omega_L. \tag{2.57}$$

In the case of a nonmonochromatic radiation consisting of many modes of a mode-locked laser, the coherence of the field in different modes is essential. The radiation of such a laser is a sequence of ultrashort high-power pulses whose duration τ_p is usually much shorter than T_2 for gas media. In this case, the interaction of a laser-light field with a two-level medium is simpler to describe by means of equations such as Eq. (2.31), i.e., in terms of time, without resorting to spectral representation.

2.1.3. Two-Level System Subject to Radiative Relaxation

When exciting isolated atoms or molecules having no surrounding particles to interact with, the only remaining mechanism responsible for level population relaxation and phase relaxation is the radiative decay of the excited state. This case can, in principle, be treated on the basis of the relations given in Section 2.1.2, assuming that both the transverse and longitudinal relaxation are governed by the same process of excited state radiative decay proceeding at a rate of Γ:

$$1/T_1 = 1/T_2 = \Gamma. \tag{2.58}$$

This holds quite true when describing the case of exact resonance ($\Omega = \omega_0$) but is entirely not so when describing that of off-resonance excitation where the detuning from resonance is

$$\Delta = \Omega - \omega_0 \gg \Gamma, \tag{2.59}$$

and, consequently, excitation takes place far on the wing of the absorption line (Makarov, 1983). This latter case is essential to the understanding of the ultimate selectivity of resonant excitation of atoms and, therefore, merits an individual discussion.

Consider the probability of excitation of a two-level medium for which the rate of radiative transition between the levels is Γ_{21} and that of the excited state radiative decay via other channels is $\tilde{\Gamma}$ (Fig. 2.6a), so that $\Gamma = \Gamma_{21} + \tilde{\Gamma}$. Let the laser pulse duration τ_p and the detuning Δ be such that the condition

$$1/\Gamma_{21} \gg \tau_p \gg 1/\Delta \tag{2.60}$$

is satisfied. The laser pulse duration is taken to be shorter than the radiative decay time in order that during the time the laser pulse lasts, the decay should affect only the amount of absorption on the line wing and not the excitation relaxation (Fig. 2.6b).

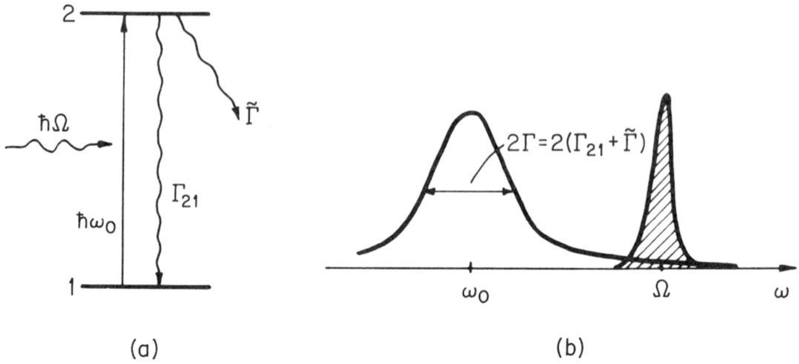

FIG. 2.6. (a) Excitation of a two-level medium with a radiative decay of the excited state and (b) the absorption line wing (Γ_{21} is the rate of radiative decay to the initial state, $\tilde{\Gamma}$ the rate of radiative decay via other channels, and Ω the laser frequency).

Based on the concept of the radiative broadening of spectral lines (Weiskopf, 1933), it would seem possible to write down at once the expression for the excitation rate,

$$W_{12} = \sigma_{12}(\Omega)I = \sigma(\omega_0)IL(\Delta/\Gamma), \tag{2.61}$$

which in the case of a large detuning from resonance has the form

$$W_{12} = \sigma(\omega_0)I(\Gamma/\Delta)^2 = \tfrac{1}{2}(\mu\mathscr{E}/\hbar\Delta)^2. \tag{2.62}$$

In fact, expression (2.62) does not describes the excitation rate of the two-level system but rather the rate of photons leaving the light wave mainly owing to the process of resonant scattering that is not accompanied by the excitation of the two-level system (Makarov, 1983). A two-level particle can get excited as a result of the sequence of quantum transition shown in Fig. 2.7. The absorption of a photon by the two-level system subject to radiative relaxation can occur only as a result of absorption of two photons at a time and be spontaneous reradiation of a photon with energy $\hbar\Omega' = 2\hbar\Omega - \hbar\omega_0 = \hbar(\Omega - \Delta)$. An approximate calculation of the excitation rate subject to condition (2.60) yields the following expression (Makarov, 1983):

$$W_{\text{exc}} = (\mu\mathscr{E}/\hbar\Delta)^4. \tag{2.63}$$

This expression differs from Eq. (2.62) commonly used for estimation purposes by a stronger detuning dependence of the excitation rate:

$$W_{\text{exc}} \sim \Delta^{-4}. \tag{2.64}$$

This difference is quite understandable from the standpoint of the two-quantum character of excitation far on the wing of the absorption line.

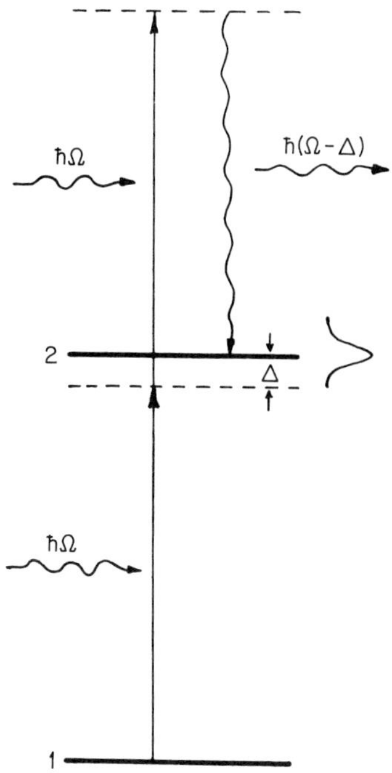

FIG. 2.7. Excitation of a two-level system subject to radiative relaxation by laser radiation whose frequency lies far on the absorption line wing.

It should be emphasized that such an abnormally low excitation on the line wing is specific only to isolated particles subject to relaxation that can compensate for the energy difference $\hbar\Delta$ only by altering the energy of spontaneous photons. The situation is, of course, trivial for a two-level system in a thermal bath (Section 2.1.2), where the possibility always exists of an energy exchange between the systems being excited and the bath. In this case, the rate of excitation far on the spectral line wing is described by the standard expression (2.62).

2.2. Two-Photon Resonant Excitation of Two- and Three-Level Systems

Let us now consider another frequently encountered case of the resonant excitation of a high-lying level through a two-photon resonant quantum transition where the energies of the two photons are summed (Goeppert-Mayer, 1931):

$$\hbar\Omega_1 + \hbar\Omega_2 = E_f - E_g. \tag{2.65}$$

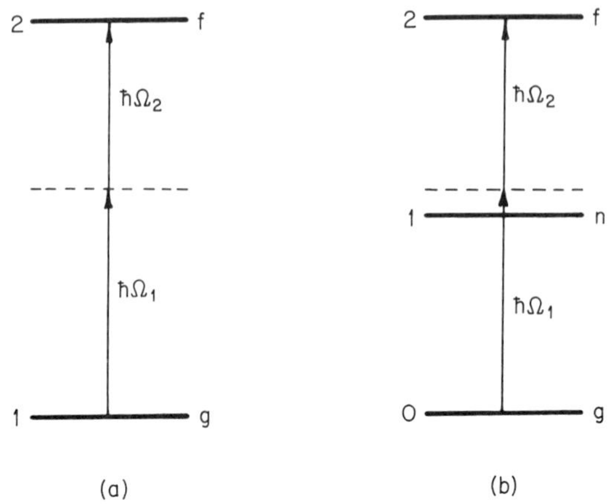

FIG. 2.8. Two-photon excitation in (a) the absence and (b) the presence of an intermediate level close to resonance.

Such an excitation may occur with an intermediate quantum state either absent or present (Fig. 2.8). In the former case (Fig. 2.8a), we speak of the two-quantum excitation of a two-level system, although the two-quantum transitions in a real atom or molecule are actually always related in some way to the presence of far intermediate levels. In the latter case (Fig. 2.8b), the presence of a close intermediate level materially influences the two-quantum excitation characteristics, and so we speak of the two-quantum excitation of a three-level system.

2.2.1. Two-Quantum Excitation Probability

2.2.1.1. Two-Frequency Field. For generality, let the light field be a sum of two waves with frequencies Ω_1 and Ω_2 and amplitudes $\mathscr{E}_1$ and $\mathscr{E}_2$:

$$\mathbf{E}(t) = \mathbf{e}_1\mathscr{E}_1 \cos(\Omega_1 t - k_1 z) + \mathbf{e}_2\mathscr{E}_2 \cos(\Omega_2 t - k_2 z). \tag{2.66}$$

Let the light-wave frequencies satisfy the two-photon resonance condition (2.65). The probability that a two-photon transition will occur in a unit of time, W_{gf}, will then be defined by the expression (Bloembergen, 1965)

$$W_{gf} = (2\hbar^4\Gamma)^{-1}|A_{gf}|^2\mathscr{E}_1^2\mathscr{E}_2^2, \tag{2.67}$$

where A_{gf} is the composite matrix element of the two-photon transition, given by

$$A_{gf} = \sum_n \left[\frac{(\boldsymbol{\mu}\mathbf{e}_1)_{gn}(\boldsymbol{\mu}\mathbf{e}_2)_{nf}}{\omega_{ng} - \Omega_1} + \frac{(\boldsymbol{\mu}\mathbf{e}_2)_{gn}(\boldsymbol{\mu}\mathbf{e}_1)_{nf}}{\omega_{ng} - \Omega_2}\right], \tag{2.68}$$

where $\boldsymbol{\mu}$ is the electric dipole moment operator, summation is extended over all the intermediate quantum states n associated with the initial state g and the final state f, and $\Gamma = \Gamma_{gf}$ is the half-width at half-weight of the $g \to f$ transition. Equations (2.67) and (2.68) can be obtained either by calculating the probability of the two-quantum transition by perturbation theory (Dirac, 1958; Cagnac *et al.*, 1973) or from the expression for the imaginary part of the third-order nonlinear susceptibility $\chi''(-\Omega_1, \Omega_1, \Omega_2, -\Omega_2)$ for a quantum system in the light field (2.66) (Bloembergen, 1965; Bloembergen and Levenson, 1976). It should be borne in mind that Eqs. (2.67) and (2.68) are written for the simplest case of nondegenerate quantum states g and f on the condition that the intermediate states are far from resonance with the fields of frequencies Ω_1 and Ω_2:

$$|\omega_{ng} - \Omega_i| \gg \Gamma, \qquad (i = 1, 2). \tag{2.69}$$

These expressions can be used to calculate the two-photon excitation rate in the simplest case where the population probability of the level f is so low that the two-photon transition saturation can be neglected.

Sometimes it proves convenient to introduce the cross section for the two-photon absorption of radiation of frequency Ω_1 in the presence of a field of frequency Ω_2 with an intensity of $I_2 = (c/8\pi)(\mathscr{E}_2^2/\hbar\Omega_2)$:

$$\sigma^2(\Omega_1) = (32\pi^3\Omega_1/\hbar^3c^2)f(\Omega_1 + \Omega_2)|A_{gf}|^2 I_2\hbar\Omega_2, \tag{2.70}$$

where $f(\Omega_1 + \Omega_2)$ is the absorption line form factor of the two-photon transition, normalized integrally to unity.

The selection rules for two-photon transitions differ from those for one-photon transitions, so that the ways of laser excitation of high-lying states are appreciably extended. In particular, for two-photon transitions, a change in the angular momentum of a particle is allowed, $\Delta L = 0 \pm 2$, and it is precisely this change that is forbidden for one-photon transitions. A systematic discussion of the selection rules for two-photon transitions can be found in a number of works (Worlock, 1972; Cagnac *et al.*, 1973; Bloembergen and Levenson, 1976).

2.2.1.2. Effect of an Intermediate State. Experimental Example. If one of the laser-radiation frequencies, Ω_1 or Ω_2, comes close to resonance with the frequency of transition to an intermediate state n, there occurs, in accordance with Eq. (2.68), a sharp increase in the two-photon transition probability. This was clearly observed in the two-photon spectroscopy of the 3S $\to$ 4D transition in the Na atom when one of the laser frequencies approached the frequency corresponding to the transition onto the intermediate level 3P. When $E(3P) - \hbar\Omega_2$ was equal to 0.1 cm^{-1}, the magnitude of the two-photon absorption cross section $\sigma^{(2)}(\Omega_1)$, in accordance with Eq. (2.70), amounted to $(5 \times 10^{-14})I_2$ [cm^2]. Even at a laser output power of $\hbar\Omega_2 I_2 = 10^3$ W/cm^2,

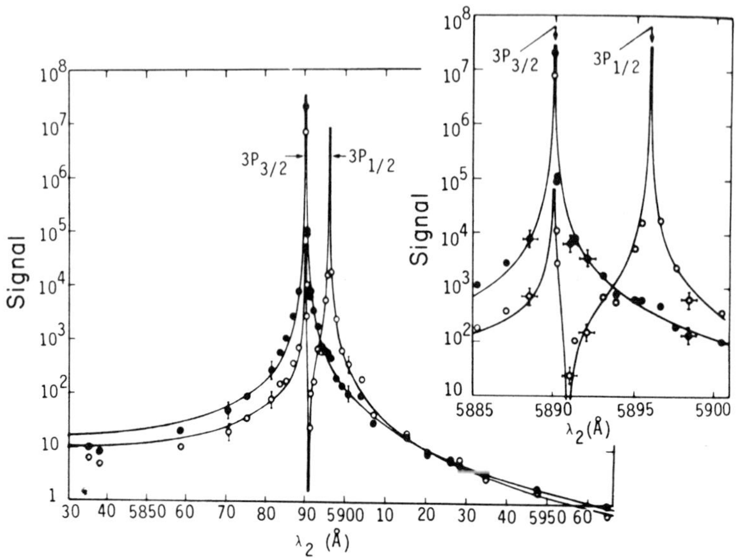

FIG. 2.9. Dependence of the rate of the $3S(F = 2) \to 4D_{5/2}$ (●) and $3S(F = 2) \to 4D_{3/2}$ (○) two-photon transitions of the Na atom on the laser wavelength λ_2 at a fixed total frequency $(\Omega_1 + \Omega_2)$ of both lasers. Inset is the region of wavelengths close to resonance with the $3P_{3/2}$ and $3P_{1/2}$ intermediate states. [From Bjorkholm and Liao (1974).]

the cross section for the two-quantum absorption of the photon $\hbar\Omega_1$ approached the typical one-quantum transition cross section. Figure 2.9 shows experimental relationships between the rates of the $3S(F = 2) \to 4D_{5/2}$ and $3S(F = 2) \to 4D_{3/2}$ two-photon transitions and the laser wavelength λ_2. Both of the two-photon transitions exhibit a strong dispersion dependence in the vicinity of the resonance with the intermediate level. For the $3S(F = 2) \to 4D_{5/2}$ transition, there is only one intermediate resonant state, because the $3P_{1/2} \to 4D_{5/2}$ one-photon transition is forbidden. At the same time, both the $3P_{3/2}$ and $3P_{1/2}$ states can be intermediate for the $3S(F = 2) \to 4D_{3/2}$ two-photon transition. In this case, the cross section $\sigma^{(2)}$ has a steep minimum when the laser frequency is exactly midway between the two intermediate states. This can be explained by the fact that the resonant contributions from each of the intermediate states to Eq. (2.68) for the two-photon transition probability are of opposite signs and, therefore, cancel out. In the vicinity of the resonances with the intermediate states, the two-photon excitation probability is increased by seven orders of magnitude. Where the probability of a stimulated two-quantum transition is greatly increased, saturation should be taken into account. The solid curves of Fig. 2.9 were calculated from Eq. (2.70) with only two intermediate states considered. Thus, the simplest estimates of the two-photon transition probability by the above formulas agree well with the experiment.

2.2.2. Power Shift and Broadening of Two-Photon Resonance

Two-photon transitions between energy levels are possible only as a result of atomic or molecular quantum state perturbations due to a light field. Such a perturbation inevitably affects both the ground and final quantum states involved in the transition, and as a result, the two-photon resonance is shifted and broadened in proportion to the laser radiation power.

2.2.2.1. Level Shift for a Two-Level System. In the general case, the shift of a level m in an external nonresonant light field $\mathbf{E} = \mathbf{e}\mathscr{E}\cos(\Omega t - kz)$ can easily be estimated using perturbation theory (Mizushima, 1964; Bonch–Bruevich and Khodovoi, 1967):

$$\Delta E_m = \frac{\mathscr{E}^2}{4}\sum_n\left[\frac{(\boldsymbol{\mu}_{mn}\mathbf{e})^2}{E_m - E_n - \hbar\Omega} + \frac{(\boldsymbol{\mu}_{mn}\mathbf{e})^2}{E_m - E_n + \hbar\Omega}\right]. \tag{2.71}$$

This formula is also valid for the case of exact resonance, but in this case the term Γ_{mn} describing the finite width of the quantum transition $m \to n$ must be added to the denominator. For example, in the simplest case of a two-level system (Fig. 2.1), Eq. (2.71) reduces to

$$\hbar\,\delta\omega = \tfrac{1}{4}\mathscr{E}^2(\boldsymbol{\mu}_{12}\mathbf{e})^2/\hbar^2(\omega_0 - \Omega)^2. \tag{2.72}$$

The shifts $\delta\omega$ of both quantum levels are equal in magnitude but opposite in sign. When $\Omega < \omega_0$, the perturbation due to the light field increases the distance between the levels, and conversely, when $\Omega > \omega_0$, this distance is reduced. In the case of exact resonance ($\Omega = \omega_0$), the shifts cancel out. It is exactly for this reason that no light shifts are observed in the case of interaction of a purely two-level system with a light field (Section 2.1.1). Note that the shift of the level m in a linearly polarized field of frequency $\Omega \to 0$ (or with $\hbar\Omega \ll |E_m - E_n|$) coincides with the quadratic shift in a dc field of the same strength.

2.2.2.2. Two-Photon Resonance Frequency Shift. Equation (2.71) also holds true for two-photon transitions. For instance, the formula for the shift of the initial level has the form

$$\Delta E_g = -\frac{\mathscr{E}_1^2}{4\hbar}\sum_n\frac{(\boldsymbol{\mu}\mathbf{e}_1)_{gn}(\boldsymbol{\mu}\mathbf{e}_2)_{ng}}{\omega_{ng} - \Omega_1}. \tag{2.73}$$

If the frequency Ω_1 comes close to resonance with ω_{ng}, the energy shift may be considerable. As is the case with the two-level system, the shift changes sign as the frequency passes through resonance. In the vicinity of resonance, $\Omega_1 \simeq \omega_{ng}$, the mixing of the states m and n, which may cause saturation of the transition, becomes important. In the case of exact resonance, we must take into consideration the finite width of the quantum transition $g \to n$ by adding the term $-i\Gamma_{gn}$ to the denominator of Eq. (2.73). In this case, the real and imaginary parts of the obtained expression describe the power-dependent shift ΔE_g and broadening $\Delta\Gamma_g$ of the ground state involved in the two-photon transition (Bloembergen and Levenson, 1976):

$$\Delta E_g = -(\mathscr{E}_1^2/4\hbar)|(\boldsymbol{\mu}\mathbf{e}_1)_{gn}|^2(\omega_{ng} - \Omega_1)/[(\omega_{ng} - \Omega_1)^2 + \Gamma_{ng}^2], \tag{2.74}$$

$$\Delta\Gamma_g = +(\mathscr{E}_1^2/4\hbar)|(\boldsymbol{\mu}\mathbf{e}_1)_{gn}|^2\Gamma_{ng}/[(\omega_{ng} - \Omega_1)^2 + \Gamma_{ng}^2]. \tag{2.75}$$

In the presence of a second, strong light field of frequency Ω_2, the shift and broadening due to this field must also be taken into account. If $\Omega_1 \simeq \omega_{ng}$ and $\Omega_1 + \Omega_2 \simeq \omega_{fg}$, the second light field is automatically in resonance with the frequency ω_{fn}. Therefore, the power shift and broadening of the level f can be described by the same expressions as Eqs. (2.74) and (2.75).

The two-photon resonance frequency shift depends on the sum of the shifts of the initial and final levels involved in the two-photon transition (Bloembergen and Levenson, 1976):

$$\begin{aligned}\Delta\omega_{fg} = &-(\mathscr{E}_1^2/4\hbar^2)|(\boldsymbol{\mu}\mathbf{e}_1)_{ng}|^2(\omega_{ng} - \Omega_1)/[(\omega_{ng} - \Omega_1)^2 + \Gamma_{ng}^2]\\ &-(\mathscr{E}_2^2/4\hbar^2)|(\boldsymbol{\mu}\mathbf{e}_2)_{fn}|^2(\omega_{fn} - \Omega_2)/[(\omega_{fn} - \Omega_2)^2 + \Gamma_{fn}^2].\end{aligned} \tag{2.76}$$

When $\Omega_1 + \Omega_2 = \omega_{fg} = \omega_0$, the detunings $(\omega_{ng} - \Omega_1)$ and $(\omega_{fn} - \Omega_2)$ become equal in magnitude but opposite in sign. In this case, with the matrix elements $(\boldsymbol{\mu}\mathbf{e}_1)_{ng}$ and $(\boldsymbol{\mu}\mathbf{e}_2)_{fn}$ being specified, the two-photon resonance frequency can be made to suffer no shift by properly selecting the ratio of the amplitudes $\mathscr{E}_1$ and $\mathscr{E}_2$ of the light fields.

2.2.2.3. Two-Photon Resonance Broadening. In quite the same way, the broadening of two-photon resonance in saturation-free conditions is determined by the total broadening of the initial and final levels (Bloembergen and Levenson, 1976):

$$\begin{aligned}\Delta\Gamma_{fg} = &(\mathscr{E}_1^2/4\hbar^2)|(\boldsymbol{\mu}\mathbf{e}_1)_{ng}|^2(\Gamma_{ng})/[(\omega_{ng} - \Omega_1)^2 + \Gamma_{ng}^2]\\ &+ (\mathscr{E}_2^2/4\hbar^2)|(\boldsymbol{\mu}\mathbf{e}_2)_{fn}|^2(\Gamma_{fn})/[(\omega_{fn} - \Omega_2)^2 + \Gamma_{fn}^2].\end{aligned} \tag{2.77}$$

Comparing Eqs. (2.67) and (2.68) for the two-photon transition rate W_{gf} with Eqs. (2.74) and (2.75) for the frequency shift, we can obtain the following relationship between these quantities for the case where the main contribution is made by a single intermediate level (Liao and Bjorkholm, 1975):

$$W_{gf} \sim \Delta E_g \cdot \Delta E_f. \tag{2.78}$$

This relationship shows that the two-photon transition rate can be increased in any way leading to an increase in the shift of the levels: either by increasing the radiation intensity or by drawing closer to resonance with the intermediate level. At the same time, if the ground level shift ΔE_g is small, a large shift of two-photon resonance frequency can be obtained in account of the final level shift ΔE_f even when the two-photon transition probability is still low. In other words, it is quite possible to observe large shifts of the two-photon resonance frequency under saturation-free conditions.

2.2.2.4. Experimental Example. Power shifts and broadenings of two-photon resonance were experimentally studied in detail (Liao and Bjorkholm, 1975; Bjorkholm and Liao, 1975). Utilized in the experiments was the 3S → 4D transition of the Na atom (Fig. 2.10). The radiation frequencies of two CW dye lasers were tuned so that the $3P_{3/2}$ and $3P_{1/2}$ levels could be intermediate virtual levels. The two-photon excitation of the atom was monitored by measuring the intensity of fluorescence on the 4P → 3S transition (330 nm), which occured in the course of the cascade decay of the 4D state. The transition probability maximum, i.e., the two-photon resonance frequency, was found by varying the sum of the laser frequencies, $\Omega_1 + \Omega_2$, in the vicinity of the resonance frequency $\omega_0 = \omega(4D \to 3S)$. To ease the task of measuring the shift and broadening values, the laser beams were made to propagate counter to each other, which reduced the Doppler broadening from 3.6 GHz down to a residual value of $(\Omega_1 - \Omega_2)v_0/c = 62$ MHz (Vasilenko *et al.*, 1970). Figure 2.10b shows the shift of the 3S($F = 2$) level, which depends mainly on the laser intensity I_2 at $\lambda_2 = 589$ nm, and Fig. 2.10c

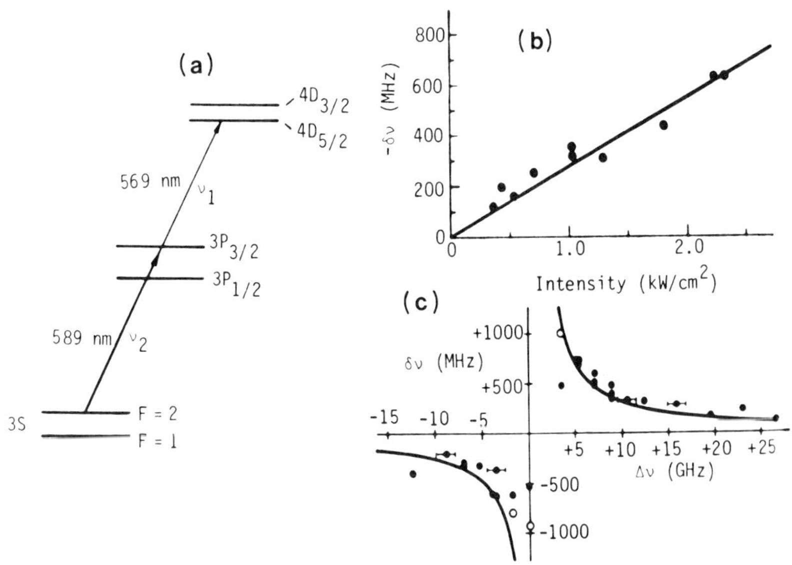

FIG. 2.10. Power shift of the two-photon resonance frequency on the 3S → 4D transition of the Na atom. (a) Diagram of the levels used, (b) dependence of the shift of the initial level 3S($F = 2$) on the radiation intensity at 589 nm; (c) dependence of the shift of the 3S($F = 2$) level on the frequency detuning from exact resonance with the 3S($F = 2$) → $3P_{3/2}$ intermediate level transition. [From Liao and Bjorkholm (1975).]

shows the relation between the shift $\delta\nu$ of the same level and the detuning of the wavelength λ_2 from resonance with the frequency of the intermediate $3S(F = 2) \rightarrow 3P_{3/2}$ transition.

Thus, the frequency and width of two-photon absorption resonance largely depend on the intensities of the light fields inducing the two-quantum transition. It is exactly this feature that differentiates between two- and one-photon excitation.

As the probability of a stimulated two-quantum transition grows higher, saturation starts gaining in importance. In this case, the two-quantum excitation must be described in terms of probability amplitudes by means of the time-dependent Schrödinger equation. The coherent excitation of multilevel systems was treated systematically in a number of works (Letokhov and Makarov, 1978; Cantrell *et al.*, 1980). Let us consider here the simplest case of a three-level system in a single-frequency field.

2.2.3. Three-level System in a Single-Frequency Field

Consider the excitation of the upper level of a three-level system (Fig. 2.11a) in a single-frequency field ($\Omega_1 = \Omega_2$), which can take place by way of two-quantum transitions only. To gain an understanding of the difference and similarity between two- and one-quantum excitation, assume that the system suffers no relaxation. In this case, by analogy with Section 2.1.1, we can use the Schrödinger equation to obtain three equations for the probability amplitudes of each of the three quantum states. With the rotating wave approximation in the case of exact resonance ($2\Omega = \omega_{10} + \omega_{12}$), these three

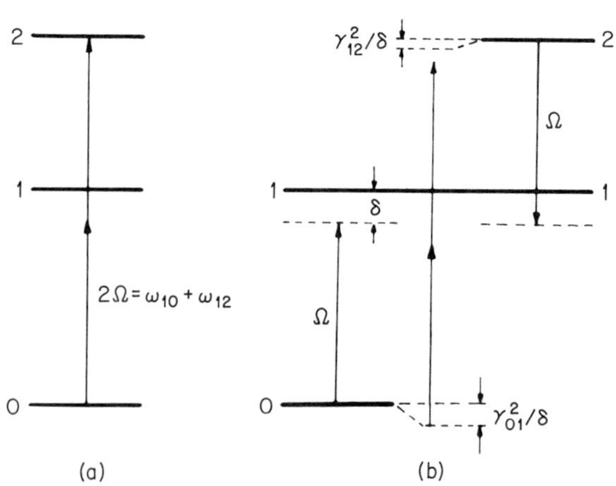

FIG. 2.11. Two-photon excitation of a three-level system by a single-frequency Ω field. (a) Level diagram, (b) effect of the difference between the level shifts due to the dynamic Stark effect on the optimum two-photon resonance frequency.

equations have the following simple form:

$$\dot{a}_0 = i\tfrac{1}{2}\gamma_{01}a_1 \exp\left(i\frac{\omega_{21}-\omega_{10}}{2}t\right),$$

$$\dot{a}_1 = i\tfrac{1}{2}\gamma_{01}a_0 \exp\left(i\frac{\omega_{10}-\omega_{21}}{2}t\right) + i\tfrac{1}{2}\gamma_{12}a_2 \exp\left(i\frac{\omega_{10}-\omega_{21}}{2}t\right), \quad (2.79)$$

$$\dot{a}_2 = i\tfrac{1}{2}\gamma_{12}a_1 \exp\left(i\frac{\omega_{21}-\omega_{10}}{2}t\right),$$

where $\omega_{21} = (E_2 - E_1)/\hbar$, $\omega_{10} = (E_1 - E_0)/\hbar$, $\gamma_{01} = (\mu_{01}\mathscr{E})/\hbar$, and $\gamma_{12} = (\mu_{12}\mathscr{E})/\hbar$. With the light-field amplitude remaining constant, the population of the upper level can easily be found (Cantrell *et al.*, 1980):

$$\rho_{22}(t) = \langle a_2^*(t)a_2(t)\rangle \simeq \frac{\gamma_{01}^2 \cdot \gamma_{12}^2}{\gamma_{01}^2 + \gamma_{02}^2} \sin^2 \frac{\gamma_{01}^2 + \gamma_{12}^2}{8\delta} t. \quad (2.80)$$

Here δ is the detuning of the field frequency Ω from resonance with the intermediate level:

$$\delta = (\omega_{21} - \omega_{10})/2 = \Omega - \omega_{10}. \quad (2.81)$$

It follows from Eq. (2.80) that in the case of exact resonance the population probability of the upper level can never reach unity unless $\gamma_{01} \neq \gamma_{12}$. Moreover, if the transition dipole moments differ greatly, the population of the upper level is negligibly small. It is exactly this fact that makes the behavior of the three-level system in the case of exact resonance different from that of the two-level system (Section 2.1.1). The reason for the absence of complete population oscillations in the case of exact two-quantum resonance is that the exact resonance condition

$$2\Omega = \omega_{10} + \omega_{21} \quad (2.82)$$

is not optimal. To explain, let us refer to the diagram of transitions in Fig. 2.11b. As discussed earlier, an off-resonance monochromatic light causes the levels to shift as a result of the dynamic Stark effect. Because of the difference in dipole moment between the $0 \to 1$ and $1 \to 2$ transitions, the shifts of the initial and final levels differ in magnitude. This causes the optimum frequency Ω_{opt} of the two-photon resonance to shift. This shift of the frequency Ω_{opt} with respect to $(\omega_{10} + \omega_{21})/2$ is small and where $\gamma_{01}, \gamma_{12} \ll 2|\delta|$, it is

$$\Omega_{\text{opt}} - (\omega_{10} + \omega_{21})/2 \simeq (\gamma_{12}^2 - \gamma_{01}^2)/8\delta. \quad (2.83)$$

As previously noted, this shift depends on the radiation intensity (Liao and Bjorkholm, 1975).

A more rigorous analysis shows that complete population oscillations between the initial (0) and final (2) levels in a three-level system are, in fact, possible, provided that the field frequency Ω is equal to the optimum two-photon resonance frequency Ω_{opt}. The level population kinetics is in this case described by the expressions

$$\rho_{00}(t) \simeq \cos^2(\gamma_{01} \cdot \gamma_{12}/4\delta)t; \qquad \rho_{22}(t) \simeq \sin^2(\gamma_{01}\gamma_{12}/4\delta)t. \tag{2.84}$$

Moreover, it is also possible to find the evolution of the level populations for the case where the field frequency Ω is detuned from the optimum two-photon resonance Ω_{opt}:

$$\rho_{00}(t) \simeq \cos^2\{[(\Omega - \Omega_{opt})^2 + \gamma_2^2]^{1/2}t\} + \frac{(\Omega - \Omega_{opt})^2}{\gamma_2^2 + (\Omega - \Omega_{opt})^2} \sin^2\{[(\Omega - \Omega_{opt})^2 + \gamma_2^2]^{1/2}\}t, \tag{2.85a}$$

$$\rho_{22}(t) \simeq \frac{\gamma_2^2}{\gamma_2^2 + (\Omega - \Omega_{opt})^2} \sin^2\{[(\Omega - \Omega_{opt})^2 + \gamma_2^2]^{1/2}\}t. \tag{2.85b}$$

Here γ_2 is the Rabi frequency of the two-photon transition:

$$\gamma_2 = (\gamma_{01}\gamma_{12}/4\delta). \tag{2.86}$$

If $\Omega = \Omega_{opt}$, Eqs. (2.85) reduce to Eqs. (2.84) describing complete population oscillations.

The half-width $(\Delta\Omega)^{(2)}$ of the two-photon resonance, hence the power broadening of the two-photon transition, is exactly equal to γ_2:

$$(\Delta\Omega)^{(2)} = \gamma_2 = (\gamma_{01}\gamma_{12}/4\delta). \tag{2.87}$$

Thus, where saturation is noticeable, we should allow for a considerable broadening of two-photon excitation resonance, which limits the excitation selectivity.

2.2.4. Doppler-Free Two-Photon Excitation

There are two reasons why two-photon resonance is of interest to photoionization spectroscopy. First, the summation of the energies of two photons makes it possible to reach high-lying states in the UV region by means of simple tunable lasers operating in the visible range. Second, the two-photon excitation of particles moving in a field produced by two counter-propagating light waves with equal or sufficiently close frequencies excludes the Doppler broadening of the excitation resonance width (Vasilenko *et al.*, 1970). This can be explained on the basis of simple qualitative considerations.

In a frame moving together with an atom, the field frequency Ω is perceived as the frequency $\Omega - kv$, where v is the projection of the atomic velocity on the propagation direction of one of the light waves. A photon from the other, counter-propagating light wave will have the frequency $\Omega + kv$ in the atom's frame. The requirement for resonance, i.e., for the equality between the total energy of the two absorbed photons and the energy of the quantum transition, leads to the condition

$$(\Omega - kv) + (\Omega + kv) = 2\Omega = \omega_0 = \omega_{12} \tag{2.88}$$

which is independent of the atomic velocity. If condition (2.88) is satisfied, all atoms interact with the field, no matter what their velocity, i.e., if two counter-propagating photons are absorbed, the Doppler shift is completely eliminated, and so the resonance width is determined by the homogeneous width of the two-photon transition.

Computation of the frequency dependence of the two-photon excitation of a particle in a standing-wave field for the case of the Doppler broadening of the absorption line yields the following expression (Vasilenko *et al.*, 1970; Bloembergen and Levenson, 1976):

$$W_{\rm st}^{(2)}(\Delta) = W_{\rm tr}^{(2)}\{\tfrac{1}{2}(\pi \ln 2)^{1/2}(\Gamma_{\rm hom}/\Delta\omega_{\rm D}) \exp[-(\Delta \ln 2/\Delta\omega_{\rm D})^2] + L(\Delta/\Gamma_{\rm hom})\}, \tag{2.89}$$

where $L(x) = (1 + x^2)^{-1}$, $\Delta = 2\Omega - \omega_0$, $W_{\rm tr}^{(2)}$ is the probability of two-photon excitation in the field of a single traveling wave in the center of the Doppler line, and $\Delta\omega_{\rm D}$ the Doppler width. The resonance shape for two-photon excitation in a standing-light-wave field is shown in Fig. 2.12a. The width of

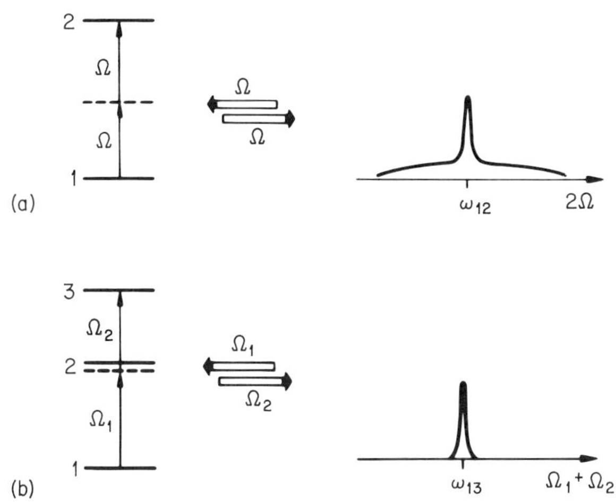

FIG. 2.12. Narrowing of a Doppler-broadened line in two-photon absorption. (a) Doppler-broadened pedestal in a standing-wave field, (b) no pedestal in the field produced by two counter-propagating waves with close but different frequencies.

the narrow two-photon excitation resonance is equal to the homogeneous width Γ_{hom} of the two-photon transition, which is determined by the total width of the initial and final transition states. The intensity ratio between the resonance and the pedestal resulting from the absorption of two photons from a single traveling wave is approximately equal to $(\Delta\omega_D/\Gamma_{\text{hom}})$.

The Doppler broadening can also be reduced in two-photon excitation with unequal frequencies of the two counter-propagating light waves (Fig. 2.12b). In this case, it proves possible to enchance both the rate and the selectivity of excitation by eliminating the Doppler-broadened pedestal. Let the frequency of one of the exciting waves be Ω_1 and that of the counter-propagating wave be Ω_2. Then, instead of Eq. (2.88), we have the following two-photon resonance condition:

$$\Omega_1 + \Omega_2 + (\Omega_1 - \Omega_2)(v/c) = \omega_0. \tag{2.90}$$

With the detuning from resonance,

$$(\Omega_1 - \Omega_2)/\omega_0 \ll \Gamma_{\text{hom}}/\Delta\omega_D, \tag{2.91}$$

the Doppler broadening will be brought down to the homogeneous broadening. If $|\Omega_1 - \Omega_2| > \omega_0\Gamma_{\text{hom}}/\Delta\omega_D$, the narrow two-photon excitation resonance will have a residual Doppler broadening of

$$\Delta\omega_{\text{res}} = \Delta\omega_D|\Omega_1 - \Omega_2|/(\sqrt{\ln 2}\,\omega_0). \tag{2.92}$$

For typical parameter values ($\Delta\omega_D/\omega_0 = 3 \times 10^{-6} \cdot \Gamma_{\text{hom}} = 10^{-4}$–$10^{-3}\ \text{cm}^{-1}$), the maximum permissible detuning $|\Omega_1 - \Omega_2| = 30$–$300\ \text{cm}^{-1}$.

An essential feature of the two-photon excitation of a Doppler-broadened transition is that all the particles are effectively excited from the lower level no matter what their velocity, the two-photon excitation linewidth being substantially smaller than the Doppler width.

2.3. Multistep Resonant Excitation of Multilevel Systems

Resonant excitation of an atom or molecule to a high-lying state forms the basis for photoionization spectroscopy. For a real atom or molecule with a large number of quantum levels, a most simple way to do this is to use multiple-frequency laser radiation to successively excite the multilevel quantum system up the ladder of discrete energy levels to some state near the ionization limit. The multistep excitation scheme used must provide for the maximum probability of exciting the atom or molecule to this final, high-lying state f.

Let us consider the simple multilevel system model shown in Fig. 2.13a. Let the system be simultaneously irradiated by laser pulses of the corresponding resonant frequencies ω_{kn}, whose duration is τ_p. It is not very difficult to formulate conditions for effective excitation of the final state f. First, the laser-pulse duration τ_p must be shorter than the population relaxation time T_1 of any of the intermediate quantum levels k.

$$\tau_p \lesssim T_1^{(k)} \qquad (k = i, \ldots, f). \tag{2.93}$$

Second, the laser-pulse fluence $\Phi(\omega_{kn})$ at the transition frequency ω_{kn} must ensure absorption saturation, i.e., the following condition similar to (1.8) must be satisfied:

$$\Phi(\Omega_{kn}) \gtrsim \Phi_{\text{sat}}^{kn} = 1/\sigma_{kn}, \tag{2.94}$$

where σ_{kn} is the cross section for the stimulated transition $k \to n$ at the frequency ω_{kn}. In the preceding expression, by the fluence $\Phi(\omega_{kn})$ is meant only the fluence within the limits of the absorption spectral line Γ_{kn} on this transition, account being taken, if necessary, of the line power broadening according to Eq. (2.49).

Under the effect of the multiple-frequency radiation subject to conditions (2.93) and (2.94), there occurs absorption saturation throughout the entire quantum transition sequence, while the excitation energy remains within the quantum system being excited. Using rate equations, we can easily find the distribution of populations n_k among the excited levels at specified $\Phi(\omega_{kn})$ and σ_{kn}. It proves most simple to do in the case where the saturation of each

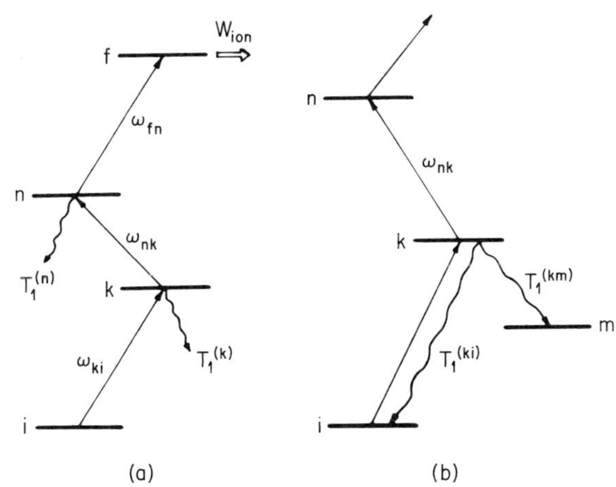

FIG. 2.13. (a) General scheme of multistep resonant excitation of a multilevel system in a multifrequency laser light field and (b) the effect of the decay of one of the intermediate states to a metastable state on the excitation efficiency.

resonant quantum transition is so strong that the population probability of the final state f reaches its maximum:

$$n_i/g_i = \cdots = n_k/g_k = \cdots = n_f/g_f, \tag{2.95}$$

where g_k is the statistical weight or degeneracy of the kth level. The partial population of the level f will then be defined by the simple relation

$$n_f = g_f/\sum_k g_k. \tag{2.96}$$

As far as the maximum population of the final state is concerned, it is advantageous to exite such a sequence of quantum levels as has its level degeneracies increasing progressively. For instance, in the case of a three-step atomic excitation, we can select a sequence of dipole quantum transitions with progressively increasing orbital moments (S → P → D → F), for which the level degeneracies grow higher as follows: $g_1 = 1$, $g_2 = 3$, $g_3 = 5$, $g_4 = 7$. In this case, the partial population of the final level will be equal to $\frac{7}{16}$ or 44%.

It frequently proves possible to effect multistep excitation where the final state suffers decay into the ionization continuum concurrently with excitation. This is, for example, the case with systems in which the final state f either is an autoionization state or lies in the continuum, so that the last quantum transition $n \to f$ is a "level" → "continuum" transition. In this case, if the ionization decay rate is high enough, i.e.,

$$W_{\text{ion}} \gtrsim 1/\tau_p, \tag{2.97}$$

there occurs depletion of the final state f and, by virtue of saturation condition (2.94), of the initial and intermediate levels as well. As a result, a 100% ionization of the multilevel quantum system is ensured.

Situations in which condition (2.93) cannot be satisfied for all intermediate states are not uncommon. For example, the first excited level often decays during the time the exciting pulse lasts, i.e., $T_1^{(k)} \lesssim \tau_p$. In this case, the saturation of the quantum transitions $k \to n$ and $i \to k$ depends on the radiation intensities on these transitions and not on the fluence. To ensure saturation, the following condition, similar to (2.94), must be satisfied:

$$I(\omega_{ik}) \gtrsim I_{\text{sat}}^{ik} = 1/\sigma_{ik} T_1^{(k)}. \tag{2.98}$$

As a rule, satisfying this condition does not impose too heavy demands on the intensity, for the rapid radiative decay $i \to k$ is accompanied by an increase in the excitation cross section σ_{ik}.

However, there are cases where the intermediate state does not decay to the initial quantum state, but rather to some longer-lived additional state (Fig. 2.13b). As a result, there is a certain probability that the system will fall out of resonance with the multiple-frequency radiation, and this limits the excitation probability of the final state f. The extent to which the population of the level f is reduced depends on the ratio between the rates of decay of the level k to the initial and metastable states:

$$\zeta_{\text{dec}} = [1/T_1^{(km)}]/[1/T_1^{(ki)} + 1/T_1^{(km)}]. \tag{2.99}$$

It often proves impossible to attain saturation at all excitation stages. Such situations may arise for various reasons. Sometimes, excitation is effected through an intercombination transition whose excitation cross section is smaller than that of the allowed transitions by two or three orders of magnitude. Because of the low pulse energy of tunable lasers with a narrow emission line, satisfying the saturation condition (2.94) for such a weak transition may turn out to be impossible. The same applies to the last transition to the continuum, whose cross section is especially small (see Chapter 3). Finally, a strong saturation of a quantum transition inevitably broadens the absorption line. This may lead to a distortion of the absorption spectrum measured on some intermediate transition $k \to n$ while varying the laser frequency Ω_{kn}. In this case, it is advisable to reduce the laser-pulse fluence or intensity to such a level as causes no strong saturation. In this way the absorption spectrum distortion can be eliminated, but at a sacrifice to the excitation probability is reduced is defined by the simple relation

$$\zeta_{\text{sat}} \simeq I(\omega_{kn})/I_{\text{sat}}^{kn} \quad \text{or} \quad \Phi(\omega_{kn})/\Phi_{\text{sat}}^{kn}, \tag{2.100}$$

where

$$I(\omega_{kn}) \ll I_{\text{sat}}^{kn} \quad \text{or} \quad \Phi(\omega_{kn}) \ll \Phi_{\text{sat}}^{kn}.$$

The physics of multistep excitation of multilevel systems is, of course, much more profound than may be inferred from the simplest qualitative considerations given above. Here we have restricted ourselves to the simplest incoherent interaction approximation. For multilevel systems with a noticeable level degeneracy, this approximation is quite acceptable for estimation purposes, because the degeneracy largely smoothes out coherent oscillations. Nevertheless, an adequate quantitative theory of multistep excitation of multilevel quantum systems in multiple-frequency fields is still to be developed.

2.4. Multiphoton Resonant Excitation of High-Lying States

For some atoms (H, He, halogens, and noble gases) and simple molecules, the first excited electronic state lies fairly high (above 6 eV). To effect resonant excitation of such atoms and molecules requires a laser radiation in the vacuum ultraviolet region, which is not quite mastered yet by laser technology. An alternative possibility is the multiphoton excitation of such high-lying states by means of longer-wave but sufficiently powerful laser radiation. The two-photon excitation was already considered in Section 2.2. What we have in mind here is the case of resonant excitation involving the absorption of three or more photons (Fig. 2.14).

Higher-order multiphoton processes were extensively studied in the works concerning the multiphoton ionization of atoms (for reviews, see Delone, 1975; Lambropoulos, 1976; Delone and Krainov, 1984). The direct multiphoton ionization on the transition from a discrete ground state to a final state in the continuum features no clearly defined resonance, and so it is of no interest to resonant photoionization spectroscopy. However, there is an important exception—multiphoton ionization via an intermediate resonance, as shown in Fig. 2.14b. In this case, the ionization process includes the resonant k-photon excitation of a high-lying discrete resonant level r and the subsequent m-photon (usually $m = 1$ or 2) nonresonant ionization of this discrete level. This process was considered in the very first works on multiphoton atomic ionization, as it lowers the order k_0 of the direct ionization of the atom to the continuum c.

The dependence of the ionization probability on the field frequency Ω and amplitude $\mathscr{E}$ in the k_0-photon ionization through the intermediate k-photon

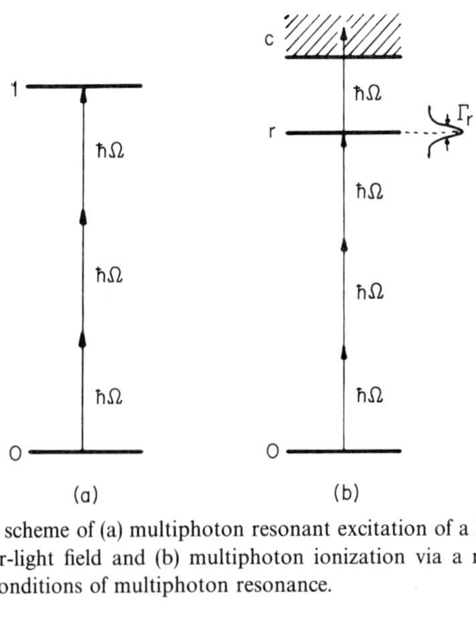

FIG. 2.14. General scheme of (a) multiphoton resonant excitation of a high-lying level 1 in a single-frequency laser-light field and (b) multiphoton ionization via a resonant intermediate state excited under conditions of multiphoton resonance.

resonance has the form (Keldysh, 1964)

$$W_{0c}^{(k_0)}(\mathscr{E}, \Omega) = \frac{|V_{0r}^{(k)}(\mathscr{E})|^2}{|\hbar\Delta_{0r}(\mathscr{E})|^2 + |\hbar\Gamma_r(\mathscr{E})|^2} W_{rc}^{(k_0-k)}, \tag{2.101}$$

where $\Delta_{0r} = |E_r(\mathscr{E}) - E_0(\mathscr{E}) - k\hbar\Omega|/\hbar$ is the detuning of the field frequency Ω from k-photon resonance and $\Gamma_r(\mathscr{E})$, the width of the excited level, which in conditions of multiphoton excitation by a strong field is determined by the broadening due to the m-photon ($m = k_0 - k$) ionization:

$$\Gamma_r(\mathscr{E}) = W_{rc}^{(k_0-k)}. \tag{2.102}$$

The quantity $V_{0r}^{(k)}(\mathscr{E})$ in Eq. (2.101) is the matrix element of the k-photon transition $0 \to r$:

$$V_{0r}^{(k)} = \sum_{s_1 \cdots s_{k-1}} \frac{V_{0s_1} V_{s_1 s_2} \cdots V_{s_{k-1} r}}{[(k-1)\Omega - \omega_{s_1 0}][(k-2)\Omega - \omega_{s_2 0}] \cdots [\Omega - \omega_{s_{k-1},0}]\hbar^{k-1}} \tag{2.103}$$

where summation is extended over $(k-1)$ intermediate states s, $V_{s_1 s_2} = \mu_{s_1 s_2}\mathscr{E}$, and $\mu_{s_1 s_2}$ is the matrix element of the transition dipole moment between the states s_1 and s_2. The k-photon transition matrix element $V_{0r}^{(k)} \sim \mathscr{E}^k$.

The probability of the k-photon transition $0 \to r$, described by the first multiplier of Eq. (2.101), may be represented in the following equivalent form:

$$W_{0r}^{(k)} = \frac{|V_{0r}^{(k)}(\mathscr{E})|^2}{|\hbar\Delta_{0r}(\mathscr{E})|^2 + |\hbar\Gamma_r(\mathscr{E})|^2} = \sigma^{(k)} I^k, \tag{2.104}$$

where I is the radiation intensity and $\sigma^{(k)}$ the cross section of the k-photon transition.

Multiphoton excitation requires a very strong field, which inevitably perturbs the atomic states. This was already discussed for two-photon excitation (Section 2.2), but for multiphoton excitation of a higher order, the perturbation is much stronger if between the levels 0 and r there are no intermediate transitions whose energy is close to $\hbar\Omega$. To illustrate, Fig. 2.15 shows spectra of the four-photon ionization of the Cs atom in the presence of a three-photon resonance (Lompre *et al.*, 1978) for several laser intensity values in the range 2×10^8 to 2×10^9 W/cm^2. The broadening of the three-photon resonance and the shift of its frequency toward the short-wavelength side are clearly seen. The broadening is mainly due to the ionization broadening of the intermediate $r = $ 6F state. At a typical photoionization cross section of $\sigma_i \simeq 10^{-17}$cm^2 on the transition to the continuum and a laser intensity of $I = 10^9$ W/cm^2, the broadening (2.102) is

$$\Gamma_r = \mathrm{W}_{rc} = \sigma_i I \simeq 6 \times 10^{10}\ \mathrm{s}^{-1}. \tag{2.105}$$

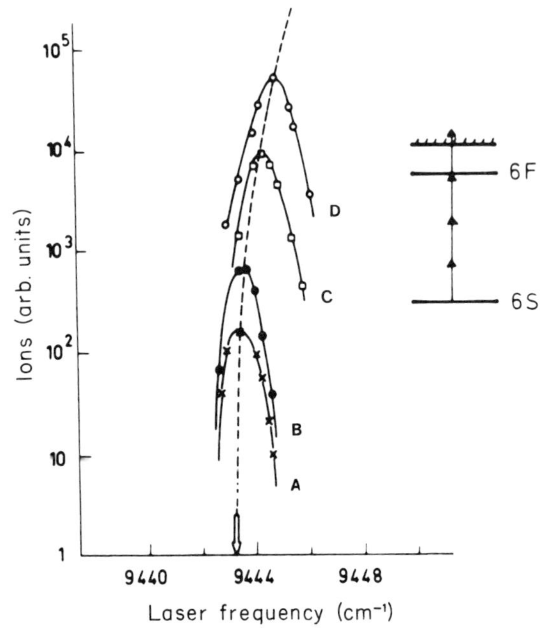

FIG. 2.15. Four-photon ionization of Cs via an intermediate resonant level. Multiphoton ionization yield as a function of laser frequency in the vicinity of the 6S → 6F three-photon resonance (the arrow indicates the exact position of the resonance in a weak field) at various laser intensity values (in 10^8 W/cm^2): (A) 1.8; (B) 5.2; (C) 11; and (D) 15.7. [From Lompre *et al.* (1978).]

This broadening limits the selectivity of resonant excitation. It becomes especially noticeable where it is necessary to ensure a high probability of excitation of a resonant state. It is apparently for this reason that it proves impossible to achieve coherent population oscillations on the resonant transition $0 \leftrightarrow r$.

Where the multiphoton excitation of an atom is effected by means of a wideband laser pulse containing many axial modes, in evidence is also an asymmetric distortion of the multiphoton resonance in the form of a long short-wavelength wing of the spectral line. Qualitatively, it is explained by the level shifts due to the Stark effect in the multimode field featuring intensity fluctuations.

3. Ionization of Excited Atoms

Several schemes may be used to carry into effect the resonant ionization of atoms. A feature common to all such schemes is the preliminary selective excitation by laser radiation of one or several intermediate atomic levels and subsequent ionization of the excited atoms alone. Resonant excitation has been dealt with in the preceding chapter. The present chapter discusses the next step: the ionization of excited particles. Let us first consider a most simple case: the ionization of atoms. The molecular case substantially differs from the atomic and will be treated individually later (Chapters 5 and 6).

Qualitatively, the basic photoionization schemes may be classified as illustrated in Fig. 3.1. The excited states that are far from the ionization limit

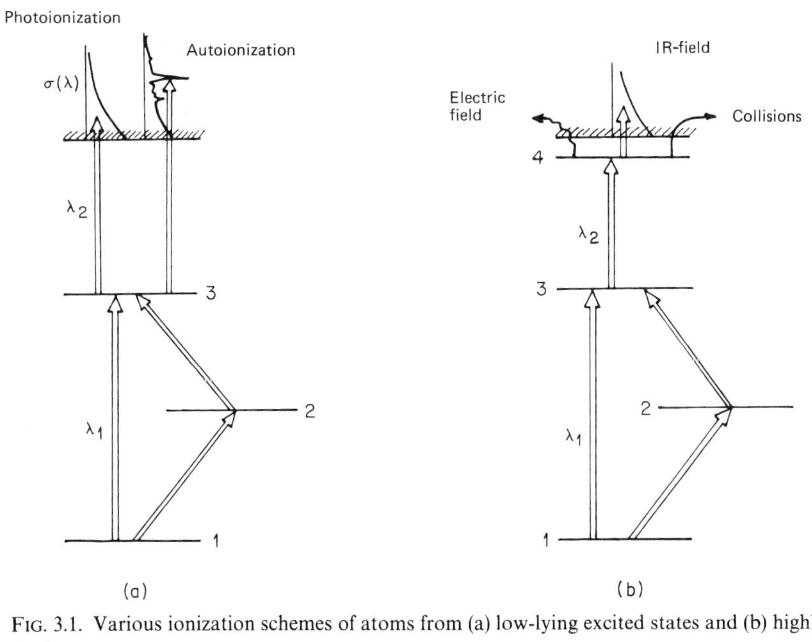

FIG. 3.1. Various ionization schemes of atoms from (a) low-lying excited states and (b) highly excited states.

(Fig. 3.1a) can be effectively ionized only by laser radiation. The following two possibilities exist in this case: direct nonresonant photoionization on transition to the continuum and resonant photoionization on transition to an autoionization state. The highly excited (Rydberg) states lying close to the ionization limit are comparatively easy to ionize with a high efficiency by one of the following means (Fig. 3.1b): an electric field pulse, an IR radiation pulse, and collisions with other particles or with walls. The ionization schemes making use of the Rydberg and autoionization states apparently involve a greater total of resonant excitation steps between the ground state and the ionized state.

All these schemes have already been tested experimentally, particularly for the isotope-selective ionization of atoms, and will be considered in detail in the subsequent sections of this chapter.

Before going into the details of atomic multistep photoionization schemes, consider some characteristics of these schemes from the viewpoint of the realization of their ultimate values. These characteristics include the required energy fluences of laser pulses, ionization yield, and selectivity of ionization.

In accordance with expressions such as Eq. (2.94), the minimum *energy fluence* at the ionization state is required for the schemes making use of those resonant transitions from the excited state upward for which the ionization cross section is a maximum. With the proper choice of a sequence of transitions and intermediate quantum states, it is possible to attain a situation in which all of the cross sections of successive transitions lie in the region $\sigma_1, \sigma_2, \ldots > 10^{-15}\ \text{cm}^2$ and the critical energy fluxes of the pulse are rather small: $\Phi_1, \Phi_2 \cdots < 10^{-5}$–$10^{-4}\ \text{J/cm}^2$. Direct photoionization of excited atoms to the continuum has a very small cross section and, from this point of view, is far from being optimal.

The *photoionization yield* reaches a maximum when all the successive transitions are simultaneously saturated by several laser pulses. If, during the laser pulse, the final excited state of a multistep process decays into a continuum, it is possible to completely deplete the ground level through excitation and thus achieve an ionization yield of 100%. Such conditions can be realized for all of the schemes illustrated in Fig. 3.1, except the last one in which the highly excited state is ionized by an electric field. If the pulse of the electric field is switched on after the laser pulses are over, only the atoms in the last excited state are ionized 100%. High saturation of all the successive quantum transitions with laser pulses equalizes the populations of all of the atomic levels (initial, intermediate, and final). Only a fraction of the atoms reaches the final state and can be ionized by the electric field. It should be noted that the use of a continuous electric field to deplete the final state is complicated by a strong Stark effect for a highly excited state.

In some cases, for example, in experiments with very rare isotopes, the *selectivity* that can be obtained in the first step is not sufficient. Here the subsequent use of resonant excitation of highly excited states and of autoionization states is of particular interest.

This short general comparison of the different schemes for multistep selective photoionization of atoms shows that, as far as the achievement of the maxima of certain characteristics is concerned, those schemes that make use of the narrow autoionization states are of particular interest. Minimum requirements on the energetics of the laser pulse make it possible to achieve a photoionization yield of 100% and a very high photoionization selectivity at the same time only for atoms with narrow autoionization states. For example, this does not apply to atoms of the alkali elements. If a lower ionization yield (30–50%) is assumed, a universal scheme for the ionization of highly excited (Rydberg) states by pulsed electric fields can be used.

Finally, in certain cases the atoms should not be photoionized in a beam but rather in a buffer gas. This allows a lower rate of laser pulses with higher energy fluxes to be used. The simplest schemes for two- and three-step photoionization to the continuum can therefore be used. In addition, it should be noted that the multiphoton resonant excitation of the last quantum state, which requires high exciting laser-pulse intensities, automatically provides for a highly effective photoionization by the same powerful laser pulse of the atoms at this level on direct transition to the continuum. This feature has already been mentioned in Section 2.4 (see Fig. 2.14).

3.1. Kinetics of Stepwise Photoionization

To realize the importance of matching the excitation and ionization rates, consider first the kinetics of the simplest case of two-step photoionization.

3.1.1. Kinetics of Two-Step Photoionization

The excitation of atoms and their photoionization can be treated by approximating the simple rate equations for the density N_1 in the ground state 1, the density N_2 in an intermediate excited state 2, and the ion density N_{i}:

$$\frac{dN_1}{dt} = -W_{12}(N_1 - N_2) + \frac{1}{\tau_{12}} N_2; \qquad N_1(t = 0) = N_0, \quad (3.1a)$$

$$\frac{dN_2}{dt} = W_{12}(N_1 - N_2) - W_{\mathrm{i}} N_2 - \frac{1}{\tau_{12}} N_2; \qquad N_2(t = 0) = 0, \quad (3.1b)$$

$$\frac{dN_{\mathrm{i}}}{dt} = W_{\mathrm{i}} N_2. \quad (3.1c)$$

For simplicity's sake, the statistical weights of the ground and excited states have been assumed to be equal, and the decay of the excited level only takes place to the ground state. When the rates W_i and W_{12} are much larger than the relaxation rate of the excited atoms $1/\tau_{12}$, and the duration of the laser pulse at the frequencies Ω_1 and Ω_2 is much shorter than the relaxation time τ_{12}, the radiative relaxation in (3.1) may be neglected. The synchronized pulses of the exciting and ionizing radiation are assumed to be rectangular and are switched on at $t = 0$. With these assumptions the following expression for the total number of ions can be derived from (3.1):

$$\frac{N_i}{N_0} = 1 - \frac{a_1 + a_2}{2a_2} \exp[-(a_1 - a_2)t] + \frac{a_1 - a_2}{2a_2} \exp[-(a_1 + a_2)t], \quad (3.2)$$

where a_1, a_2 are given by the equations

$$a_1 = W_{12}[1 + (W_i/2W_{12}); \qquad a_2 = W_{12}[1 + (W_i/2W_{12})^2]^{1/2}. \quad (3.3)$$

The exponents in Eq. (3.2) differ greatly. The maximum of the ratio $(a_1 - a_2)/(a_1 + a_2)$ can be obtained with $W_i/W_{12} = 2$ and equals 0.18. The coefficients in Eq. (3.2) just slightly depend on the ratio W_i/W_{12}, and the maximum of $(a_1 + a_2)/2a_2$ (approximately equal to 1.2) can also be obtained with $W_i/W_{12} = 2$.

Thus, the characteristic time for the ionization of most atoms is given by the following equation (Ambartzumian *et al.*, 1976a):

$$\tau_i = \frac{1}{a_1 - a_2} = \frac{1}{W_{12}} \left\{ 1 + \frac{W_i}{2W_{12}} - \left[1 + \left(\frac{W_i}{2W_{12}} \right)^2 \right]^{1/2} \right\}^{-1}. \quad (3.4)$$

A maximum rate for two-step photoionization or a minimum value for τ_i can be obtained with $W_2 = 2W_{12}$. Hence, the conversion of each excited atom into an ion will be optimal under the following conditions:

(1) Negligible relaxation of atoms from excited states:

$$W_{12}, W_i \gg 1/\tau_{12}. \quad (3.5)$$

(2) Matching of the excitation and ionization rates:

$$\frac{W_i}{2W_{12}} = \frac{\sigma_{2i} I_2}{2\sigma_{12} I_1} = 1 \quad \text{or} \quad I_2/I_1 = 2\sigma_{12}/\sigma_{2i}. \quad (3.6)$$

To ionize all of the excited atoms, the intensity of the ionizing radiation I_2 must be approximately $(2\sigma_{12}/\sigma_{2i})$ times higher that of the exciting radiation.

It is possible to achieve two-step ionization using short radiation pulses whose durations $\tau_p^{(1)}, \tau_p^{(2)}$ are shorter than the relaxation time τ_{12} of the population of the intermediate state. These pulses should be accurately synchronized in time to within $\Delta\tau$, which should also be shorter than τ_{12}. To get a considerable portion of the atoms excited and subsequently photoionized, the energy flux of the pulse Φ_1, Φ_2 (photons/cm^2) must comply with conditions similar to Eqs. (3.6) and (1.8).

$$\Phi_1 \gtrsim \Phi_{\text{sat}}^{(1)} = (2\sigma_{12})^{-1}, \tag{3.7}$$

$$\Phi_2 \gtrsim \Phi_{\text{sat}}^{(2)} = (\sigma_{2i})^{-1}. \tag{3.8}$$

3.1.2. Coherent Effects in Photoionization

Strictly speaking, the simple calculations based on rate equation approximation hold true only if the effects of coherent interaction between the laser-light field and the atom can be neglected. The conditions of coherent and incoherent interaction for the simplest case of a two-level system have been presented earlier in the text (Section 2.1.2). The validity of the rate equation approximation for the interaction of the field with multilevel systems has been considered in many papers; see, for example, the treatment of a three-level system (Letokhov and Chebotayev, 1977, Chapter 5) and a multistep excitation (Ackerhalt and Eberly, 1976; Salomaa and Stenholm, 1978).

This approximation describes the atom–field interaction well enough since either the atom loses coherency during its interaction with the field (for example, due to level degeneracy or a rapid decay of the final state) or the field of the light is multimode and hence its time of coherency is shorter than the duration of the laser pulse. If more precise calculations are required, the distribution of atoms over the many sublevels, particularly degenerate sublevels, and the atomic velocity distribution within the region of interaction with the field should be taken into account. Attempts are being made in the literature (Chang, 1980) to improve the accuracy of the calculations. Only when these have been done will it be constructive to allow for the more delicate effects of the coherent radiation–atom interaction. Some of the possible effects of this kind are discussed below.

There are two possibilities for the occurrence of coherence effects in stepwise atomic photoionization. First, the last excited energy level from which the atoms become photoionized (for example, state 3 in Fig. 3.1) may consist of several overlapping sublevels. In principle, during the pulsed excitation of such a level the atom may be in a so-called superposition state which causes "quantum beats". This may result in an oscillating time dependence of the photoionization yield, i.e., bring about quantum beats in the photocurrent. Such a possibility was already discussed (Letokhov, 1966) and has been realized in some experiments with Rydberg Na atoms (Leuchs and Walther, 1979; Leuchs *et al.*, 1979). The coherent beats observed in the yield and angular distribution of photoelectrons agree well with the theory (Georges and Lambropoulos, 1978; Chien *et al.*, 1983). The additional capabilities of laser photoionization spectroscopy, associated with measurements of photoelectron parameters (angular distribution, energy), will be considered in Chapter 11.

The second cause for the occurrence of coherence effects lies in the fact that the last photoionized quantum state may have a population oscillating in time because of the oscillations taking place at the Rabi frequency between this level and the initial state. This is possible for two-level and multilevel systems with discrete quantum levels if the Rabi frequency γ is much larger than the rate of phase relaxation in each quantum state (see Chapter 2). A simple case of a two-step photoionization is considered in accordance with the scheme in Fig. 3.1a in which atomic levels 1 and 2 are equally degenerate. Let us assume that the amplitude of the single-mode coherent field $\mathscr{E}$ is large enough to induce oscillations with a Rabi frequency γ between levels 1 and 2, i.e., condition (2.34) is valid.

The probability of the atoms being in the intermediate state, averaged over the oscillation period γ^{-1}, is equal to $\frac{1}{2}$, which corresponds to a strong saturation of the transition. This effect provides nothing new as far as the probability of photoionization, averaged over an oscillation period, is concerned. However, the probability of photoionization and hence the photocurrent will oscillate at the frequency γ. This oscillation is only observable when

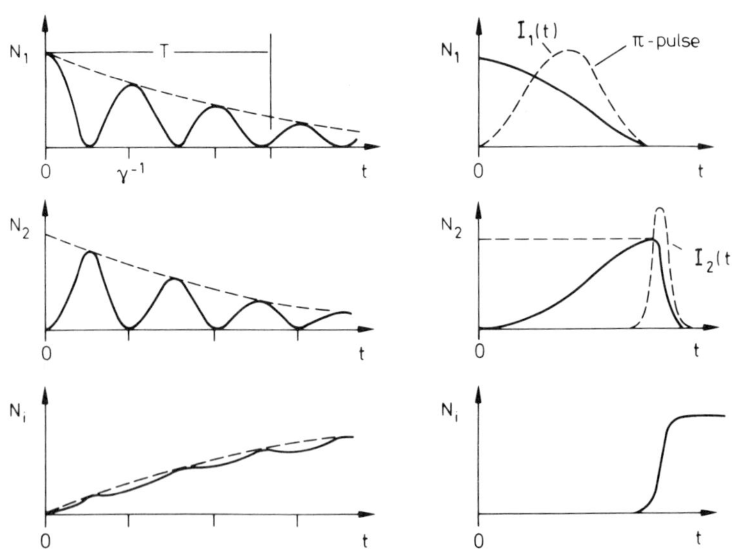

FIG. 3.2. Coherent effects of excitation of the intermediate state in two-step atomic photoionization. (a) Photocurrent oscillations with $W_{ion} \ll \gamma$ and (b) the possibility of complete atomic photoionization with an intermediate state excited by a π pulse. [From Letokhov (1977b).]

the photoionization rate W_i is smaller than the oscillation frequency γ (Fig. 3.2). Otherwise, the superposition atomic state decays quickly because of the high probability of a transition into the continuum. In this case the oscillations decay with the characteristic time $\tau_i = 1/W_i$.

There is another limiting case of some interest (Fig. 3.2b). Assume that the atomic transition $1 \rightarrow 2$ is acted on by a short "π pulse" that inverts the population of the level. If, after full inversion of the population, the atom is irradiated with a powerful pulse of photoionizing radiation, all of the atoms of a particular sort can be photoionized (Letokhov, 1977b). Numerical calculations show the usefulness of the coherent excitation by short laser pulses compared with the incoherent one (Diels, 1976). With photoionization, the coherent oscillations of the final discrete level population manifest themselves in the photoelectron spectrum as well (Rzazewski, 1983).

3.2. Photoionization of Excited Atoms

Let us consider now the various methods for ionization of excited levels in greater detail and direct our attention to the schemes of photoionization from those excited levels that are relatively far from the ionization limit and cannot be considered as being highly excited (Fig. 3.1).

3.2.1. Photoionization Cross Section from Excited States to the Continuum

Cross sections for photoionization from the ground state are well known for many atoms (Marr, 1967; Berkowitz, 1979), but there has been comparatively little information in the literature about the cross sections for photoionization from excited states, until investigations were started into the multistep photoionization of atoms by laser radiation. The output intensity of lasers and their narrow tunable radiation spectrum are quite adequate for selective excitation of any atomic level, and subsequent photoionization from a given excited state can help to measure the cross section of this process.

3.2.1.1. Measurement of Photoionization Cross Section by the Method by Quantum State Depletion. The method of multistep photoionization and, particularly, the observation of the dependence of the photoionization yield on the energy flux of the photoionizing laser pulse simplify the absolute measurement of the photoionization cross section. This method of measurement has been developed and used with success by several authors (Zeman, 1976; Heinzmann *et al.*, 1977; Nygaard *et al.*, 1975; Ambartzumian *et al.*, 1976a,b; Bradley *et al.*, 1976).

With a sufficiently high rate of excitation, $\sigma_{12}\Phi_1 \gg 1$, the transition that is being excited from $1 \rightarrow 2$ will be saturated, and, as seen from (3.2), the density of ions will be determined by the expression

$$N_i \simeq N_0[1 - \exp(-\tfrac{1}{2}\sigma_{2i}\Phi_2)]. \tag{3.9}$$

It is seen from (3.9) that the density of the atoms and the energy density of the ionizing radiation must be known in order to determine the photoionization cross section σ_{2i} of the excited state 2. This method does not require that the density of excited atoms be known. The ionization cross section is determined by a comparison of the theoretical dependence (3.9) and the experimental curve $N_i(\Phi_2)$.

This method was used to measure the photoionization cross section of Cs atoms in the excited states $7^2P_{3/2}, 7^2P_{1/2}$ (Zeman, 1976) and $6^2P_{3/2}, 6^2P_{1/2}$ (Nygaard *et al.*, 1975), and of Rb atoms in the states $6^2P_{3/2}, 6^2P_{1/2}$ (Ambartzumian *et al.*, 1976a,b), as well as the cross section of the autoionization resonance of Mg for the transition $3s3p(^1P_1^0) \to 3p^2(^1S_0)$ (Bradley *et al.*, 1976).

Let this method be illustrated for Rb atoms. The energy levels of the Rb atom are shown schematically in Fig. 3.3a. The ionization energy for Rb is $E_i = 4.18$ eV. Rubidium was excited to the 6^2p state by a tunable dye laser that was pumped by the second-harmonic radiation of a *Q*-switched ruby laser. The output energy of the dye laser was 2×10^{-5} J with a pulse duration of 20 ns and a spectral width of 0.5 cm^{-1}. The exciting radiation ν_1 was tuned to the transition $5^2S_{1/2} \to 6^2P_{3/2}, 6^2P_{1/2}$, whose frequencies are 23,799 and 23,725 cm^{-1}, respectively. The subsequent ionization of the excited atoms was done either by a ruby laser ($\nu_2 = 14{,}403$ cm^{-1}) or by its second harmonic. The output energy of the ruby laser was 0.5 J, with a pulse duration of 20 ns. Figure 3.3b shows the dependence of the total ion yield on the energy flux of the ionizing radiation energy Φ_2. The intersection of the linear part of this curve with the plateau (point A in Fig. 3.3b) corresponds to the energy flux Φ_2 for which the ionization of almost every excited atom occurs. The photoionization cross-section for the 6^2P states was around $(1.5\text{–}1.7) \times 10^{-17}$ cm^2 at $\lambda_2 = 6943$ Å and decreased by approximately an order of magnitude at the wavelength of the second harmonic. These numerical values agree well with the results of calculations by the Burgess–Seaton formulas (Burgess and Seaton, 1960).

3.2.1.2. Frequency Dependence of the Photoionization Cross Section. The atomic photoionization cross section drops drastically when the photoionization energy $\hbar\omega_2$ considerably exceeds the ionization limit. By changing the photon energy $\hbar\omega_2$, we can measure the frequency dependence of the cross section. Such measurements were carried out on Cs atoms in the lowest excited states $6^2P_{3/2}$ and $6^2P_{1/2}$ (Nygaard *et al.*, 1975). The ionization potential of Cs equals 3.89 eV. The exciting radiation was tuned to be in resonance with the transitions $6^2S_{1/2} \to 6^2P_{3/2}, 6^2P_{1/2}$ ($\lambda_1 = 8521$ and 8944 Å,

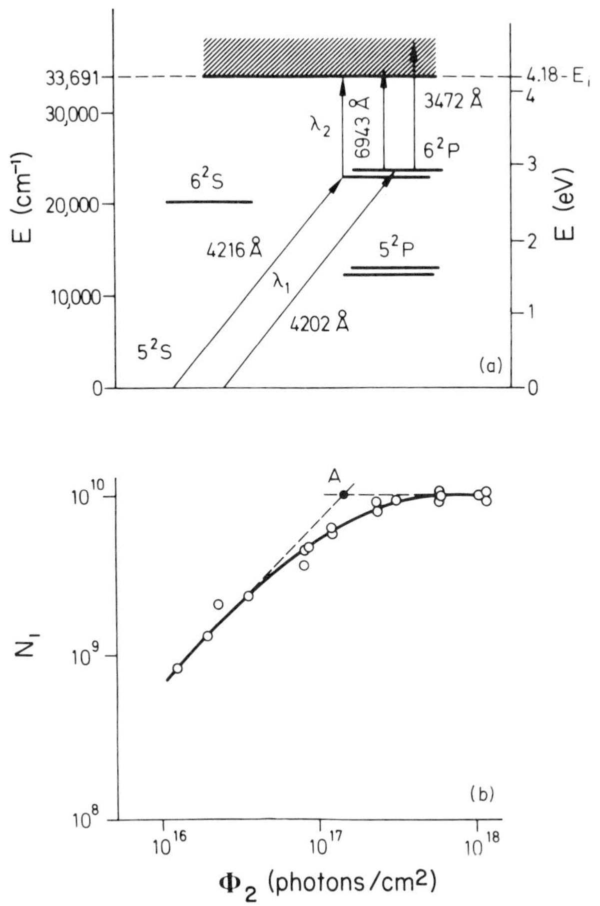

FIG. 3.3. Measurement of the photoionization cross section of Rb atoms through the laser-induced depletion of the excited states. (a) Partial energy level diagram of rubidium with transitions for two-step photoionization and (b) the total photoion signal as a function of the energy flux of the ionizing radiation of the ruby laser. [From Ambartzumian *et al.* (1976a).]

respectively). The subsequent ionization was caused by radiation in the 2500–4944-Å range (λ_2). A cesium lamp was used as the source of the exciting radiation. A mercury–xenon lamp with a grating monochromator was used for the subsequent ionization. Figure 3.4 shows the experimental and theoretical dependence of the photoionization cross section of Cs in the $6^2P_{3/2}$ state on the ionizing radiation.

A smooth frequency dependence, such as the one illustrated in Fig. 3.4 is found only for atoms with one valence electron. For atoms with several valence electrons, the pattern changes greatly. Instead of a smooth frequency

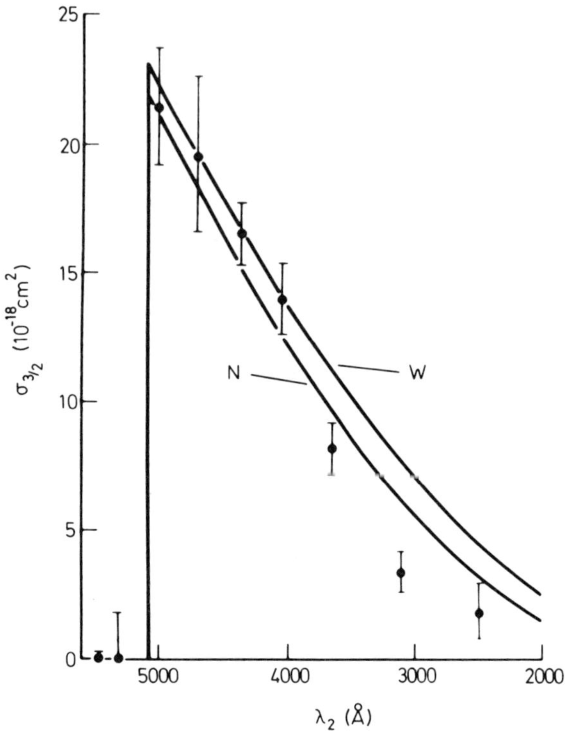

FIG. 3.4. Photoionization cross section from the $6^2P_{3/2}$ state of Cs σ_{2i} in units of 10^{-18} cm^2 as a function of wavelength; (N) Norcrosss and Stone (1966) (W), Wiesheit (1972). [From Nygaard *et al.* (1975).]

dependence, sharp resonances and minima are observed. As an illustration, Fig. 3.5 shows the frequency dependence of the photoionization cross section of the uranium atom in a certain excited state (Janes *et al.*, 1976). The exciting radiation (λ_1) was generated by a narrowband (spectral width $\simeq 0.02$ Å) dye laser pumped by a nitrogen laser. It was tuned to the 4266.324-Å resonance line of the uranium atom, which corresponds to a photon energy $\hbar\omega_1 = 2.905$ eV. The excited atoms were then ionized by a similar dye laser whose wavelength (λ_2) varied from 3605 to 3830 Å. The experimental results are given in Fig. 3.5. The dashed curve corresponds to the spectral width of the ionizing radiation, equal to 2 Å. The ionization limit of the excited state (Fig. 3.5) is 3777 ± 2 Å, which corresponds to a photon energy $\hbar\omega_2 = 3.282 \pm 0.002$ eV. The ionization potential of uranium was found to be $E_i = 6.187 \pm 0.002$ eV. The solid curve (Fig. 3.5) corresponds to a 0.6-Å spectral width for the ionizing radiation. There is a sharp peak at the ionization

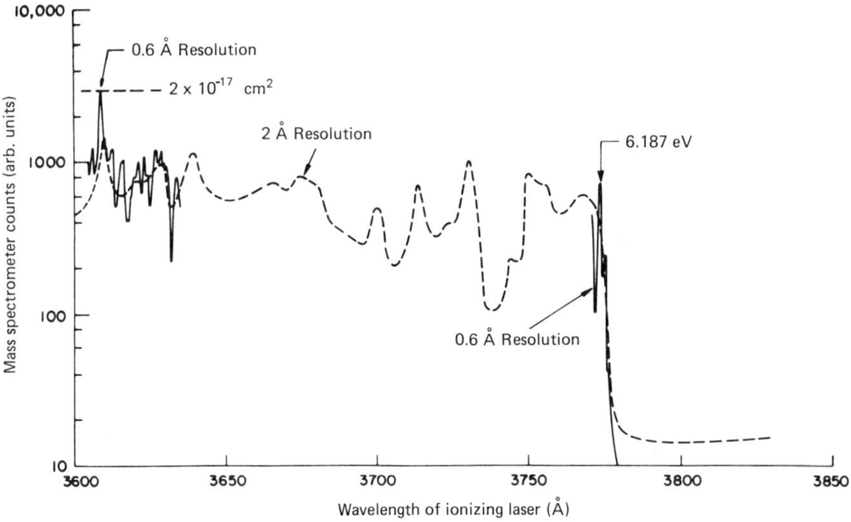

FIG. 3.5. Photoionization spectrum of uranium from a state excited by laser radiation at the 4266.324-Å uranium resonance line. [From Janes *et al.* (1976).]

wavelength 3609 Å. Its maximum corresponds to an ionization cross section of 2×10^{-17} cm$^2(\Delta\omega_L/\Delta\omega_a)$, where $\Delta\omega_L$ is the laser linewidth and $\Delta\omega_a$ is the linewidth of the optical transition. The presence of sharp resonances in the ionization spectrum of uranium shows that autoionization often takes place.

Thus, the preceding experiments have shown that the ionization cross section ($\sigma_{2i} = 10^{-17}$ cm^2) is several orders of magnitude smaller than the excitation cross section ($\sigma_{12} = 10^{-11}$–10^{-13} cm^2) (Corliss and Bozman, 1962). It follows that the criterion for efficient two-step ionization (3.6) is different to fulfill for direct transition to continuum and that a search for ways of increasing ionization cross sections should be made. Autoionization of excited atoms is one of these ways.

3.2.2. Autoionization Resonances of Excited States

Autoionization states are bound atomic states whose energy levels relative to the outer valence electron lie above the ionization limit. Such states usually arise when the internal shell is excited. In multielectron atoms we can observe so-called shifted terms that are formed by the simultaneous excitation of two or more valence electrons. With rather strong excitation, we can observe configurations whose energy levels are above the ionization limit for one-electron excitation, i.e., such levels also become autoionized. Since there are continuous states above the ionization limit, it is probable that an atom excited to a discrete state, whose energy is higher than the ionization potential, will transfer to the continuum and be ionized. This is the essence of autoionization.

It is obvious that there are two ways for the decay of autoionization states: radiative and autoionization, i.e., nonradioative decay with a loss of electrons. The autoionization channel reduces the lifetime of the discrete state and causes its broadening. Since autoionization can be considered to be the result of the superposition of continuous and discrete states, the probability of decay increases with an increase in the coupling of the autoionization state and the continuum.

The autoionization resonances that occur when atoms are excited from their ground state into the continuum were studied both theoretically and experimentally. A rigorous theory of such autoionization resonances was proposed by Fano (1961). The presence of a discrete level in the continuum causes interference, the shape of the autoionization resonance becomes asymmetric, an absorption minimum occurring in the low-frequency wing of the resonance (Fano and Cooper, 1965; Fano, 1969). Where a strong light field that induces transitions to the continuum is present, the shape of autoionization resonances alters, and new autoionization-like resonances of the same interference origin appear (Geller and Popov, 1981).

The development of the methods of laser spectroscopy, particularly the multistep photoionization of atoms with lasers, has allowed systematic studies to be made of the autoionization states of multielectron atoms whose bound states have a complicated spectrum. Having a choice of different exciting transitions makes it possible to study autoionization states of different parity. At the same time, with the use of atomic absorption spectra, only those autoionization states that combine with the ground state can be studied, i.e., those states that have a parity opposite to that of the ground state and a total momentum that is in tune with the selection rules for this optical transition. Until recently, comparatively narrow autoionization states have been observed in the atoms of such noble gases as Ar, Kr, and Xe ($\Gamma_{a.i.} \simeq 10\ cm^{-1}$) (Stebbings, *et al.*, 1975a), the alkaline-earth elements—for example, the spectrum of Ba ($\Gamma_{a.i.} \simeq 10\ cm^{-1}$) (Cooke and Gallagher, 1978; Wynne and Herman, 1979), the atomic spectrum of uranium ($\Gamma_{a.i.} \simeq 3\ cm^{-1}$) (Janes *et al.*, 1976), and some lanthanides ($\Gamma_{a.i.} \simeq 1\text{–}2\ cm^{-1}$) (Worden *et al.*, 1978b). The widths of these autoionization states may vary within 1 or 2 orders of magnitude, and at the same time they tend to decrease for heavier atoms.

The methods of laser spectroscopy substantially widen the scope of the research of very narrow autoionization states. This can be illustrated by the long-lived autoionization state (its lifetime is about 0.5 ns, $\Gamma_{a.i.} \simeq 0.05\ cm^{-1}$) of the Gd atom (Bekov *et al.*, 1978b). Besides being of interest in the physics of multielectron atoms, very narrow autoionization states are also of great interest in the selective multistep photoionization of atoms by lasers. The cross section for the excitation to such a narrow autoionization state may be as large as 10^{-15} cm^2. This value is several orders of magnitude higher than that for the cross section for the direct photoionization to the continuum. This makes a 100% absolute yield possible.

Before considering the experiments on very narrow autoionization states, the physics of their formation in multielectron atoms such as Yb will be discussed.

The relevant electronic configuration of Yb is $4f^{14}6s^2 (Z = 70)$. The 6s electrons are valence electrons while the rest, including those in the 4f shell, form the core. There are two types of autoionization states for this system of electrons. The first type can be obtained from the ground state by exciting the outer shell of the core. This causes an electron from the adjacent shell to occupy the vacant place and, at the same time, another electron to break away (Auger effect). The second type results from the double excitation of the valence shell. As a result of the autoionization decay, one of these electrons breaks away from the atom. The resultant ion turns out to be in the ground state.

Figure 3.6 shows schematically the excitation for all types of autoionization states. The ordinate indicates the energy of the electrons to a one-particle approximation. The ionization limit is indicated by a 0. The excitation energy

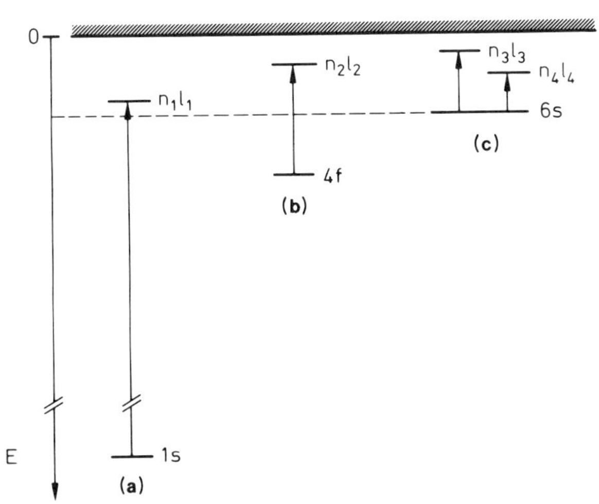

FIG. 3.6. Formation of various autoionization states for Yb ($1s^2 \ldots 4f^{14}5s^2 5p^6 6s^2$). (a) Excitation of the k shell, $|E(n_1l_1) - E(1s)| > |E(6s)|$; (b) excitation of the 4f shell, $|E(n_2l_2) - E(4f)| > |E(6s)|$; (c) double excitation of the valence shell, $|E(n_3l_3) - E(6s)| + |E(n_4l_4) - E(6s)| > |E(6s)|$. [From Bekov *et al.* (1981a).]

for autoionization states with excited K shells (Fig. 3.6a) or with excited 4f shells (Fig. 3.6b) should be sufficient for the 6s electron to break away. With a double excitation of the valence shell (Fig. 3.6c) autoionization states may arise if the total energy of valence electrons exceeds the ionization energy.

To a one-particle approximation, the autoionization decay can be represented by a two-electron transition $n_1 n_2 \to n_3 k$ ($n_i, i = 1, 2, 3$ are bound one-electron states; k is the state of the free electron). The width of the autoionization level depends on its coupling with the continuum in the following way (in terms of Coulomb units):

$$\Gamma_{\text{a.i.}} = 2\pi |\langle i|V|f\rangle|^2 = 2\pi |V(n_1 n_2, n_3 k)|^2, \tag{3.10}$$

where $|i\rangle$ is the initial and $|f\rangle$ is the final state of the system, V is the inter-electron interaction, and $V(n_1 n_2, n_3 k)$ is the matrix element of this interaction. The order of magnitude of $V(n_1 n_2, n_3 k)$ is determined by the radii and the location of the nodes of the orbitals n_1, n_2, n_3, k.

Simple estimates and numerical calculations have shown that the autoionization states obtained through double excitation (Fig. 3.6c) may be rather narrow in width (Bekov *et al.*, 1980a, 1981a). The numerical calculation by the model potential method yields the predicted positions and widths of such

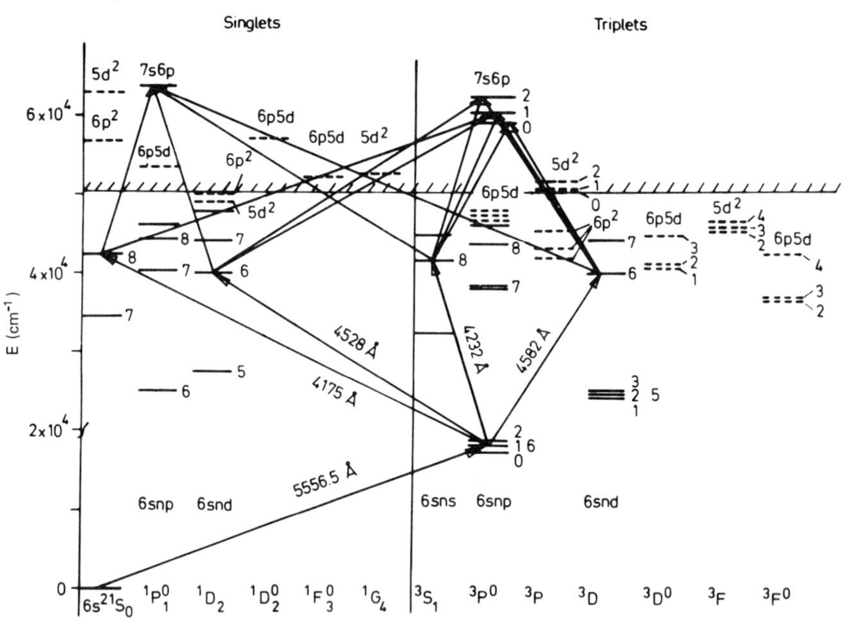

FIG. 3.7. Energy level diagram for Yb and the quantum states used for the excitation of autoionization states. [From Bekov *et al.* (1981a).]

autoionization levels. This enables us to find and study these states through stepwise excitation and ionization.

Figure 3.7 shows some of the excitation schemes of Yb to autoionization states (Bekov *et al.*, 1980a, 1981a). In the first step ($\lambda_1 = 5556.5$ Å) a dye laser brings about the transition $6s^2(^1S_0) \rightarrow 6s6p(^3P_1^0)$. Then a second laser excites the atoms to different intermediate states. The wavelength of the third laser was tuned over the range 4100–7100 Å. This made it possible to search for autoionization states in the region from 3700 to 14,000 cm^{-1} over the ionization limit, where the predicted configurations can be found. To identify the detected autoionization states, the spectra produced through different methods of excitation were compared when the intermediate states were $6s6d(^1D_2)$, $6s6d(^3D_1)$, $6s6d(^3D_2)$, $6s8s(^3S_1)$, and $6s8s(^1S_0)$. Such a choice of lower states allowed a clear classification of the observed autoionization states.

Several narrow autoionization resonances belonging to the 7s6p configuration have been discovered and identified in this way. Table 3.1 lists the calculated and experimental values of the width and energy of the detected autoionization states. Note that there is good agreement between theory and experiment.

The widths ($\Gamma_{a.i.} \simeq 1$ cm^{-1}) found experimentally and those predicted theoretically are far from being the lowest possible. The experiments with Gd atom, which has a more complex valence shell, have led to the detection of narrower autoionization states. Unlike those of the Yb atom, these narrow states have been found near the ionization limit. A simplified scheme of energy levels of gadolinium is shown in Fig. 3.8a. The dye laser in the first excitation step with $\lambda_1 = 5618$ Å transfers the gadolinium atoms from the ground state $4f^75d6s^2(^9D_2^0)$ into the $4f^75d6s6p(^9D_3^0)$ state. The second laser with $\lambda_2 = 6351.7$ Å excites the atoms into the $4f^75d6s7s(^9D_4^0)$ state. The third laser is tuned within 6300 to 6100 Å and causes the excitation of Rydberg states and those states that are within 300 cm^{-1} above the ionization limit.

TABLE 3.1. Energies E and Widths $\Gamma_{a.i.}$ of Autoionization States for the 7s6p Configuration of Yb[a]

	Calculated		Experimental	
Term	E [cm^{-1}]	$\Gamma_{a.i.}$ [cm^{-1}]	E [cm^{-1}]	$\Gamma_{a.i.}$ [cm^{-1}]
$^3P_0^0$	59,800	0.70	59,130.5	1.2
$^3P_1^0$	60,000	3.00	60,428.7	0.95
$^3P_2^0$	62,600	0.70	62,529.1	1.6
$^1P_1^0$	63,600	1.80	63,655.8	2.6

[a] From Bekov *et al.* (1981a).

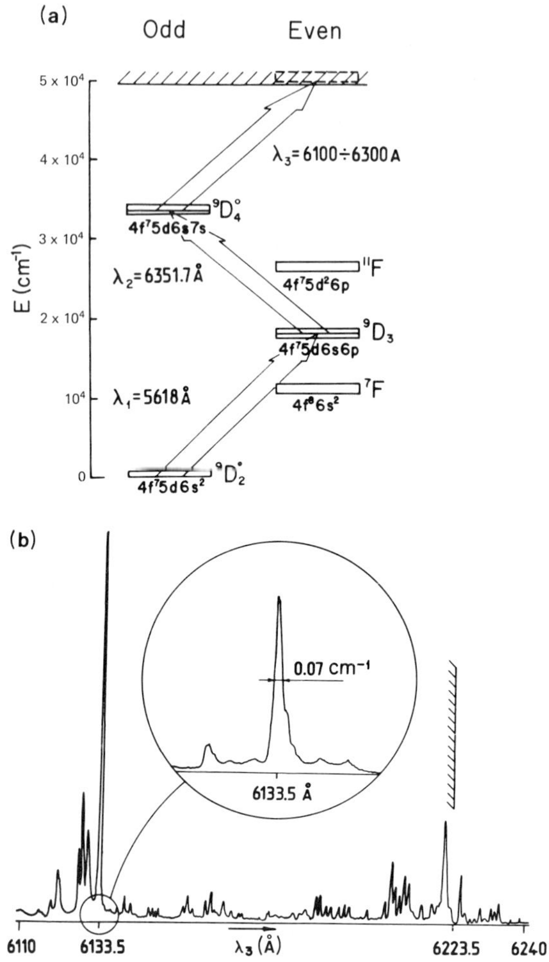

FIG. 3.8. (a) Energy levels of Gd involved in the three-step excitation of the autoionization state and the relevant quantum transitions. (b) The narrow autoionization resonance for Gd: ion current as a function of the laser wavelength λ_3 in the third excitation step (laser bandwidth $\Delta\nu_3 = 1\ \mathrm{cm}^{-1}$) and the same dependence in the vicinity of autoionization resonance $\lambda_3 = 6133.5$ Å with higher resolution (laser bandwidth $\Delta\nu_3 = 0.03\ \mathrm{cm}^{-1}$). [From Bekov *et al.* (1978b).]

Laser beams intersect the atomic beam between the electrodes, where an electric field pulse of rectangular shape is sent in 20 ns after the laser pulses. The ions pass through a slit in one of the electrodes and are registered by a secondary electron multiplier.

Figure 3.8b shows the dependence of the ion signal on λ_3 for a gadolinium atom with fine tuning of the wavelength to within 6110 to 6240 Å. A detailed investigation of the behavior of this dependence in the 6220–6225-Å range has made it possible to determine the ionization limit of gadolinium. For the ionization energy, the value $E_i = 49{,}588 \pm 5\ \text{cm}^{-1}$ has been obtained. It agrees well with the data of other workers (Martin *et al.*, 1978). The intense autoionization resonance at $\lambda_3 = 6133.5$ Å is of particular interest. The high-resolution record of this region is shown in the upper part of Fig. 3.8b. In this case the excitation in the third step was carried out by a pressure-tunable dye laser with an oscillation linewidth $\Delta\nu_3 = 0.03\ \text{cm}^{-1}$. The autoionization resonance half-width is about $0.07\ \text{cm}^{-1}$, which yields an estimated lifetime of 0.5 ns. Since the detected state is only $230\ \text{cm}^{-1}$ above the ionization limit, its decay into the state "electron plus gadolinium ion in the ground state" is most probable. Evidently the selection rules forbid a decay that leads to an increase in the lifetime of a state. Such a restriction can be removed, for example, by an electric field. A broadening of up to $0.35\ \text{cm}^{-1}$ was registered in a relatively weak electronic field ($E = 100$ V/cm).

The cross section of this autoionization transition was measured by the method of the saturation of the photoionization transition (Section 3.2.1). A value for the cross-section ($0.8 \times 10^{-15}\ \text{cm}^2$) was obtained from the dependence of the ion yield in the region of resonance maximum on the energy flux of the laser pulse Φ_3 in the third step.

This experiment proved that very narrow autoionization states with lifetimes of about 10^{-9} s are possible for multielectron atoms. The excitation cross sections of such states can be compared with the excitation cross sections of high-lying bound states. These long-lived autoionization states can be efficiently used in the stepwise ionization of atoms by laser radiation to radically increase the photoionization cross sections of atoms.

In conclusion, we present a classification of narrow and abnormally narrow autoionization resonances in the spectra of many-electron atoms (Table 3.2) (Vidolova-Angelova *et al.*, 1981a; Ivanov and Letokhov, 1985).

With the decay into the continuum being forbidden strongly enough, we can expect the occurrence of very narrow resonances that are still to be bound experimentally. Multistep excitation techniques are suitable for this purpose, for such states in most cases cannot be excited from the ground state by radiation means.

3.3. Ionization of Highly Excited (Rydberg) Atomic States

An effective and universal approach to the resonant ionization of atoms is the one in which the multistep excitation of high-lying states still below the ionization limit and the subsequent ionization either by an electric field, IR radiation, or collisions is used (Fig. 3.1b). This approach is based on the

TABLE 3.2. Classification of Narrow and Abnormally Narrow Autoionization Resonances of a Many-Electron Atom

Decay type / Excitation type	Subject to no selection rules		Subjection to selection rules	
	Width (cm^{-1}) for $Z \simeq 70$	Resonance narrowing effect	Width (cm^{-1}) for $Z \simeq 70$	Forbiddenness cause
K-shell excitation	10^3, irrespective of Z	None		None
Inner-shell excitation	10	None		
Valence-shell excitation	10	Valence-electron	1	No spin and orbital moment conservation
	$\sim Z^{-2}$	wave function	0.1	Two-electron decay channel closed
		oscillation	0.01	No Coulomb coupling between autoionization level and continuum

unique properties of highly excited atoms in states near the ionization limit (often called Rydberg states) (Stebbings and Dunning, 1983). They will now be considered briefly.

3.3.1. Properties of Highly Excited Atoms

For atoms with one optical electron (hydrogen atom or alkali atoms), the position of the energy levels in the vicinity of the ionization limit E_{i} is described by a simple expression:

$$E_{\mathrm{n}} = E_{\mathrm{i}} - \mathrm{Ry}/(n^*)^2, \tag{3.11}$$

where $\mathrm{Ry} = 13.6$ eV is the Rydberg constant, and n^* is the effective principal quantum number (for hydrogen, n^* is an integer, while for alkali atoms, $n^* = n - \Delta$, where Δ is the quantum defect). With an increase in the atomic excitation, that is, with increasing n^*, atomic characteristics and properties such as the binding energy, atomic size, lifetimes of the excited states, and polarizability are drastically changed. These changes are due to the large orbit of the electron in the highly excited state. Table 3.3 lists the properties of Rydberg states as a function of the principal quantum number (n).

The binding energy of the electrons in Rydberg states is proportional to $1/n^{*2}$, and for $n^* = 20$, the binding energy equals 30 meV, which under normal conditions is comparable to the energy of thermal motion. This property can be used to ionize highly excited atoms with continuous or pulsed electric fields, or with electromagnetic radiation in the microwave and IR range.

TABLE 3.3. Properties of Rydberg States as a Function of the Principal Quantum Number n with Numerical Data for the Hydrogen Atom[a]

Property	$n = 1$	$n = 100$	Arbitrary n (scaling)
Radius of Bohr orbit a_0 (cm)	$0.53 \cdot 10^{-8}$	$5.3 \cdot 10^{-5}$	n^2
Geometrical cross-section πa_0^2 (cm^2)	$8.8 \cdot 10^{-17}$	$8.8 \cdot 10^{-9}$	n^4
Binding energy E_n (eV)	13.6	$1.36 \cdot 10^{-3}$	$1/n^2$
Root-mean-square velocity of electron v_e (cm/s)	$2.2 \cdot 10^8$	$2.2 \cdot 10^6$	$1/n$
Separation of adjacent levels (eV)		$2.7 \cdot 10^{-5}$	$2/n^3$
Optical cross-section ($n \to n + 1$)			n^4
Quadratic Stark coefficient (shift $\ll$ smallest energy difference)			n^7
Linear Stark coefficient (shift $\simeq$ smallest energy difference)			n^2
Radiative lifetime			n^3

[a] From Stebbings (1976).

With an increase in n^*, the atomic collision cross section is increased, since the atomic radius $\langle r \rangle$ increases with n^{*2} ($\langle r \rangle = 3/2n^2 a_0$ for the hydrogen atom with a small orbital momentum, where a_0 is the Bohr radius), and its cross-sectional area is proportional to n^{*4}. The atomic diameter can reach large values—for example, 1 μm for states with $n^* \simeq 85$. Taking into account the weak bond of an excited electron, we can expect large values for the cross sections of collisions with other molecules or atoms that result in ionization.

The fact that the lifetime of a highly excited atom increases with n^{*3} (Bethe and Salpeter, 1957) is of importance since this promotes its ionization. The lifetimes of the d states of the sodium atom, for instance, are given by $\tau_d = 0.95n^{*3}$ ns for $n^* \gtrsim 4$ (Gallagher *et al.*, 1975).

The excitation cross section of a high-energy state depends on n^* and decreases with increasing n^* as n^{*-3}. On the other hand, atomic ionization becomes easier the greater the value of n^*. Therefore, an optimal value of n^* can be chosen for each ionization scheme and each particular case. For example, ionization induced through an electric field (10 kV/cm) can easily be realized with a maximum yield for states with $n^* \simeq 15$. The excitation cross sections of such states are about 10^{-14} cm^2; for alkali atoms they are two or three orders of magnitude higher.

The energy structure of the highly excited states, and hence the optical properties of the atom, can be controlled over wide limits with an external electric field. This is due to the fact that the quadratic Stark shift for states with large effective quantum numbers changes to a large linear shift similar to the shift for hydrogen atoms. This is caused by the presence of a great number of closely spaced states with different orbital angular momenta easily mixed in comparatively weak electric fields and possessing a very strong Stark effect. In addition, under such conditions the selection rules related to the change in parity of states break down, and transitions from states with small values of orbital momentum, for example from s, p, or d states, into those with any orbital momentum for a given n become possible. As a result, the number of potentially useful spectral lines in the absorption spectrum of an atom increases drastically.

The shift of a high-lying atomic level in a dc electric field may be very large. Linear changes in the energy of the d states of the sodium atom have been observed up to 40 cm^{-1} in fields of 0.3–3 kV/cm for $n^* = 15$ (Littman *et al.*, 1976a). A similar shift and splitting of the 9D state of Na with the principal quantum number $n = 9$ has been observed (Apatin *et al.*, 1978). Here, the transformation of the quadratic effect into a linear one has been studied. The general shift of the components in a 2-kV/cm field reaches 20 cm^{-1}. Thus, such a strong Stark effect makes it possible to tune the last excitation step to the oscillator frequency of a narrowband untunable laser, for instance, a gas discharge laser. It becomes possible to use them for the excitation, which increases the efficiency of the whole process of selective atomic photoionization.

The above simple picture is, strictly speaking, true for the Rydberg states of simple atoms with one valence electron. Of course, the Rydberg state exists and, what is more, is under intense study for complex atoms; see, for example, the works on the noble atoms (Stebbings *et al.*, 1975b), the alkaline-earth atoms (Wynne and Armstrong, 1979), and the rare-earth elements (Bekov *et al.*, 1980b; Camus *et al.*, 1980). The simple formula (3.11) does not adequately describe the positions of the Rydberg levels. There are two theoretical approaches used to describe them. The first is based on the assumption that the excited valence electron in a state with a large value of n is acted on by the Coulomb field of the atomic core. To calculate the arrangement of the Rydberg levels, the multichannel quantum defect method is used (Fano, 1970). In this method, information about the whole series can be obtained from the experimental data on the series of relatively low-lying Rydberg states. This enables us to extrapolate the experimental data and qualitatively study the level structure of the whole series. This method has been applied in the analysis of the Rydberg states of noble gases (Lee, 1971) and alkaline-earth elements (Lee, 1974).

The second approach is based on the theoretical calculation of the characteristics of the electronic shell of multielectron atoms using a model potential. In this method it is possible to take into account the deviation of the field from the Coulomb character. In this case, for large n, the singlet–triplet splitting may exceed the spacing between the adjacent levels of a certain Rydberg series. Such a method has been used with success in calculating the Rydberg states of the Yb atom (Vidolova-Angelova *et al.*, 1981b), with a good agreement between theory and experiment (Bekov *et al.*, 1980b).

3.3.2. Electric Field Ionization

An electric field changes the electronic spectrum of atoms so that some of the discrete levels nearest the ionization limit fall into the continuum. The other levels become autoionized (Fig. 3.9), and the probability of decay through ionization increases quickly with an increase in the principal quantum number.

3.3.2.1. Theoretical Treatment. It is easy to estimate the electric field strength at which a state with the principal quantum number n will fall into the continuum (Bethe and Salpeter, 1957). The potential energy of a Rydberg-state electron in an external electric field is $V(r) = -(1/r) - Ez$ (in atomic units), where E is the electric field strength along the z axis, and r is the distance from the center of mass. The potential energy is a maximum at the

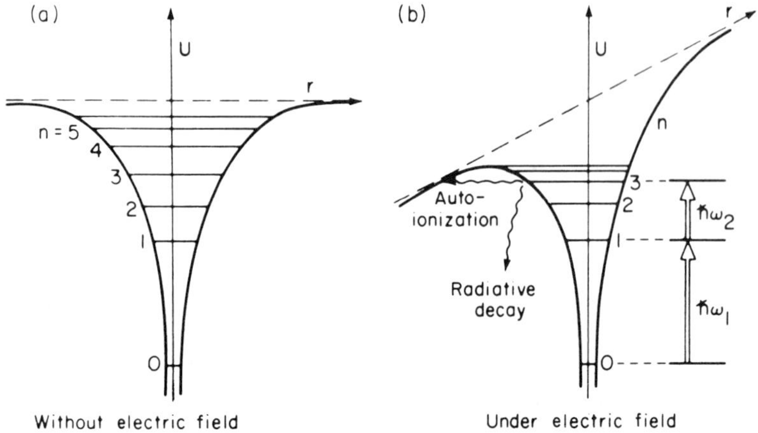

FIG. 3.9. Ionization of a highly excited atom in an external electric field: (a) optical electron levels and (b) modification of the potential in an external electric field and the formation of autoionization levels.

distance $r_{max} = E^{-1/2}$ along the z-axis and equals $V(r_{max}) = -E^{1/2}$. The energy of the electron in a state with the principal quantum number n is given by $E_n = -\frac{1}{2}n^{-2}$ (neglecting the Stark shift and level splitting). A state with a given n will fall into the continuum if $E_n \geqslant V(r_{max})$, or

$$E \geqslant 1/(16n^4) = E_{cr} \qquad (1 \text{ atomic unit} = 5 \times 10^9 \text{ V/cm}). \qquad (3.12)$$

Using several tunable lasers we can excite an atom to a state from which the probability of yield ionization is η_i. Let the ionization cross section for the combined action of a laser and an electric field be given by $\sigma_i = \sigma_R \eta_i$, where σ_R denotes the cross section for absorption of radiation in the last excitation step. The total yield of atomic ionization η_i can rather easily be made equal to unity by choosing a certain value of the electric field strength. As a result, the ionization cross section will mainly depend on the excitation cross section of the radiation in the last step, which equals 10^{-12}–10^{-15} cm^2, depending on the width of the absorption line and the oscillator strength of transitions with large changes in the principal quantum number n. It is evident that the optimal case occurs when an atom is excited by laser radiation to a state whose ionization rate in the electric field is higher than that of radiative decay.

Highly excited atomic states become hydrogenlike in fields where the Stark shift is larger than, or of the order of, the change in the Coulomb energy of the optical electron due to the electronic screening of the nuclear charge. This takes place for $E \geqslant \frac{2}{3}\Delta/n^5$ a.u., where Δ is the quantum defect.

It is not difficult to calculate the autoionization probability of high-lying states in an electric field for a hydrogenlike atom (Ivanov and Letokhov,

1975). Consider the hydrogen atom in a state with the principal quantum number n and an electric field directed along the z axis with strength E. The variables of the Schrödinger equations are separable in the parabolic coordinates ξ and η ($\xi = r + z$, $\eta = r - z$), so that the required wave function has the form $\Psi(r) = f_{n_1m}(\xi)f_{n_2m}(\eta)e^{im\varphi}$, where m is the magnetic quantum number, and n_1 and n_2 are the parabolic quantum numbers replacing n and l in spherical coordinates (Bethe and Salpeter, 1957). The functions $f_{n_1m}(\xi)$ and $f_{n_2m}(\eta)$ have the form of one-dimensional Schrödinger equations with the potentials

$$U_1(\xi) = -b_1/2\xi + (m^2 - 1)/8\xi^2 + E\xi/8, \tag{3.13a}$$

$$U_2(\eta) = -b_2/2\eta + (m^2 - 1)/8\eta^2 + E\eta/8, \tag{3.13b}$$

where $b_1 = n_i + |m|/2 + \frac{1}{2}$. The potential $U_1(\xi)$ has the form of an infinitely deep well, while $U_2(\eta)$ has the form of a barrier. Hencc, the electron may tunnel in the direction $\eta \to \infty$.

The asymptotic behavior of the function $f_{n_2m}(\eta)$ with $\eta \to \infty$ will briefly be described to a quasi-classical approximation. The autoionization decay rate is given by the total probability flux through a plane perpendicular to the z-axis,

$$W = \int_0^\infty |\Psi|^2 v_z 2\pi r \, dr, \tag{3.14}$$

where v_z is the velocity of the electron along the z-axis. The asymptotic form of the function $f_{n_2m}(\eta)$ is substituted into (3.14), and instead of $f_{n_1m}(\xi)$ the exact solution for hydrogen may be used, since this function exponentially decreases as $\xi \to \infty$ and contributions to the integral are made only in the region of small ξ.

The calculation being considered is valid as long as the potential barrier is broad enough, that is, $\eta_2 \gg \eta_1$, where η_1 and η_2 are the two turning points. The equivalent condition in atomic units has the form $|\max U_2(\eta)| \ll 1/2n^2$, which means that the electric field must only slightly influence the ionization energy of an atom in the state under consideration. Otherwise the state is absorbed into the continuum, and the problem is reduced to one of direct ionization through radiation or one in which the width of the level becomes comparable to the energy of the electron in this state. Again we are, in fact, dealing with transitions to the continuum. The condition of the field strength may be written

$$En^3 \ll \tfrac{1}{4}(n + |m|/2 + \tfrac{1}{2})^{-1}. \tag{3.15}$$

The autoionization probability W can be accurately calculated analytically up to the terms proportional to En^3. Finally, the following result for the rate of field ionization is obtained (Ivanov and Letokhov, 1975):

$$W = \frac{1}{n^2}\left(\frac{4}{En^3}\right)^{1+|m|+2n_2}[n_2!(n_2+|m|)!]^{-1} \cdot \exp\left[-\frac{2}{3n^3E} + 3(n - 2n_2 - |m| - 1)\right]. \tag{3.16}$$

Formula (3.16) has been obtained in a quasi-classical manner with consideration given only for the first-order terms in the small parameter En^3. Gallas *et al.* (1982) made a more exact numerical calculation within the framework of the quasi-classical approximation. The cumbersome exact quantum-mechanical calculation (Damburg and Kolosov, 1976) yields practically the same result. Equation (3.16) gives the field-strength dependence of the decay rate, which is shifted towards lower E with respect to the more exact dependence. For example, formula (3.16) gives a 20% overestimate for the critical field strength.

According to (3.16), the state with a minimum value for n_2 has the highest rate of autoionization. This is to be expected, since this state corresponds to an orbit that is extended the most in the opposite direction to the field. A change in n_2 of unity changes W by about one order of magnitude. The increase in autoionization probability as m decreases is explained by the localization of the electron near the z-axis.

In a pulsed electric field with $E > E_{cr}$ the highly excited state decays instantly to the continuum. The magnitude of the field strength E_{cr} at which this takes place can be roughly estimated from (3.15). A more relevant condition is that the levels with successive values of n not overlap because of broadening. If this condition is not met, all of the levels, beginning with a particular $n_0(E)$, merge into the continuum. This occurs for $W = 1/n_0^3$ a.u. or $5 \times 10^{16}/n_0^3\ s^{-1}$. The corresponding values of E for $n = 8, \ldots, 16$ can be obtained from the curves in Fig. 3.10. From the calculations and Fig. 3.10 it follows that the decay rate of the field ionization for states with $n > 10$ is higher than that of the radiation decay even when $E \gtrsim 5 \times 10^4$ V/cm (1 a.u. $= 5 \times 10^9$ V/cm).

3.3.2.2. Experimental Results. Experiments confirmed the high efficiency of field-ionization of sodium atoms in highly excited states with a dc electric field (Ambartzumian *et al.*, 1975a). The experiment employed two tunable dye lasers pumped by a pulsed N_2 laser. The parameters of the first laser resulted in the saturation of the resonance transition $3^2S_{1/2} \rightarrow 3^2P_{1/2}$. The second dye laser was tuned in the range 4210–4140 Å which allowed for the excitation of the S and D states of sodium with principal quantum numbers varying from 13 to 18. The light beams entered a cell containing sodium vapors heated to 450 K in the area of the electrodes. A sharp resonant

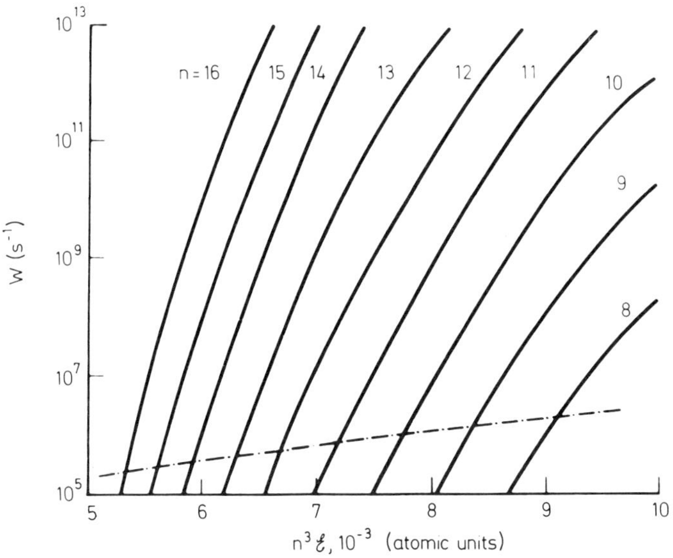

FIG. 3.10. Ionization probability of a highly excited atom as a function of the electric field strength. The quantum numbers $n = 8 \ldots, 16$; $m = 0$; $n_2 = n - 1$. The dashed line represents the probability of radiative decay (1 a.u. $= 5 \times 10^9$ V/cm).

increase in the photocurrent occurs in fields with $E = 12$ kV/cm, which agrees well with the calculations done for the levels with $n = 15$. The 15D level lies 488 cm^{-1} below the ionization limit. A similar resonance photoionization has been observed in the excitation of any of the S and D states of Na with a principal quantum number $13 \leqslant n \leqslant 18$.

The maximum in the cross section of atomic photoionization in an electric field is determined by the cross section of the autoionization excitation level, that is, by the cross section of the transition 3P → 15D. For the transition $3^2P_{1/2} \rightarrow 15^2D_{3/2}$ the excitation cross section equals $\sigma_R = 0.7 \times 10^{-14}$ cm^2. Thus, in electric fields with $E \simeq 12$ kV/cm the photoionization cross section of Na is $\sigma_R/\sigma_i \simeq 10^5$ times that of the ionization from the ground state ($\sigma_i \simeq 10^{-19}$ cm^2), and about 10^3 times that of the ionization from the excited intermediate 3P state to the continuum ($\sigma_{2i} \simeq 10^{-17}$ cm^2).

The cross section of ionization through autoionization states can be considerably increased provided that the levels with smaller principal quantum numbers ($n \sim 10$) are excited, stronger fields are used, and the atoms are excited in a beam. As can be seen in Fig. 3.10, the state with $n = 10$ decays in a 32-kV/cm field within the lifetime of that state. It is easy to form an atomic beam with a divergence of 0.1 rad. In this case, the absorption cross section of the radiation in the second step and, hence, the cross section for atomic ionization will equal $\sigma_R \simeq 5 \times 10^{-13}$ cm^2.

The structure of levels of highly excited Na, in particular of the level with $n = 15$, in weak and strong electric fields and measurements of the tunneling rate of the optical electron in a static electric field have been performed in a number of works (Littman *et al.*, 1976a,b). The Na atoms in the beam were excited by two pulsed dye lasers. The ionization rate of a single Stark state of Na with $n = 12$–14 was determined from the time dependence of the number of ions formed through excitation with laser pulses of about 5 ns duration. The method made it possible to measure the ionization rate in the 10^6–10^8-s^{-1} range. The measured and estimated values are in excellent agreement. The authors emphasize that the autoionization of hydrogenlike atoms in the lower states of high-lying terms can be successfully calculated in a hydrogenlike approximation. The behavior of the upper Stark components differs considerably from that of the lower ones. This is due to an intersection and an interaction of the upper Stark components of one term with the lower ones of a higher term. The lower states of this term are autoionized. Such behavior of the upper Stark components makes it possible to autoionize atoms in an electric field whose magnitude is less than the critical value estimated for the hydrogen atom.

The ionization of highly excited atoms in a pulsed electric field has been investigated for Na (Ducas *et al.*, 1975; Bekov *et al.*, 1977a), Xe (Stebbings *et al.*, 1975b), U (Solarz *et al.*, 1976), and Rb (Bekov *et al.*, 1977b). The results of the critical field measurements for sodium, rubidium, and uranium in states with different principal quantum numbers are shown in Fig. 3.11. The lower line shows the classical dependence (3.12). In this equation, a shift in the level due to the Stark effect and the electron tunneling were not taken into consideration. The upper line was computed using (3.16) for a pulsed electric field with a duration equal to the lifetime of a high-lying state. The experimental values for all of the works differ from the theoretical dependencies. The difference in the magnitude of the experimental and theoretical results is outside of the experimental accuracy and is, on the average, 15%. The approximated dependence (dashed line) drawn through the experimental values is parallel to the classical dependence and can be described by an empirical formula $E_{cr} = 0.07(n^*)^{-4}$.

3.3.2.3. Total Ionization Yield. Multistep ionization schemes assisted by a static pulsed electric field have their pros and cons. A choice between them should be made according to the requirements of a concrete application. New autoionization states whose decay rate is determined by the ratio given by (3.16) can be used with an external static electric field. The continuous decay of the final state makes it possible to entirely ionize all the atoms interacting with the laser radiation. However, the dc field splits and shifts the quantum levels so that the experiment is hindered. When (3.12) is satisfied, the electric field that is applied after exciting the atom into a high-lying state causes its

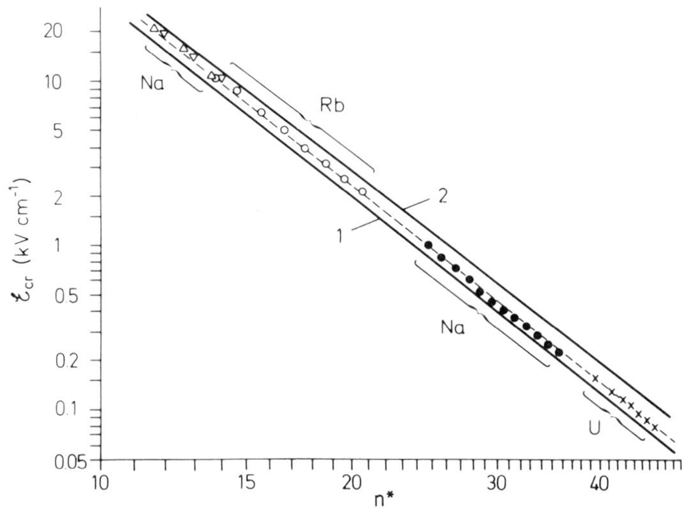

FIG. 3.11. Experimental dependence (dashed line) of critical ionization field as a function of effective principal quantum number n^* for Na, Rb, and U and its comparisons with theoretical relations (3.12) and (3.16) (solid lines 1 and 2, respectively). $\mathcal{E} = (16n^{*4})^{-1}$.

rapid decay due to the transition into the continuum. However, under such conditions, the yield is less than 100% because the atoms are distributed between the initial and intermediate states. Indeed, with multistep excitation, the population of the final discrete level is defined by expression (2.96) on condition that all transitions are saturated. Therefore, the total yield of ionization under the action of the laser and subsequent electric-field pulses is defined by the expression

$$Y = N_f/\sum_k N_k = g_f/\sum_k g_k. \tag{3.17}$$

Such situations can be realized if the Rydberg atoms are ionized by an electric field pulse switched on after the laser pulses are over. Figure 3.12 illustrates the dependence of the Yb ion yield on the energy flux of the laser pulse used in the first, second, and third excitation steps as well as on the strength of the pulsed electric field (Bekov *et al.*, 1978c). From these dependencies it is possible to find the energy flux Φ_{kn} of the laser pulse needed to saturate the transition $k \to n$. The value of Φ_{kn} is related to the parameters of the laser pulse at the frequency of the transition and the transition itself:

$$\Phi_{kn} = \Phi_s^{(kn)} \Delta\omega_L^{kn}/\Delta\omega_a^{kn}, \tag{3.18}$$

where $\Phi_s^{(kn)}$ is the saturation energy of the transition $k \to n$ determined from

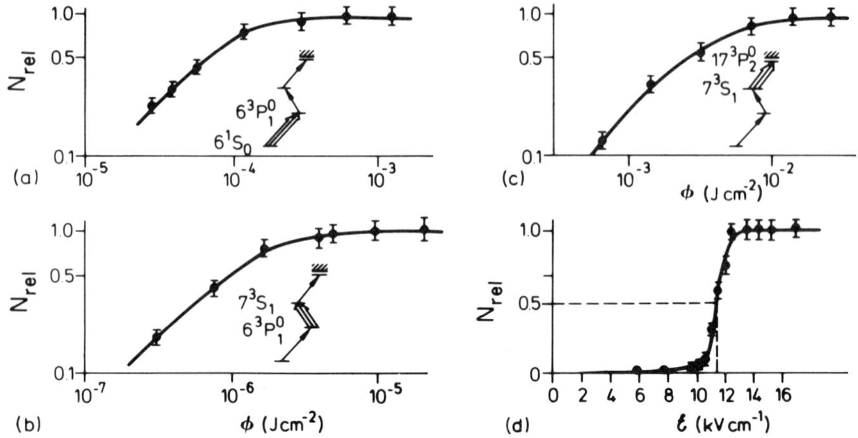

FIG. 3.12. Dependence of Yb ion yield on the laser pulse energy flux of (a) the first ($\lambda_1 = 555.6$ nm), (b) second ($\lambda_2 = 680$ nm) and (c) third ($\lambda_3 = 587.3$ nm) excitation steps and (d) on the strength of the pulsed electric field $\mathscr{E}_{\mathrm{cr}} = 11.5$ kV/cm; $17^3\mathrm{P}_2^0$. [From Bekov *et al.* (1978c).]

the expression

$$\Phi_{\mathrm{s}}^{(kn)} = \frac{\hbar\omega_{kn}}{2\sigma_{kn}}, \qquad \sigma_{kn} = \frac{g_n}{g_k}\frac{\lambda_{kn}^2}{2\pi}\frac{A_{nk}}{\Delta\omega_a^{kn}}. \tag{3.19}$$

$\Delta\omega_{\mathrm{L}}^{kn}$ is the spectral width of the laser pulse at the frequency ω_{kn}; $\Delta\omega_a^{kn}$ is the linewidth (for a beam or gas) of the transition $k \to n$; and A_{nk} is the Einstein coefficient. When all three transitions are saturated, $\frac{5}{12}$ or about 40% of the Yb atoms in the region of excitation will simultaneously be excited to the last state $17^3\mathrm{P}_2^0$. Figure 3.12d shows the dependence of the Yb ion yield on the strength of the pulsed electric field for a three-step excitation of the $17^3\mathrm{P}_2^0$ state. The ion signal becomes saturated because of the field ionization of all of the atoms in the last state. Thus, even with optimal conditions, it is impossible to attain a 100% yield because of the distribution of atoms over the intermediate levels.

From this point of view it also seems rather interesting to compare the electric field ionization involving Rydberg states with the ionization involving autoionization states. Figure 3.13 shows the results of experiments on Gd according to the excitation scheme given in Fig. 3.7. In the top part of the figure the ion yield as a function of the wavelength λ_3 of the laser in the third step is shown for the case in which the electric field pulse was applied just after the laser pulses. In the bottom part of the figure the same ion-yield dependence in the weak electric field, unable to ionize highly excited atoms, is shown. That is the reason why only the autoionization resonances can be seen in the bottom spectrum. The autoionization peak with the maxi-

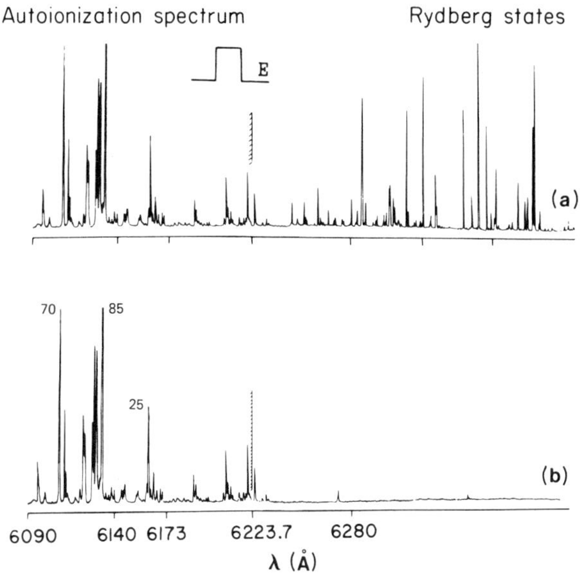

FIG. 3.13. Yield of Gd atomic ions as a function of the laser wavelength at the third stage of excitation of Rydberg and autoionization states from the intermediate quantum state $^9D_4^0$ (ionization energy limit $E_i = 6.145 \pm 10^{-3}$ eV): (a) ionization with a pulsed electric field (E = 13 kV/cm); (b) ionization with a weak dc field (E = 30 V/cm = const). [From Bekov *et al.* (1979).]

mum amplitude (its relative amplitude equals 85 and is reduced in the figure) is very narrow. It is shown separately with high resolution in the upper part of Fig. 3.8b. In the top spectrum the Rydberg states can be seen below the ionization limit (to the right) while the autoionized ones can be seen above the ionization limit (to the left). We can see that they have amplitudes of the same order of magnitude.

At first glance it seems that the existence of narrow autoionization states, which provide an optimal ionization with 100% yield, is quite a rare event. Experiments performed recently with heavy atoms have shown that this is probably not so. For a heavy atom with many valence electrons, we can always choose the last intermediate state for the transition to a narrow autoionization state.

3.3.3. Photoionization through Infrared Radiation

The quasi-classical formula may be used to estimate the effective cross sections for the atomic ionization from highly excited states through IR or microwave fields. The effective cross section for the atomic ionization from a Rydberg state with a principal quantum number n by electromagnetic radiation with a frequency ω equals (Sobel'man, 1979, Chapter 6):

$$\sigma_{ni} = \frac{64\pi}{3\sqrt{3}} \alpha \left(\frac{E_i}{\hbar\omega}\right)^3 \frac{a_0^2}{n^5}, \tag{3.20}$$

where α is the fine structure constant. The photoionization limit from the level n is given by $\hbar\omega \geqslant E_i/n^2$. As the frequency increases, the ionization cross section decreases. Its maximum value is attained when $\hbar\omega = E_i/n^2$ and equals $\sigma_{ni}^{max} = 0.34na_0^2$. The ionization cross section grows with n up to value about $5 \times 10^{-16}\,cm^2$.

At present the CO_2 laser is the most powerful and efficient source of ionizing radiation. The ionization cross section for the 10.6 μ band is $\sigma_i = 10^{-16}\ cm^2$. The intensity of the radiation whose ionization probability during the laser pulse ($\tau_p = 200$ ns) is about equal to unity, is given by $I = \hbar\omega/\sigma_i\tau_p = 10^3\ W/cm^2$. With a pulse rate repetition of 20 kHz and a laser beam cross section of about 1 cm^2, the average laser power is about 4 W. Thus, the use of CO_2 lasers in selective ionization proves to be very promising. Its application to the ionization of uranium has been used as an experiment to detect Rydberg states using time-resolved, stepwise laser photoionization (Solarz *et al.*, 1976).

3.3.4. Collisional Ionization of Excited Atoms

Let the collisions between excited atoms A* and the atoms or molecules of M that result in the loss of a weakly bound electron from the selectively excited atom A* be considered. The ionization of an excited atom A* may proceed by way of one of the following processes:

$$A^* + M \rightarrow (A + e^-) + M^+ \rightarrow A^- + M^+ \tag{3.21}$$

$$A^* + M \rightarrow AM^+ + e^- \tag{3.22}$$

$$A^* + M \rightarrow A^+ + (M + e^-) \rightarrow A^+ + M^- \tag{3.23}$$

3.3.4.1. Electron Capture. In (3.21) the negative ions A^- are formed when the atom A possesses a significant electron affinity φ_A. The reaction becomes energetically possible if

$$E(A^*) + \varphi_A \gtrsim E_i(M^+), \tag{3.24}$$

where $E(A^*)$ is the excitation energy of the atom A, and $E_i(M^+)$ is the ionization potential of particle M. Process (3.21) was observed for slow, highly excited atoms of oxygen (Kuprianov, 1968):

$$O^* + CO \rightarrow O^- + CO^+. \tag{3.25}$$

In these experiments highly excited oxygen atoms were produced in electron–molecule collisions. The excitation energy of the atom O* was much lower than the ionization potential of CO. The energy difference was compensated by the large electron affinity of the oxygen atom, which equals 1.46 eV.

At low values of the relative translational energy of the colliding particles, the maximum value of the cross section of process (3.21) depends on the capture cross section of the atom or molecule by another atom, which in turn depends on the attractive potential acting between the colliding particles.

If the collision is governed by a long-range van der Waals potential of the form $U = -C/R^6$, where R is the internuclear distance and C is the van der Waals constant, the capture cross section is given by

$$\sigma_{\text{capt}} = \tfrac{2}{3}\pi(2C/E_{\text{kin}})^{1/3}, \tag{3.26}$$

where E_{kin} is the relative kinetic energy of the atoms. Taking into account the fact that one of the atoms is excited, $C = \beta e^2\langle r^2\rangle$, where β is the polarizability of the ground state atom, $\langle r^2\rangle$ is the square of the radius of the excited electron, and e is the charge of the electron. Averaging over the Maxwell velocity distribution function, an expression for the cross section is obtained (Smirnov, 1974):

$$\sigma_{\text{capt}}(T) = 5.36(\beta e^2\langle r^2\rangle/k_{\text{B}}T)^{1/3}. \tag{3.27}$$

A typical value for σ_{capt} is 10^{-14} cm^2 when the polarizability of one of the atoms is about $10^2 a_0^3$.

Equation (3.27) also holds for the case where A is selectively excited while M is also excited (Harris and Lidow, 1976). Collisional ionization takes place if

$$E(\text{A}^*) + E(\text{M}^*) \geqslant E_{\text{i}}(\text{M}^+), \tag{3.28}$$

where $E(\text{M}^*)$ is the excitation energy of M.

3.3.4.2. Associative Ionization. For the associative ionization process (3.22), the cross sections are of the same order of magnitude as the gas-kinetic ones as long as the excitation energy of A is smaller than the ionization potential of M. For instance, for the process

$$\text{Cs}(8^2\text{P}_{1/2}) + \text{Cs}(6^2\text{S}_{1/2}) \rightarrow \text{Cs}_2^+ + e^- \tag{3.29}$$

the average cross section $\langle\sigma(v)\rangle_v = 0.33 \pm 0.09$ (Å)^2 was measured (Williams and Naiditch, 1967). The total ionization rate of the Cs atoms in the excited states $9^2\text{P} \rightarrow 15^2\text{P}$ colliding with ground state Cs atoms was measured (Devdariani *et al.*, 1978). This total rate includes the associative ionization process. The ionization rate k_{i} equals $\langle\sigma_{\text{i}}v\rangle = (4\text{–}7) \times 10^{-11}$ cm^{3}s^{-1}. A theoretical calculation using a simple model also gives the same order of magnitude.

The multistep excitation of highly excited atomic states with lasers allows the systematic study of the associative homonuclear ionization of Rydberg states. Using mass-spectrometers, it has been confirmed that Sr_2^+ ions are produced from laser populated Rydberg states by the associative ionization process (Worden *et al.*, 1978a):

$$\mathrm{Sr}(5s\,n1) + \mathrm{Sr}(5s^2) \rightarrow \mathrm{Sr}_2^+ + e^-. \tag{3.30}$$

An ion signal for the dimeric cation was detected for the $5s6d\,^3D_2$ level indicating that the binding energy of Sr_2^+ exceeds 0.77 eV. The process (3.30) was studied under conditions of low ($<10^{12}$ Rydberg atoms/cm^3) and high ($>10^{12}$ Rydberg atoms/cm^3) excitation density. Worden *et al.* (1978a) found strong evidence that the electrons initially produced by the photoionization play an important role in ionizing the high Rydberg states through an avalanche mechanism that takes place before an associative ionization collision can occur. This process effects the chemistry of these Rydberg states at high densities of excited atoms.

3.3.4.3. Electron Escape. When the atom A is highly excited, the most efficient collisional ionization process takes place (3.23). In this process the ionization can be considered to be the result of the interaction between a weakly bound electron and an incident particle. The problem may be solved by classical methods since the characteristic distance at which an interaction occurs is much less than the distance between the electron and the nucleus. Hence, the influence of the ion core on the collision may be neglected (Matsuzawa, 1971; Smirnov, 1974). Process (3.23) may proceed in the following ways: (i) the weakly bound electron collides with an atom and undergoes elastic scattering; (ii) the electron collides with a molecule and undergoes inelastic scattering; and (iii) the electron attaches itself to an atom or molecule that has a large electron affinity.

The ionization cross section of an excited atom colliding with another atom, which results in elastic electron scattering, is similar to the cross section of low-speed electron scattering from an atom and is small (10^{-17}–10^{-19} cm^2) (Smirnov, 1974). The ionization cross section of a highly excited atom during its collision with a molecule, as a result of which the molecule passes into a lower rotational state, is much larger (Hotop and Niehaus, 1968).

The cross section and rate constants for the collisional ionization of Xe atoms in definite high Rydberg states have been measured (West *et al.*, 1976). A beam of Xe atoms, excited to the metastable states $^3P_{0,2}$ through electron impact, was passed into SF_6 at a pressure of $(1–2) \times 10^{-6}$ Torr. Some of the Xe ($^3P_0$) atoms were then excited by a dye laser to a particular Rydberg state. The ions were detected by an electron multiplier, and the highly excited atoms were detected through ionization by an electric field. The results of the measurements of the ionization rate constant k_i for Xe in states with n ranging from 25 to 40 are given in Fig. 3.14. This figure also shows the results of Tehsenfeld (1970) and the theoretical values of k_i. The rate constant depends very little on n at the electron energies used in the experiment. Its average value is 3.8×10^{-7} cm^3/s. Thus, the ionization cross section of highly

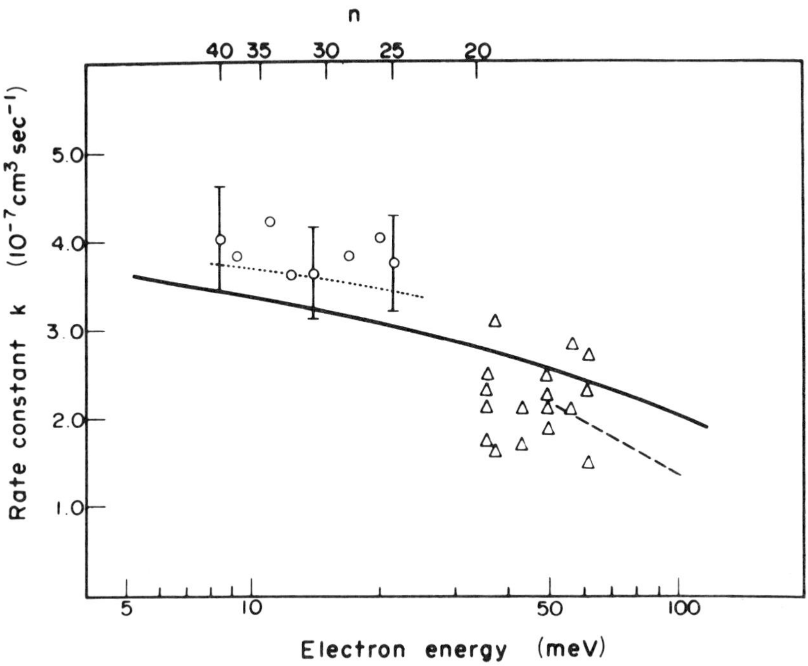

FIG. 3.14. Rate constant for electron attachment to SR_6 as a function of the kinetic of the electron bound to the Xe atom in high-lying states with the quantum number n. [From West *et al.* (1976).]

excited atoms that collide with molecules that effectively accept thermal electrons, is really very large. For states whose principal quantum number n varies from 20 to 40, $\sigma_{ni} \simeq 10^{-11}$ cm^2.

Mixed with molecules such as SF_6 selectively excited atoms can be ionized with yields approximating unity if the molecular concentration is around $(k_i\tau_{nl})^{-1}$, where τ_{nl} is the lifetime of a highly excited atom which, for example, for the d states of hydrogenlike atoms will be given by $\tau_{nd} \simeq n^3$ (ns). Thus, collisional ionization becomes effective at rather low molecular concentrations. When SF_6 is used, the concentration will be 10^{12} mol/cm^3 if $n \gtrsim 20$.

Collision ionization appears to be a useful method for obtaining ions from highly excited atoms. In particular, this method has been successfully used to detect highly excited atoms in a thermionic diode (Niemax, 1983). To illustrate, Fig. 3.15 shows the dependence of the ionization probability for Rb atoms in the n^2P_J states on the principal quantum number n, which has been measured in this work. As is clear from the graph, at a buffer gas pressure of 0.1 Torr the ionization probability for $n \gtrsim 25$–30 is close to unity.

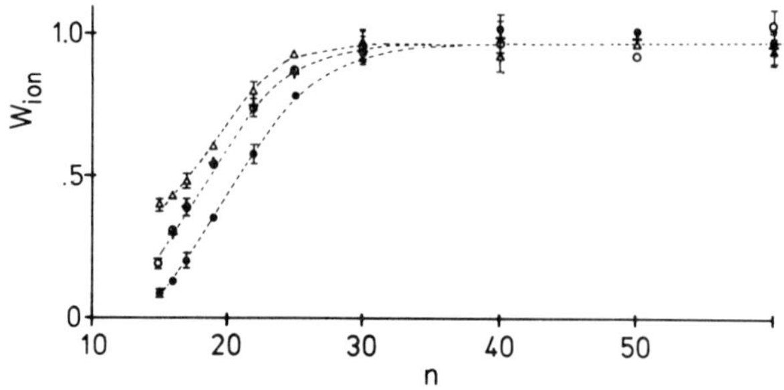

FIG. 3.15. Probability of ionization of the Rb atom in the n^2P_J state upon collisions with Ne atoms at various pressures (●, 0.12; ○, +, 1.2; and Δ 12 Torr) as a function of the principal quantum number n. [From Niemax (1983).]

3.4. Schemes for Stepwise Resonance Ionization of Various Atoms

Combining the various atomic resonance excitation techniques considered in Chapter 2 with the methods for ionizing the excited atoms, which have been dealt with earlier in this chapter, we can obtain a host of different schemes for resonance ionization of atoms. All these schemes can easily be classified, and it is quite possible to indicate optimal schemes of two- or multiple-photon resonance ionization for every atom in the periodic table.

3.4.1. Classification of Ionization Schemes

All the possible resonance ionization schemes can be classified by several characteristic features:

(1) the number of the independent radiation frequencies (colors) used;

(2) the number of excitation stages, including both resonance and nonresonance transitions;

(3) the type of the multiphoton resonance and nonresonance transitions used;

(4) the total number of the absorbed photons, and so forth.

The resonance ionization of any atom can, in principle, be effected by a two-step excitation scheme. All the *two-step* ionization schemes can be divided into three types (Fig. 3.16):

(a) ionization in a *single-frequency* field ω, with only one excitation stage being resonant:

$$A(\omega, \tilde{\omega}) \rightarrow A^+, \qquad \text{(Ia)}$$

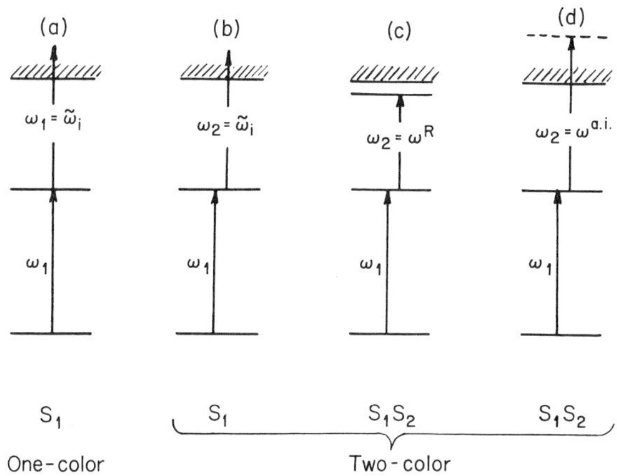

FIG. 3.16. Various two-step resonance atomic ionization schemes (type I): (a) with a single one-photon resonant stage in a single-frequency field; (b) with a single one-photon resonant stage in a two-frequency field; (c) with two one-photon resonance ionization stages involving a Rydberg state in a two-frequency field; (d) the same involving an autoionization state.

where the wavy line over the frequency symbol ω denotes the absorption of a photon in the nonresonance excitation stage as a result of a direct transition to the continuum. Incidentally, with this notation, a simple single-step nonresonance photoionization process can be denoted $A(\tilde{\omega}) \to A^+$;

(b) ionization in a *two-frequency* field, with only one excitation stage being resonant:

$$A(\omega_1, \tilde{\omega}_2) \to A^+; \tag{Ib}$$

(c) ionization in a *two-frequency field*, with *both* excitation stages being resonant:

$$A(\omega_1, \omega_2^R) \to A^{**}, \tag{Ic}$$

$$A(\omega_1, \omega_2^{a.i.}) \to A^+. \tag{Id}$$

In this case, the resonant absorption of the second photon of frequency ω_2 can take place either on a transition to a highly excited (Rydberg) state (this being denoted by the superscript R on the frequency symbol ω_2) or on a transition to a narrow autoionization state (indicated by the superscript a.i. at the frequency symbol ω_2). The highly excited state A^{**} can be ionized either by an electric field or by collisions (see Fig. 3.1.b).

For many atoms, the two-step ionization is not optimal for various reasons. For example, in many practical applications it is desirable that use should be made of a laser radiation in the well-developed visible and ultraviolet regions. In this case, it is advantageous to employ a three-step ionization scheme. All the *three-step* ionization schemes can likewise be divided into three types (Fig. 3.17):

(a) ionization in a *two-frequency* field, with only two excitation stages being resonant:

$$A(\omega_1, \omega_2, \tilde{\omega}_1 \text{ or } \tilde{\omega}_2) \rightarrow A^+; \tag{IIa}$$

(b) ionization in a *three-frequency* field, also with only two excitation stages being resonant:

$$A(\omega_1, \omega_2, \tilde{\omega}_3) \rightarrow A^+; \tag{IIb}$$

(c) ionization in a *three-frequency* field, with *all* the three excitation stages being resonant:

$$A(\omega_1, \omega_2, \omega_3^{R}) \rightarrow A^{**} \quad \text{or} \quad A^+, \tag{IIc}$$

or

$$A(\omega_1, \omega_2, \omega_3^{a.i.}) \rightarrow A^+. \tag{IId}$$

Many atoms have their first resonant transitions in the vacuum ultraviolet, which, with VUV lasers lacking, can be excited on account of two- or three-photon resonance processes. In this case, the first excited state lies much closer to the ionization threshold than to the ground state. Therefore, two-step excitation schemes are sufficient for ionization involving multiphoton

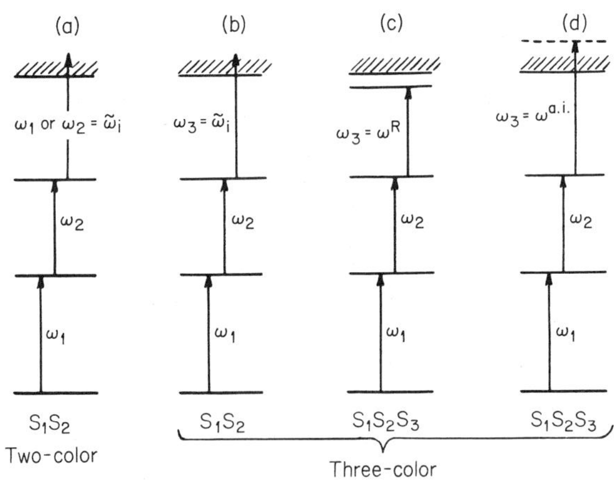

FIG. 3.17. Various two-step resonance atomic ionization schemes (type II): (a) with two one-photon resonant stages in a two-frequency field; (b) with two resonant stages in a three-frequency field, (c) with three resonance excitation stages involving a Rydberg state in a three-frequency field; (d) the same involving a narrow autoionization state.

processes. All ionization schemes involving a *two-photon* resonance transition in the first stage can be divided as follows (Fig. 3.18):

(a) *single-step* ionization in a single-frequency field, involving a *single* two-photon resonance excitation stage on a transition to a narrow autoionization or a Rydberg state:

$$A(2\omega) \rightarrow A^{+} \quad \text{or} \quad A^{**}. \tag{IIIa$'$}$$

Sometimes, to increase the ionization probability, use is made of two resonant photon frequencies to tune to quasi-resonance with some intermediate level:

$$A(\omega_1 + \omega_2) \rightarrow A^{+} \quad \text{or} \quad A^{**} \tag{IIIa$''$}$$

(b) two-step ionization in a *single-frequency* field, involving a single two-photon resonance excitation stage:

$$A(2\omega_1, \tilde{\omega}_1) \rightarrow A^{+} \tag{IIIb}$$

(c) two-step ionization in a *two-frequency* field, involving a single two-photon resonance excitation stage:

$$A(2\omega_1, \tilde{\omega}_2) \rightarrow A^{+} \tag{IIIc}$$

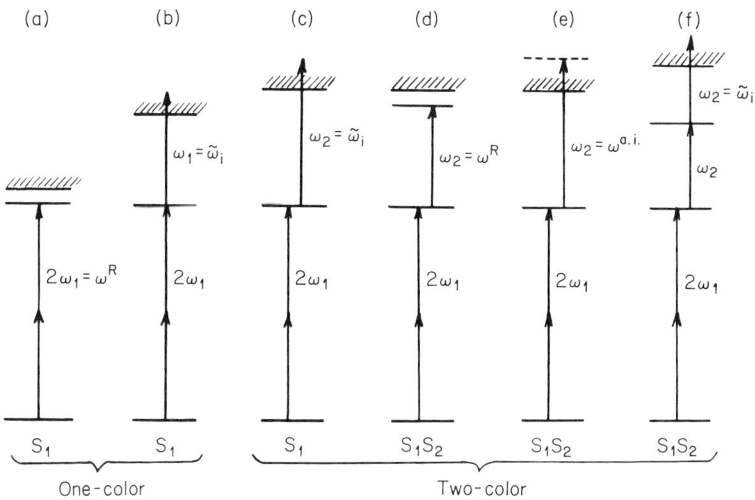

FIG. 3.18. Various multistep ionization schemes (type III) with two-photon resonance excitation at the first stage: (a) involving the excitation of a Rydberg state in a single-frequency field; (b) with a single resonant stage in a single-frequency field; (c) with a single resonant stage in a two-frequency field; (d) with two resonance excitation stages involving a Rydberg state; (e) the same involving an autoionization state; (f) three-step ionization with two resonant (two- and one-photon) excitation stages.

(d) two-step ionization in a *two-frequency* field, with the first (two-photon) and the second (one-photon) excitation stages being resonant:

$$A(2\omega_1, \omega_2^{R}) \rightarrow A^{**} \quad \text{(IIId)}$$

$$A(2\omega_1, \omega_2^{a.i.}) \rightarrow A^{+} \quad \text{(IIIe)}$$

(e) three-step ionization in a *two-frequency* field, with the first (*two-photon*) and the second (*one-photon*) excitation stages being resonant:

$$A(2\omega_1, \omega_2, \tilde{\omega}_2) \rightarrow A^{+} \quad \text{(IIIf)}$$

Finally, to effect resonance ionization of those elements that are especially difficult to excite, e.g., halogens and noble gases, use can be made of three-photon resonance excitation. The corresponding ionization schemes can be divided, by analogy with case III, as follows (Fig. 3.19):

(a) two-step ionization in a *single-frequency* field, involving a single three-photon resonance excitation stage:

$$A(3\omega, \tilde{\omega}) \rightarrow A^{+} \quad \text{(IVa)}$$

(b) two-step ionization in a *single-frequency* field, involving one three-photon resonance and one two-photon nonresonance excitation stages:

$$A(3\omega, 2\tilde{\omega}) \rightarrow A^{+} \quad \text{(IVb)}$$

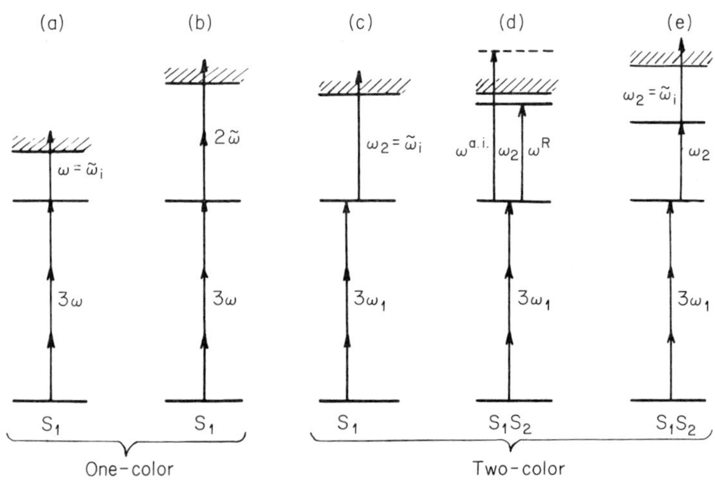

FIG. 3.19. Various multistep ionization schemes (stage IV) with three-photon resonance excitation at the first stage: (a) with a single resonant stage in a single-frequency field; (b) with a single resonant stage and two-photon excitation at the second stage in a single-frequency field; (c) with a single resonant stage in a two-frequency field; (d) with two resonance excitation stages; (e) three-step scheme with a two-frequency field.

(c) two-step ionization in a *two-frequency* field, involving one three-photon resonance and a one-photon nonresonance transitions:

$$A(3\omega_1, \tilde{\omega}_2) \rightarrow A^+ \tag{IVc}$$

(d) two-step ionization in a *two-frequency* field, involving one three-photon resonance and a one-photon resonance transitions:

$$A(3\omega_1, \omega_2) \rightarrow A^{**} \text{ or } A^+ \tag{IVd}$$

(e) three-step ionization in a *two-frequency* field, involving one (the first) three-photon resonance excitation stage and two one-photon stages of resonance and nonresonance excitation:

$$A(3\omega_1, \omega_2, \tilde{\omega}_2) \rightarrow A^+. \tag{IVe}$$

Of course, some other combinations of multiphoton and multistep transitions are also possible.

Note that for the realization of all potentialities of atomic photoionization spectroscopy involving multistep excitation, of great importance is the number of resonance excitation stages that affects the total selectivity of multistep ionization.

In the general case, the selectivity of multistep photoionization may be expressed as

$$S = S_{\text{exc}} S_{\text{ion}}, \tag{3.31}$$

where S_{exc} is the selectivity of excitation of the atomic stage being ionized and S_{ion} the selectivity of the ionizing stage. The excitation selectivity S_{exc} is in turn the product of the selectivities of all the excitation stages involved:

$$S_{\text{exc}} = \prod_k S_k, \tag{3.32}$$

where k is the number of the excitation stages. The selectivity of the ionizing stage depends on the method of ionization. For instance, where ionization is effected on transitions to the continuum, it is safe to assume, because of the great photoabsorption bandwidth, that $S_{\text{ion}} \simeq 1$, i.e., there is no selectivity at all. A similar situation is with the ionization of Rydberg atoms with an electric field, for the field ionizes any highly excited atoms. And it is only with photoionization through narrow autoionization states that a high selectivity can be realized at the last, ionizing stage. Indicated in Figs. 3.16–3.19 is the selectivity multiplication effect for various multistep excitation and ionization schemes. As can be seen, the simplest two-step photoionization scheme involving transitions to the continuum provides for no selectivity multiplication effect, while in three-step excitation schemes for an autoionization or a Rydberg state, it is at a maximum. Besides, the selectivity of any excitation stage depends on the number of the photons being absorbed in the course of the excitation process at this stage. The selectivity is apparently at a maximum for the one-photon resonance absorption process and is lower for the two- and especially three-photon resonance processes because of the broadening and shifting of the levels in the intense light field required to ensure the necessary high probability of excitation (see Section 2.4 and Fig. 2.15).

3.4.2. Optimal Atomic Ionization Schemes

The choice of ionization scheme (I–IV) is governed by the ionization potential of the given atom and the arrangement of its energy levels. Of no small importance is, of course, the available selection of lasers to carry out the resonance photoionization of the atom. As a rule, the atoms of any element are amenable to ionization by various types of schemes.

The possible resonance ionization schemes for each element in the periodic table have already been analyzed (Hurst *et al.*, 1979a). Our analysis is supplemented with consideration of possibility of utilizing three-step excitation (scheme II) and determination of the number of resonance excitation stages.

Appendix A presents Tables 1–5 listing transition wavelengths for the ionization schemes of various atoms, according to their position in the periodic table. To determine the radiation wavelength necessary to ionize the excited atom, we have calculated the boundary wavelength λ_{lim} corresponding to the atomic transition from the excited level to the ionization limit, as radiation with $\lambda < \lambda_{\text{lim}}$ is required to ionize the atom. Since there are several methods to ionize an atom from a high-lying Rydberg state, a transition to such a state can be chosen for the last stage of laser excitation. The Rydberg levels of various atoms are as yet not adequately studied. The tables therefore indicate approximate wavelength values for the transition to the level with the principal quantum number $n^* = 15$, which lies near the first ionization limit ($E_{\text{i}} - E_{n^*=15} = 487.5\ \text{cm}^{-1}$). According to the estimate (3.12), the electric field strength required to effectively ionize an atom from this state is $E > E_{\text{cr}} = 6.2\ \text{kV/cm}$.

The tables also give the values of the wavelengths $\lambda_{\text{i}}^{\text{exp}}$ and $\lambda_{\text{R}}^{\text{exp}}$ used to ionize atoms or to excite their Rydberg state in the experiments utilizing the indicated transition scheme. In some of the experiments, use was made of several similar transition sequences. In such cases, the tables indicate, as a rule, only a single characteristic sequence of the transitions employed. However, to list the experimental studies most fully, the tables sometimes (e.g., in the case of sodium) present several schemes that differ but slightly.

3.4.2.1. Alkali Metals. Table 1a (Appendix) lists data on the ionization of alkali metal atoms by the two-step scheme type I which is most common to these elements. The three-step scheme type III is more suitable for lithium, but can be used for the other atoms of this group as well. However, in so far as the ionization potential of alkali metals decreases with increasing serial number, the use of the three-step scheme for potassium, rubidium, and cesium is somewhat complicated by virtue of the fact that their resonance transition wavelengths lie in the far-red region. The table gives no ionization scheme for francium because the necessary spectral information for this element is lacking. But the presence of the known transition from the ground state FrI with $\lambda_1 = 7177$ Å allows one to use for this element the ionization scheme type Ib (Fig. 3.16).

The d-shell elements of the first group (Table 1b, Appendix) possess a higher ionization potential. They can also be ionized by the two- and three-step schemes. Their resonance transition wavelengths lie in the ultraviolet region of the spectrum that can be reached by doubling the frequency of a dye laser.

The same holds true also for the lightest alkali-earth metals, beryllium and magnesium (Table 2a, Appendix), whose ionization potentials are high. The other elements of this subgroup are conveniently handled by schemes type I and II: two- and three-step ionization without the use of frequency doubling.

The d-shell elements of the second subgroup (Table 3a, Appendix) have not very high ionization potentials, but their first resonance transitions lie in the near ultraviolet range. Therefore they are amenable only to the two-step atomic photoionization schemes, with use being made in most cases of the second-harmonic output of a dye laser or two-photon resonance excitation.

However, for the elements of group III with an incomplete d electron shell (Table 3b,c, Appendix), use can be made not only of two-step schemes but also of three-step ones. In the latter case, the first resonance transition is an intercombination type, i.e., such a scheme is less advantageous as far as energy is concerned. Especially numerous are experiments on the multistep photoionization of rare earth elements (Table 3c, Appendix) that require no frequency doubling.

Halogen atoms (Table 4a, Appendix) have high ionization potentials, and their resonance ground-state transitions lie in the VUV region. For this reason, their laser photoionization can only be effected by schemes involving multiphoton transitions under the action of an intense laser light. The use of three-photon resonance in this case requires so intense a radiation that it causes the two-photon ionization of the excited atom (scheme IV). A process with two-photon resonance (scheme III) is more advantageous here.

Ionization potentials are especially high in noble gases (Table 5, Appendix). Nevertheless, there are a number of experiments in which they have been ionized by means of a stepwise laser excitation in conjunction with a nonlaser excitation of the atom to a metastable state. Some works suggest that use should be made for the purpose of two- and three-step laser atomic ionization from the ground state, involving the generation of a VUV radiation by the four-wave nonlinear mixing of the output frequencies of several lasers (e.g., an excimer, neodymium, and dye lasers).

4. Photoionization Detection of Single Atoms

4.1. Problems of Single-Atom Detection

The ultimate goal of photoionization laser spectroscopy is to detect single atoms in a sample. Since only free atoms can be subjected to selective resonance photoionization, the process of detection consists of several successive stages. First, it is necessary to release the atoms from the sample; second, to deliver them into the region of irradiation and to irradiate all the atoms. Then, it is necessary to provide effective photoionization of the maximum fraction of the exposed atoms and, finally, to detect the resulting photoions. Each stage is characterized by its efficiency: η_{rel} is the probability of release of atoms from the sample to free space, η_{del} the efficiency of delivery of atoms to the irradiation volume, η_{irr} the probability of irradiation of atoms during their flight through the irradiation volume, Y_{ion} the yield or efficiency of resonance photoionization of atoms by laser light in the irradiation volume, and η_{det} the efficiency of collection and detection of photoions. The total probability of detection of a single atom in the sample η can be expressed as

$$\eta = \eta_{\text{rel}}\eta_{\text{del}}\eta_{\text{irr}}Y_{\text{ion}}\eta_{\text{det}}. \tag{4.1}$$

In Chapter 3 different methods and schemes of highly effective atomic photoionization (that is, the conditions when $Y_{\text{ion}} \simeq 1$) are considered. Under this condition it is possible to detect *single atoms in the zone of irradiation* which is typical of most of the experiments. When $\eta \simeq 1$, ultimate sensitivity, that is, the detection of *single atoms in the sample*, can be obtained. The attainment of this ultimate sensitivity essentially depends on specific conditions of experiment.

An atom in a sample is rather an indefinite concept, which may, in fact, have at least three versions: (1) the atom is inside the sample; (2) the atom is adsorbed on the sample; and (3) the atom is in the buffer gas. In the first case, the sample must be strongly heated or even completely evaporated so that the atom can be released. In the second case, it is sufficient to desorb the atoms being detected without heating or evaporating the sample. In both of these cases, the processes of release and delivery of atoms to the region under irradiation can be realized most effectively in a vacuum or in a low-pressure gas. In the third case, the atoms can be photoionized directly in the buffer.

The resonance photoionization of atoms and the detection of photoions in a buffer gas differ essentially from those in the vacuum. For example, the atoms in the vacuum quickly cross the region of irradiation, and it is easy to accelerate and detect the photoions by an electron multiplier. On the other hand, in the buffer the atoms stay long in the irradiated region, and the resulting photoelectrons can be detected by means of the electron avalanche in the gas. Because of this difference the numerical values of efficiencies at different stages in relation (4.1) differ greatly for these cases.

The detection of atoms on the sample surface is an intermediate case that compensates for the shortcomings of the two previous cases. The particles to be detected can accumulate on the surface for a long time and desorb to the vacuum just before laser irradiation. Therefore all three cases of photoionization detection of atoms—inside the sample, on its surface, and in the buffer gas—should be considered separately.

4.2. Detection of Atoms in a Vacuum

There are at least two types of detectors that can be used in this case. They differ in the motion of the atoms being detected: directed motion in a beam and random motion at a low pressure when collisions are negligible and the atoms move about freely (Fig. 4.1). The most widely used case is the first one, in which the atoms are ionized in the beam formed by the atoms released from the sample (Fig. 4.1a). The secondary electron multiplier is usually used as a detector of the ions deflected from the atomic beam by the electric field. In the second case, the resulting ions are directed to the channeltron-type electron multiplier (Fig. 4.1b). An important feature of this type of photoionization detector is that it reduces the contribution of the η_{del} geometrical factor which greatly restricts the detection sensitivity with an atomic beam. It is also possible to withdraw the resulting photoions through a hole in a cavity and then detect or use them in a more complicated experiment (acceleration, mass separation, etc.). Such a potential detector is schematically shown in Fig. 4.1c and discussed in more detail in Section 4.2.4.

The beamless detectors (Fig. 4.1b,c) are usable only in experiments in which the atoms are at a low pressure and there are no other charged particles but photoions. In real experiments, when the sample is heated to a

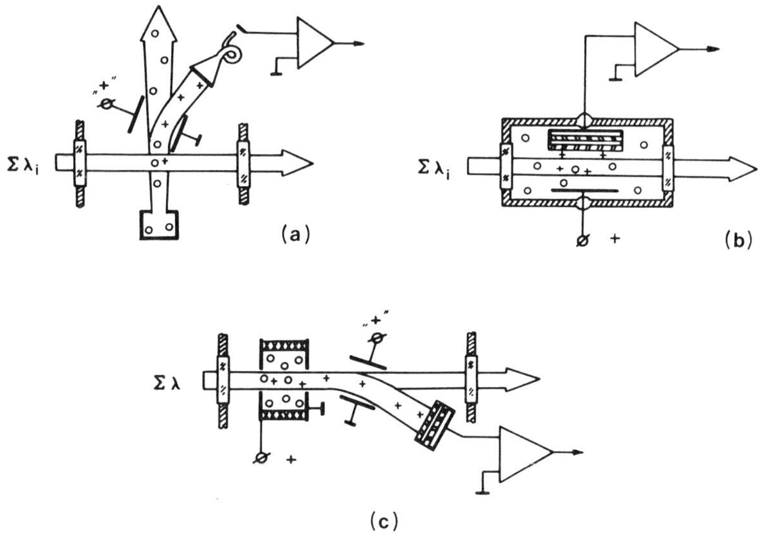

FIG. 4.1. Various types of atomic laser ionization detector. (a) Atomic-beam detector wherein photoions are registered by means of a secondary electron multiplier; (b) low-pressure gas cell in which photoions are registered by means of a multiplier in the form of a microchannel plate; and (c) enclosed low-pressure cavity wherein photoions are produced which are then extracted outward by an electric field.

high temperature, the detector's sensitivity is limited by the background of the thermal ions, which cannot be removed in a beamless detector. The spectral line broadening caused by the Doppler effect in a beamless detector, however, can be overcome by means of two-photon excitation of atoms in counter-running laser beams (see Section 2.2.4).

The photoionization detection of atoms in a beam was studied in detail in many experiments with Na atoms (Bekov *et al.*, 1978a,d), Yb atoms (Bekov *et al.*, 1978c,d), long-lived isotopes of Yb in metal foil (Bekov *et al.*, 1979), long-lived isotopes of Eu in foil ("off-line" operation with an accelerator) (Alkhazov *et al.*, 1983a,b; Zherikhin *et al.*, 1984a), and short-lived isotopes of Eu ("on-line" operation with an accelerator) (Alkhazov *et al.*, 1984; Fedoseyev *et al.*, 1984). We now discuss the characteristics of photoionization detection of atoms in a beam released from the sample into the vacuum.

4.2.1. Release of Atoms from the Sample and the Formation of an Atomic Beam

The typical features of the accumulation of atoms in the sample and their release into the vacuum are studied in the above-mentioned experiments with radioactive isotopes of Yb and Eu. These isotopes are produced in nuclear reactions under the action of a proton beam, through their subsequent separation by a mass separator and implantation into the sample (Bekov *et al.*, 1979; Alkhazov *et al.*, 1983a,b; Zherikhin *et al.*, 1984a).

According to the half-decay time of radioactive isotopes, there are three modes of operations with them. First, it is possible to implant mass-separated isotopes into different parts of the foil sample and then release them by heating in an oven, like the one shown in Fig. 4.2. This "off-line" mode of operation with isotope accumulation is possible only for comparatively long-lived isotopes ($T_{1/2}$ is several days or more). Second, the separated isotopes can be implanted and simultaneously evaporated in the form of an atomic beam. This method is suitable for isotopes with lifetimes of several seconds limited by the diffusion release time of implanted atoms. Third, it is possible to ionize selectively certain isotopes at the mass separator inlet similarly by laser radiation. This enables the separation of isotopes, that is, nuclei with the same mass at the inlet (not the outlet) of the mass separator.

For all of the operation modes, a tantalum target heated to 2000–3000°C is exposed to an intense beam of protons. The reactions between the nuclei

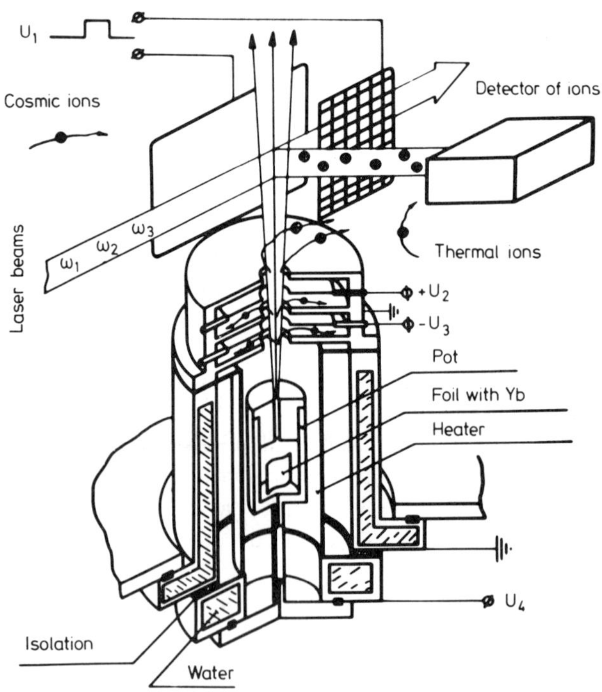

FIG. 4.2. High-temperature oven designed to produce an atomic beam of the trace element implanted into foil. [From Bekov *et al.* (1979).]

of the target and the protons give rise to almost the whole series of stable and unstable isotopes of various elements. In the source of ions, the reaction products are ionized; and after they are collimated, they enter the mass separator where the isotopes undergo spatial mass separation, and the ions of different isotopes with energies of about 30 keV fall onto different sections of tantalum foil (50 μm thick). The ions penetrate the foil to about 100 Å and are neutralized. The atoms turn out to be isolated from the outside, and under normal conditions they stay in the foil for a long time. Thus, the implantation of ions into foil is a convenient method of accumulating, storing, and transporting ultrasmall amounts of elements (Bekov *et al.*, 1979).

In the experiments, tantalum foil was used for implantation. As the foil was heated to over 1200°C, one could observe the diffisional release of Yb atoms from tantalum. In the experiments with Yb, the lower estimate for the probability of atomic release from heated foil was $\eta_{rel} > 0.07$. The dynamics of the release of implanted atoms from heated tantalum foil was investigated by using radioactive isotopes of Eu (Alkhazov *et al.*, 1983b). The authors disclosed several temperature zones in which the release of atoms from heated foil was most effective. In the first zone (600–800°C) about 40% of the implanted isotopes are released. In the second zone (1200–1500°C) the amount of atoms is about 30%. At 1310°C the half-release time of atoms is 10 s. In the third temperature zone, the rest of the isotopes (30%) are released. It becomes significant at a temperature over 1800°C. It is not clear yet why the temperature dependence of atomic release is so complex. The presence of several temperature zones governs the procedure of temperature increase in the course of the experiment. Since with increasing temperature the number of thermal ions rises, the low-temperature zones are the most efficient and convenient for experiments. In this case the atomic release time can be controlled from several to tens of minutes.

A beam of free atoms (Yb, Eu) was formed in the following way. The foil with implanted ions was placed in a tantalum crucible with a narrow cylindrical channel (Fig. 4.2). The crucible was located inside a tantalum tube that was heated electrically. The design of the oven, with its small dimensions, allowed the crucible temperature to be brought up to 1800°C. Upon being collimated by the narrow channel, the free atoms formed a well-directed atomic beam.

The inevitable thermal ions are reflected by the electrostatic protection system shown in Fig. 4.2. The five diaphragms placed in series along the path of the atomic beam have potentials: zero, $U-$, zero, $U+$, zero. The diaphragms, spaced at 5 mm, had holes 6 mm in diameter. At 1900°C the protection reduces the number of background thermal ions (without gating) to 3–4 s^{-1} with potentials $U-$ and $U+$ of several tens of volts. When $U-$ is reduced to below 20 V, the background increases drastically. For example, as $U-$ changes from -20 V to -12 V, the thermal ion background increases 10 times. A change in $U+$ does not lead to such a sharp dependence. Variations in $U-$ and $U+$ make it possible to find the minimum of the thermal background ions.

According to the degree of collimation of the atomic beam, 1–10% of the atoms released from the target can be delivered to the region of irradiation, that is, $\eta_{del} \ll 1$. This is the most serious limitation on the probability of detection of a single atom in the sample. Thus, much attention is being given to the development of schemes for photoionization detection that are free of this shortcoming. In Section 4.2.4, consideration is given to one such scheme based on photoionizing the atoms directly in the hot space of the atomic beam source (Andreyev *et al.*, 1985, 1986).

4.2.2. Irradiation of Atoms in a Beam

The average transit time τ_{tr} of atoms with velocity v_0 through the region of laser irradiation of size a is

$$\tau_{tr} = a/v_0, \tag{4.2}$$

where $v_0 = (2\kappa T/M)^{1/2}$ is the average thermal velocity of atoms at the evaporation temperature in the source T.

The laser radiation must irradiate all the atoms in a continuous atomic beam so that each atom will be detected. Two irradiation modes are possible: pulsed-periodic and continuous.

The probability of irradiation of an atom during its flight through the irradiated region is given by the relation

$$\eta_{irr} = f\tau_{tr}, \qquad f \leqslant 1/\tau_{tr}, \tag{4.3}$$

where f is the pulse repetition frequency. In order that all the atoms be intercepted, that is, the condition $\eta_{irr} \simeq 1$ be realized, the pulse repetition frequency f must satisfy the condition

$$f \gtrsim 1/\tau_{tr}. \tag{4.4}$$

In the typical experiment, $a \simeq 0.1$–1.0 cm and $v_0 = 10^4$ cm/s. Thus a high pulse-repetition frequency $f = 10$–100 kHz is required.

As far as the realization of condition (4.4) is concerned, it is dye lasers pumped by metal-vapor (Cu, Pb, Au) and excimer lasers with a high pulse-repetition frequency that are most suitable today. In the USSR, for example, a commercial sealed-off Cu-vapor laser (model ILGL-101) is available (Burmakin *et al.*, 1978). Its parameters are: average generation power is 4 W(510.6 nm—70%; 578 nm—30%), pulse-repetition frequency is 10 kHz (variable), laser pulse duration is 18 to 20 ns, radiation divergence is

10^{-3} rad, and guaranteed lifetime is 1000 h. This laser served as the basis for designing dye lasers with the following parameters (Zherikhin *et al.*, 1981): generation in the region 530–720 nm on alcoholic solutions of 110- , 6G- , and B-rhodamine and 17-oxazine, the 6G-rhodamine amplifier efficiency is 30% with a laser spectrum width of 0.8 cm^{-1}; the average generation power is up to 0.6 W; the efficiency of radiation transformation to the second harmonic in the region 265–360 nm is 5%. The power of sealed-off Cu lasers with a long lifetime can be increase by several times (Burmakin *et al.*, 1979). The development of similar lasers on the vapor of other metals—Au ($\lambda = 628.2$ nm) (Markova *et al.*, 1978) and Pb ($\lambda = 723$ nm) (Kirillov *et al.*, 1979)—will make it possible to widen the tuning range of dye lasers towards the red side (to 850–900 nm) and towards the blue side (to 340 nm) by generating the second harmonic.

The irradiation efficiency that can be obtained in experiments with a dye laser pumped by a Cu laser with $f = 10$ kHz is usually $\eta_{irr} \simeq 0.05$–0.1 (Zherikhin *et al.*, 1984a). It is possible, however, to increase η_{irr} to 0.3–1.0 by using a multipass irradiation geometry (Alkhazov *et al.*, 1984).

As the atomic beam is exposed to CW laser radiation, it is easy to attain $\eta_{irr} \simeq 1$. But in this case the multistep photoionization efficiency falls rapidly, since it is difficult to excite the atoms at the W_{exc} rate that exceeds the rate of radiative decay of the kth intermediate state $1/\tau_{dec}^{(k)}$. Second, it is very difficult to use efficiently the entire radiation of a laser beam with its cross-sectional dimension a, because the relaxation time of intermediate levels is usually much shorter than the time of flight of atoms through the laser beam, that is,

$$\tau_{dec}^{(k)} \ll \tau_{tr} \simeq a/v_0. \tag{4.5}$$

Therefore CW irradiation can be applied only when there is no need to reach ultimate values of Y_{ion}.

4.2.3. Effective Photoionization of Atoms in a Beam

In the method of multistep photoionization, the atoms are excited by laser radiation to an intermediate state in one or several steps; then the excited atoms only are photoionized (see Chapter 3).

According to the method of ionization from the intermediate state there may be nonresonance photoionization of excited atoms and resonance ionization of excited atoms. In the first case, the excited atoms are ionized by additional or the same radiation. This case is characterized by a low value of photoionization cross section ($\sigma_{ki} = 10^{-17}$–10^{19} cm^2) and accordingly by a high value of saturation energy fluence ($\Phi_{sat}^{(i)} = 0.01$–1 J/cm^2). In the second case, the excited atoms are transferred either to an autoionization state, which decays to the continuum in a short time, or to a high-lying Rydberg state with subsequent ionization by electric field pulses. The resonance excitation cross sections of autoionization or high-lying states exceed the nonresonance photoionization cross section by several orders.

In order that the maximum fraction of atoms should be intercepted, the repetition rate of laser pulses f must be sufficiently high (10^4–10^5 Hz). Under these conditions the average laser radiation intensity essential for ionization to the continuum must amount to several kilowatts per square centimeters:

$$I_{av} = f\Phi_{sat}^{(i)} = f/\sigma_{ki} = (10^4\text{–}10^5\ \text{Hz})(0.01\text{–}1.0\ \text{J/cm}^2) \\ = 10^2\text{–}10^5\ \text{W/cm}^2. \tag{4.6}$$

It is much more efficient to ionize the atoms through Rydberg or autoionization states for which $\Phi_{sat}^{(i)} \simeq 10^{-6}$–$10^{-4}$ J/cm^2. Therefore, even when every atom moving with a thermal velocity is exposed to radiation, the average intensity of ionizing radiation will be

$$I_{av} = (10^4\text{–}10^5\ \text{Hz})(10^{-6}\text{–}10^{-4}\ \text{J/cm}^2) = 10^{-2}\text{–}10\ \text{W/cm}^2. \tag{4.7}$$

Thus, in experiments with atomic beams it is advisable to use the method of resonance ionization from excited states through Rydberg or autoionization states. It is in this case that a high pulse-repetition frequency can be obtained, due to moderate requirements on their energy.

In Chapter 3 (Fig. 3.12) the ion yield for a three-step excitation of the Rydberg state of an atom and subsequent ionization of highly excited atoms by an electric field pulse is given as a function of the laser-pulse energy fluence at each step and the electric field strength. Under optimal experimental conditions (all quantum transitions are saturated), about one-half the atoms in the zone of detection at the instant the laser pulses arrive are excited to a high-lying state and every highly excited atom becomes ionized.

Under such conditions the dependence of ion signals on the temperature of an oven with pure ytterbium was studied (Fig. 4.3a). As the temperature was decreased, at first the amplitude of the signal gradually dropped; then it became highly unstable, which was conditioned by the fluctuations of the number of atoms in the excitation volume. At low beam density, such fluctuations may be of the same order as the average number of atoms in the irradiation volume.

The atomic beam density can be made so low that in most cases there will be no more than one atom in the irradiation volume. The statistics of appearence of atoms in the zone of detection was studied under these conditions by ion counting (Fig. 4.3b,c). The distribution obtained was similar to the Poisson distribution.

The three-step photoionization of atoms in a beam through autoionization states under the action of laser pulses with a high repetition rate was studied

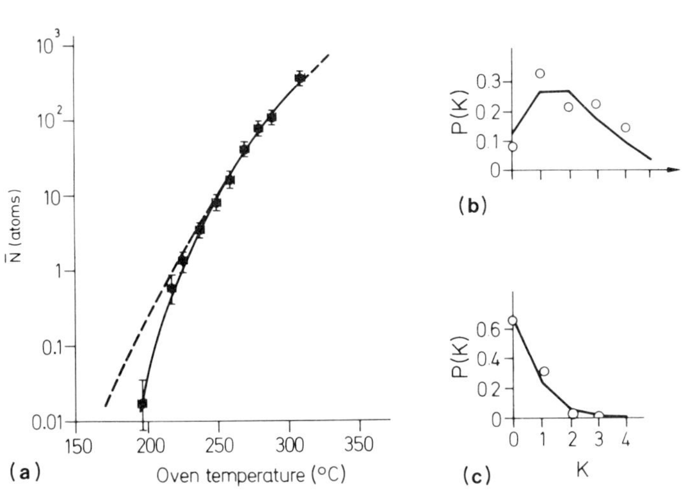

FIG. 4.3. Temperature dependence of the ion signal of Yb and the fluctuation of the number of atoms in the detection volume. (a) Dependence of the Yb ion yield on the temperature in the oven at maximum ion yield (the dashed curve is extrapolation; the solid curve is experiment); (b) the distribution of the number of Yb ions detected in the time $T_{obs} = 10$ s with the average number of ions $\bar{N}_i = 3$ and the average number of atoms in the excitation space $\bar{N} = 0.04$; and (c) the same as (b) with $T_{obs} = 20$ s, $\bar{N}_i = 0.35$, $\bar{N} = 0.003$. [From Bekov *et al.* (1978c).]

in detail for Eu atoms (Zherikin *et al.*, 1984a). This study was necessary to reach a maximum measurement sensitivity of the hyperfine and isotopic structures of short-lived isotopes of Eu accessible in very small amounts. Figure 4.4 shows a simplified scheme of this experiment. The isotopes were formed in the tantalum target (20 g) of the mass separator irradiated by a beam of protons with an energy of about 1 GeV and an intensity of about 10^{12} s^{-1}. An ion beam with a separated isotope energy of 30 keV was directed from the mass separator into a sealed tantalum tube (50 mm long and 5 mm in diameter). When it was heated to 1500°C, the implanted ions escaped in the opposite direction as a collimated beam of neutral atoms. A system of diaphragms at different potentials cut off the thermal ions and electrons. The atoms of Eu were excited to an autoionization state with its energy of 53271 cm^{-1} in a three-step process ($\lambda_1 = 576.520$ nm, $\lambda_2 = 557.266$ nm, and $\lambda_3 = 555.7$ nm) by three laser beams at a right angle. The laser beams were passed through the atomic beam 15 times in order that the fraction of exposed atoms (η_{irr}) and the ionization efficiency could be increased (Fig. 4.4) (Alkhazov *et al.*, 1984; Fedoseyev *et al.*, 1984). The average radiation powers of the dye lasers excited by a Cu laser ($f = 10$ kHz) were $P_1 = 0.04$ W, $P_2 = 0.2$ W, and $P_3 = 0.3$ W, respectively, at the first, second, and third steps. At such powers it was possible to provide saturation at the first and the second transitions; but because of a low value of cross section of

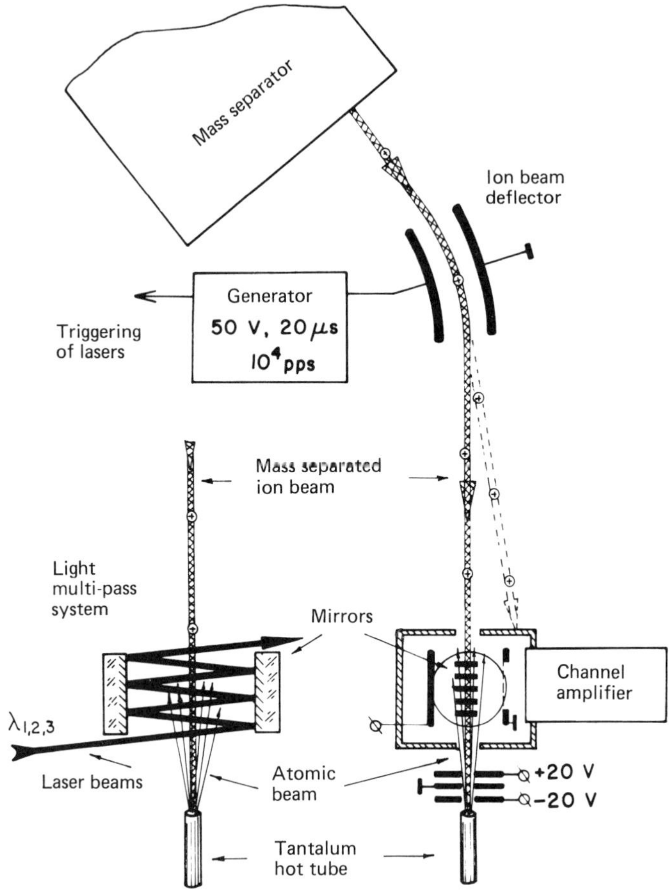

FIG. 4.4. Arrangement of ion, atomic, and laser beams during the photoionization of radioactive isotopes on-line with a proton accelerator. [After Alkhazov *et al.* (1984); Fedoseyev *et al.* (1984).]

the transition to an autoionization state ($\sigma_{\text{a.i.}} = 5 \cdot 10^{-17}$ cm^2), the intensity at the third step was about 100 times lower than the saturation intensity.

The total efficiency of detection of Eu atoms, that is, the ratio of the flow of detected photoions to the flow of radioactive isotopes implanted in the heated oven, was $\eta = 3 \times 10^{-4}$. Figure 4.5 shows the origin of this value, which according to (4.1) was caused by several effects. The efficiency of release of implanted ions into the vacuum in the form of neutral atoms $\eta_{\text{rel}} \simeq 1$. This was proved by the direct measurements described in Section 4.2.1. Considerable losses occur when the atomic beam is formed. As a result, only 2.5% of

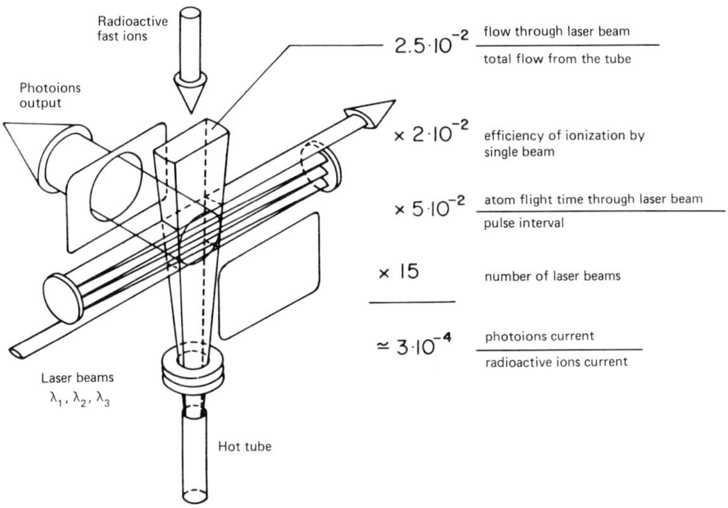

FIG. 4.5. Total efficiency of photoionization detection of Eu atoms in a sample.

atoms are delivered into the region under radiation, that is, $\eta_{del} = 2.5 \times 10^{-2}$. The efficiency of ionization of the atoms in this region by three laser pulses at the same time $Y_{ion} = 2 \times 10^{-2}$. As noted above, this low value is conditioned by using the red wavelength transition to an autoionization state at the third step, which has comparatively a small excitation cross section. Finally, the time of flight of atoms through one light beam, the diameter of which is 20 times smaller than the interval between the laser pulses, $T = (1/f) = 10^{-4}$ s. Therefore, only a small fraction of atoms (5×10^{-2}) is irradiated by one laser beam. In order that the coefficient of interception of atoms by laser radiation be increased (by up to 15 times), the beam must pass through different nonoverlapping region of atomic beam. This provides an increase of η_{irr} by 15 times. As a result, in accordance with (4.1), the total efficiency of photoionization detection of one radioactive ion implanted into a neutralizing hot tube is $\eta = (1)(2.5 \times 10^{-2})(5 \times 10^{-2})(2 \cdot 10^{-2})(15) = 3 \cdot 10^{-4}$. This is rather a high value of atomic detection in a sample. Moreover, if the multistep photoionization scheme is improved by exciting a Rydberg state, we can expect Y_{ion} to be 0.5, and accordingly, the detection efficiency can be increased to $\eta \simeq 10^{-2}$.

4.2.4. Method of Enhancement the Total Ionization Efficiency

The analysis shows that it is rather easy to detect single atoms in the region of irradiation but much more difficult to do it in a sample. The main obstacles are, first, the delivery of atoms into the irradiated region in the case of an atomic beam is not efficient, and second, there is partial interception of the atoms because of their rapid flight through the irradiation volume. So, in order that these shortcomings should be overcome, a method based on multiple flight of atoms through the region under irradiation is being developed. It is considered in more detail below.

The conventional method of photoionization of atoms in a beam, even when lasers with a high pulse-repetition rate (up to 10^4 Hz) are used, makes it possible to intercept just several percent of all the atoms with the laser beam diameter $a \simeq 1$ mm. The new approach allows increasing the efficiency of interception of atoms of the oven and subsequent extraction of the resulting ions by electric field through a small hole in the wall (Andreyev *et al.*, 1985, 1986). The small hole serves both to admit the laser beam and extract the photoions. Every chaotically moving atom inside the cavity has two possibilities: it either comes into the region of the laser beam during pulsed irradiation and is ionized, or it escapes through the hole in the wall. Let us introduce the ratio between the probabilities of these two processes: the probability that the atom will be irradiated upon entering the laser beam region, P_{irr}, and the probability that the neutral atom will exit through the hole in the cavity wall, P_{exit}. In the case of a simple cavity model with an average size of l and two holes (Fig. 4.6), the following estimate is valid for this ratio:

$$P_{irr}/P_{exit} = 2fl/v_0 k_{wall}, \tag{4.8}$$

where f is the laser pulse repetition frequency, v_0 the mean atomic velocity, and k_{wall} the transmission efficiency of the channel hole in the cavity wall.

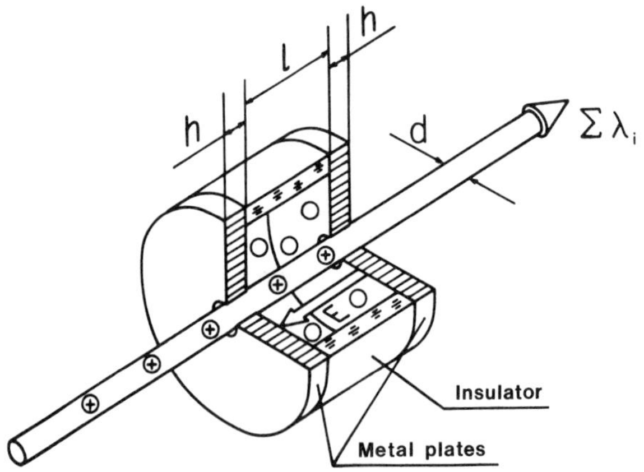

FIG. 4.6. Scheme of atomic photoionization in an oven cavity and the extraction of ions from it. [From Andreyev *et al.* (1986).]

Consider a typical experiment with a Cu laser ($f = 10^4$ Hz) and a small cavity ($l = 1$ cm) with $k_{wall} \simeq 0.3$ (at $h/d = 3$, h being the wall thickness and d the hole diameter) (Ramsey, 1966). At $v_0 = 5 \times 10^4$ cm/s, $P_{irr}/P_{exit} \simeq 1.5$, i.e., the atom has a 60% probability of being irradiated and a 40% probability of leaving the cavity without suffering irradiation. Expression (4.8) does not contain the hole diameter d, which is assumed to be equal to the laser beam diameter a. So the value of d can be arbitrary within certain limits, but for the multistep photoionization to be saturated, it is advisable to increase the laser radiation intensity by decreasing d.

We can also obtain a simple expression similar to Eq. (4.1) for the ionization detection probability, which is determined by the probability of the atom being ionized in the cavity, $\eta_{ion} = Y_i P_{irr}/(Y_i P_{irr} + P_{exit})$, the probability of the ion being extracted from the cavity, η_{extr}, and the probability of the ion being detected, η_{det}:

$$\eta = \eta_{ion}\eta_{extr}\eta_{det} = [Y_i P_{irr}/(Y_i P_{irr} + P_{exit})]\eta_{extr}\eta_{det}$$

$$= \frac{Y_i}{Y_i + (v_0 K_{wall})/(2f\,l)}\eta_{ext}\eta_{det}. \tag{4.9}$$

If the condition $P_{irr}/P_{exit} \gg Y_i^{-1}$ is satisfied, almost a 100% probability of atomic ionization can be attained even at a low per-pulse ionization yield Y_i, because the atom crosses the irradiation region many times.

To extract the resulting photoions we can use a flat condenser formed with two walls of the cell (Fig. 4.6). With $d \ll l$ the hole essentially does not distort the electric field uniformity. The basic cause that does not enable the ions to be pulsed out through the hole is that they have a transverse velocity. For effective extraction of photoions their transverse shift during pushing out must not exceed the dimension of the hole. From this the value of thc ejecting voltage U can be estimated. For example, when $l = 1$ cm, $d = 1$ mm, $h = 3$ mm, and $T = 800$ K, then $U \simeq 100$ V.

The efficiency of photoionization and the extraction of photoions in this way was investigated experimentally for Na (Andreyev *et al.*, 1985) and Sr (Andreyev *et al.*, 1986b) atoms. They demonstrated the possibility of extracting almost all the resultant photoions and attaining a total cfficiency of ionization of atoms in the cavity of $\eta \gtrsim 50\%$.

This method is particularly convenient for ionizing atoms through Rydberg states. In this case, the extraction electric field can serve also for the ionization of the atoms in the Rydberg states.

The loss of resolution due to the rejection of the atomic-beam geometry can be eliminated by using the two-photon excitation of the atom in a standing wave, which excludes the Doppler broadening (Vasilenko *et al.*, 1970) (see Section 2.2.4).

4.3. Detection of Atoms in a Buffer Gas

In use are two types of atomic detector in which the ionization and detection of an ion takes place in the presence and on account of collisional effects in a buffer gas. These are the thermionic diode and the proportional counter. Both types have long been known in experimental physics. The laser light makes it possible to excite and ionize in a resonance fashion the atoms being detected. Collisions in the thermionic diode cause ionization of highly excited atoms, and those in the proportional counter result in the development of an electron avalanche in the presence of an electric field. Let us discuss in short both types of detector.

4.3.1. Atomic Detection in Thermionic Diodes

The hot-cathode gas diode (Hertz, 1923; Kingdon, 1923) is not as popular as the proportional counter, and so its principle of operation needs to be explained. This device uses the phenomenon of very effective collisional ionization of highly excited atoms. Such an ionization detector of excited atoms operates in the pressure range 10^{-3}–1 Torr, a range intermediate between the pressures required for vacuum detectors (less than 10^{-5} Torr) and the proportional counter (1–10^3 Torr) described in Section 4.3.2. Although the hot-cathode gas diode detector has long been used in photoionization studies (Foote and Mohler, 1925; Lawrence and Edlefson, 1929), a new interest in this instrument has been aroused in connection with the experiments on the laser ionization of alkali metal vapors (Collins *et al.*, 1974; Curry *et al.*, 1976; Harvey and Stoicheff, 1977). The use of the hot-cathode gas diode for the laser spectroscopy of the Rydberg levels of atoms (Niemax, 1982) proved especially effective. Special modifications of the hot-cathode gas diode were developed for the purpose, in which the effect of the electric field on the Rydberg atoms was minimized (Harvey, 1981; Thompson and Stoicheff, 1982). Following these works, we shall consider briefly the principle of operation of the laser hot-cathode gas diode detector.

Figure 4.7 shows a schematic diagram of a typical thermionic diode detector using an electrostatic shield to minimize the effect of the electric field. The lower part of the detector is a diode with a tungsten thermionic cathode surrounded by a cylindrical anode. The work function of tungsten is reduced in the presence of alkali metal vapors, thus causing thermionic emission and a flow of current through the diode. When the electron emission is high enough, the electron cloud around the cathode forms a toroidal potential barrier, the negative space charge limiting the current through the diode. The positive ions formed in the detector enter the potential minimum of toroidal shape where they can circulate for a long time, compensating the

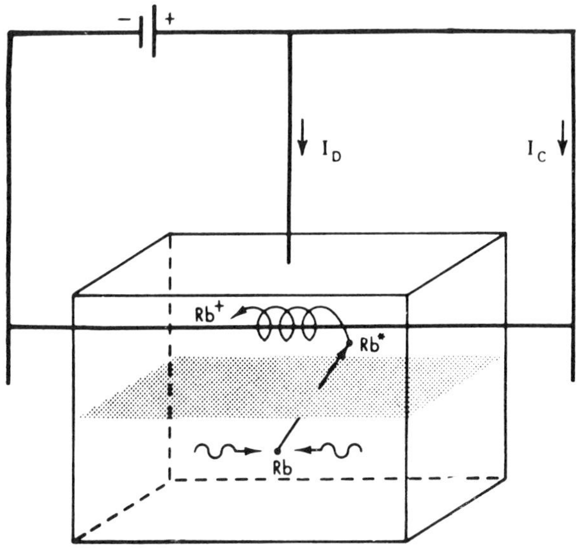

FIG. 4.7. Schematic diagram of a thermoionic diode detector with an electrostatically screened region (upper section). In the lower section, the highly excited atom A* produced as a result of laser resonance excitation are ionized by collision to yield the ions A^+ which drift into the detection (diode) section. [From Thompson and Stoicheff (1982).]

space charge. The sensitivity of the detector is high because a positive ion of charge e compensates a space charge equal to the product of the current i through the diode by the lifetime τ_i of the ion in the potential well. It is for this reason that the sensitivity of the detector is increased by $(i\tau_i/e) \simeq 10^4$–10^6 times (the ratio of the change in the diode current to the ion current).

To eliminate the effect of the electric field, use is made of laser radiation to generate highly excited atoms in the upper part of the detector, which is separated from the diode by a fine wire mesh. Such screening is necessary because even a weak electric field strongly affects the Rydberg atoms (the polarizability of a Rydberg atom is $\sim n^7$, see Section 3.3.1 and Table 3.3). The positive ions are produced as a result of collisions between the highly excited atoms and the atoms of a buffer gas filling the detector at a moderate pressure (e.g., Ne at 0.1 Torr). The efficiency of collision ionization of highly excited atoms distant from the ionization threshold by an amount $\delta E \gtrsim kT$ is close to 100% (see Section 3.3.4 and Fig. 3.15). The ions produced diffuse through the mesh and into the diode where they enter the potential minimum and reduce the space charge.

Another modification of the hot-cathode gas diode detector with a reduced effect of the electric field on Rydberg atoms has a ring-shaped structure (Beigang *et al.*, 1984). In such a ring-shaped diode, the cathode comprises several electrically connected wires arranged axisymmetrically round a tubular anode. Because of the symmetry, the electric field at the diode axis, in the laser irradiation region, is cancelled. Still another modification enables us to experiment with elements having very low vapor pressures, for which it is difficult to ensure the establishment of a negative space charge. In this case, use can be made of activation of the cathode with another element (Lorenzen *et al.*, 1984).

The optimum operating conditions and characteristics of the laser hot-cathode gas diode detector were investigated in a number of works (Harvey, 1981; Thompson and Stoicheff, 1982; Niemax, 1983). With the detection conditions being optimized, the sensitivity of the detector can reach as high a value as 1–10 ions/s, although it is, in principle, limited by the presence of charged particles in the detector volume and cannot provide for the detection of single atoms. Nevertheless, the hot-cathode gas diode detector is fairly convenient for experiments with highly excited atoms available in sufficient amounts. A series of such experiments are discussed in Chapter 7.

4.3.2. Laser Detection in the Proportional Counter

In the proportional counter, the buffer gas pressure and the applied voltage are high enough to provide for the development of an electron avalanche every time an electron–ion pair appears as a result of the laser resonance ionization of an atom subject to detection.

The photoionization detection of atoms in buffer gases was studied in detail in many experiments with Cs atoms (Hurst *et al.*, 1977a,b,c) and Cs atoms formed by the nuclear fission reaction of ^{252}Cf (Kramer *et al.*, 1978). Also, the process taking place in detecting the photoion–photoelectron pair in a proportional counter were studied (Hurst *et al.*, 1978a, 1979a). The characteristics of photoionization detection of atoms in buffer gases taken from these works are discussed below.

4.3.2.1. Photoionization Efficiency in Buffers. As atoms are detected in a buffer gas, the time of their stay in the exposed region is defined by their diffusion in the gas. The time of diffusion of atoms from the spherical region with its diameter a in this case is

$$\tau_{\rm dif} = (a^2/16D)(p/760), \tag{4.10}$$

where D is the diffusion coefficient at a pressure of 1 atm and p is the buffer gas pressure (Torr). In the case of Rb atoms in argon (1 atm), for example, $D = 0.25\ \text{cm}^2\ \text{s}^{-1}$ (Franzen, 1959). When $a = 0.1$–1 cm and $p = 760$ Torr, $\tau_{\rm dif}$ ranges from 0.25 to 2.5×10^{-3} s. In order that all the atoms in this region should be irradiated, the laser pulse repetition rate must satisfy the condition:

$$f \gtrsim 1/\tau_{\text{dif}}, \tag{4.11}$$

which is not so rigid as (4.4). Therefore, it is possible to apply even non-resonance photoionization of excited atoms that requires a high energy fluence of the ionizing pulse; nevertheless, we can restrict ourselves to a moderate average laser radiation power

$$I_{\text{av}} = f\Phi_{\text{sat}}^{(\text{i})} = (10^2\text{–}1\ \text{Hz})(0.01\text{–}1\ \text{J/cm}^2) = 0.01\text{–}10^2\ \text{W/cm}^2. \tag{4.12}$$

The first successful experiment on the detection of single atoms by means of two-step photoionization in buffer gases was carried out under such conditions (Hurst *et al.*, 1977a,b). The Cs atoms were excited in the buffer to the $7^2P_{3/2}$ state by the radiation ($\lambda_1 = 4593$ Å) of a lamp-pumped pulsed dye laser. The same pulse ($\lambda_2 = \lambda_1$) photoionized the excited atoms at the transition to the continuum, as had been done for atoms at two-color excitation (Ambartzumian *et al.*, 1971, 1976a; Ambartzumian and Letokhov, 1972). The laser radiation fluence ($\Phi_1 = \Phi_2 = 0.2$ J/cm^2) during focusing was sufficient to saturate both transitions, which ensured 100% conversion of the Cs atoms to photoions in the irradiated volume (0.05 cm^3).

Compared with the previous experiments with Rb atoms, the new feature of this experiment was a buffer gas used as a working mixture (argon and methane mixture at 100–200 Torr) in the ionization chamber of the proportional counter. This enabled increasing the signal from one photoelectron by about 10^4 times. Figure 4.8 shows the distribution of counter pulse amplitude for the case when a Cs atom is detected in 20 laser pulses. For comparison, it also shows a single-electron distribution attained with a mercury lamp, the light of which interacts with the counter walls. It is well-known that one electron in the proportional counter gives an exponential-like amplitude distribution of pulses. The fact that these distributions are similar points to the sensitivity of the counter being sufficient for stable detection of one Cs atom in the volume of laser irradiation.

Photoionization detection of single atoms in buffer gases is of great interest for detecting nuclear reaction products. The first experiment aimed at detecting single daughter atoms formed several seconds after the nuclear fragment stops was carried out in Oak Ridge (Kramer *et al.*, 1978). The authors investigated the reaction of spontaneous fission of ^{252}Cf into non-symmetric nuclear fragments, one of which was the nucleus of Cs. They applied the photoionization technique to detect Cs atoms in the buffer gas. The laser detector of Cs atoms is schematically shown in Fig. 4.9. A piece of foil with implanted Cf ions was placed in front of the inlet of a surface-barrier detector responsive to nuclear decay fragments. The fission intensity

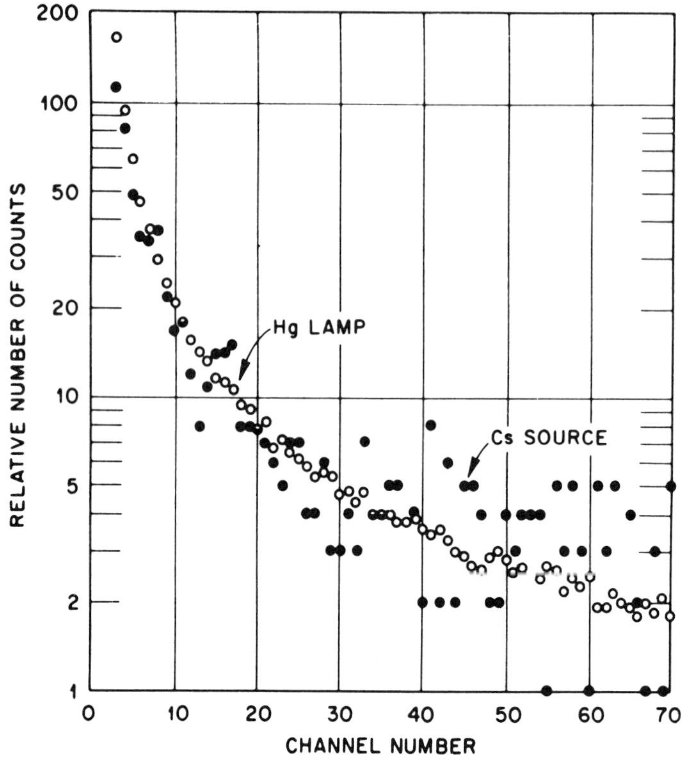

FIG. 4.8. Distribution of the pulse amplitude of a proportional counter in the case of detection of a single cesium atom in the irradiation region during 20 laser pulses. Shown for comparison is the one-electron distribution obtained with the incoherent light source, a Hg lamp. [From Hurst *et al.* (1977a,b).]

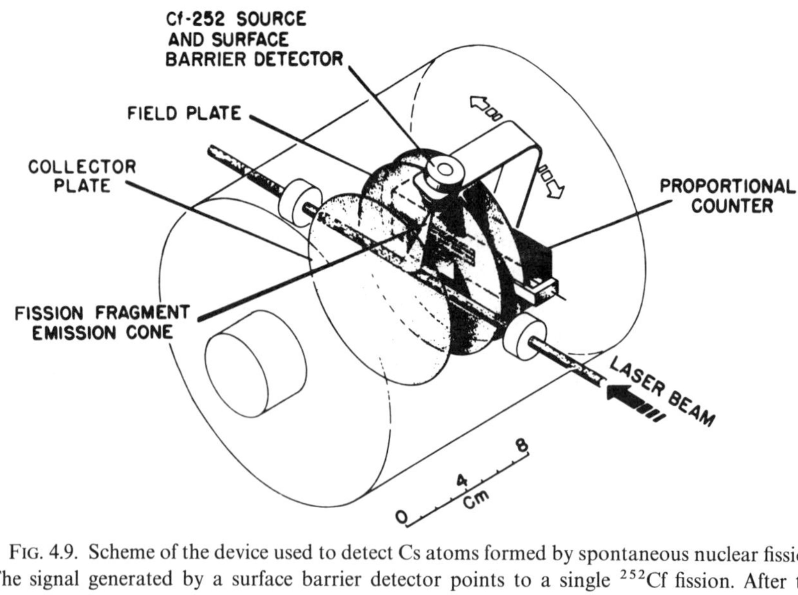

FIG. 4.9. Scheme of the device used to detect Cs atoms formed by spontaneous nuclear fission. The signal generated by a surface barrier detector points to a single ^{252}Cf fission. After the initial ionization has been swept onto the collector plate, a dye laser pulse is used to photoionize the selected single atom. The released photoelectron drifts in the proportional counter and is detected as a single event. [From Kramer *et al.* (1978).]

was about one event per second. The laser beam was cut by the axis of the nuclear fragment collector at 5 cm from the source. Deceleration was performed in the gas mixture (90% Ar + 10% CH_4) at a pressure of around 300 Torr.

A heavy fragment of Cf nuclear fission produced 10^6 electrons in the ionization track, which were collected on the plate in 2 μs. The collecting plate was charged negatively 40 μs after the nuclear decay and 10 μs before the laser pulse. The photoelectrons formed by selective two-step photoionization drifted through the hole into the proportional counter that detected them. Resonance excitation was performed by a lamp-pumped pulsed dye laser, the spectrum of which covered the wavelengths of two transitions: $6^2S_{1/2} \rightarrow 7^2P_{3/2}$ (4555 Å) and $6^3S_{1/2} \rightarrow 7^2P_{1/2}$ (4593 Å). The same pulse performed at the same time the nonresonant photoionization of excited Cs atoms.

The experiment demonstrated that a considerable fraction of the Cs nuclei was thermalized to neutral atoms. Thus, when the buffer gas composition is optimal, it is possible to realize conditions when almost all high-speed ions are neutralized and, hence, to apply laser methods for the detection of single atoms to the particles produced initially as high-speed ions.

Controllable laser generation of photoelectron–photoion pairs with thermal energy in the gas medium of the proportional counter made it possible to investigate the fluctuations in proportional counter gain (Hurst *et al.*, 1978a). On this basis they proposed to use the difference in the amplitude distributions of the signals, as the electron avalanche starts from one to two photoelectrons, in order to discriminate the cases of detection of one or two atoms in the region of irradiation (Hurst *et al.*, 1979b).

4.3.2.2. Advantages and Disadvantages of Photoionization in Buffer Gases. The photoionization detection of atoms in a buffer gas has definite pros and cons compared with the detection in a vacuum. Some of the pros can be summarized as follows:

(1) When the buffer gas composition and pressure are chosen and a potential is applied across the gas cell, it is possible to detect photoelectrons with a proportional counter. This makes it possible to detect one atom in the irradiated volume (Hurst *et al.*, 1977a,b).

(2) Due to collisions, the buffer gas can ionize highly excited atoms. This effect is used to detect Rydberg atoms in the thermionic diode, although its sensitivity (10^3 atoms) is far from the limit required, that is, a single atom (Niemax, 1983).

(3) The buffer gas limits the free escape of atoms from the irradiated volume during the diffusion time τ_{dif} expressed by (4.10). This property is useful for detecting atoms in the absence of lasers with a sufficiently high pulse-repetition rate.

(4) The buffer gas provides neutralization of the nuclear reaction products initially formed as ions with a high kinetic energy. This property was demonstrated in the above-described experiment aimed at detecting the Cs atoms formed by the fission of ^{252}Cf atoms (Kramer *et al.*, 1978). The positive ions lost their kinetic energy in the buffer gas; then, cooled to a thermal energy, they captured an electron and were neutralized. After that, they could be detected by photoionization.

The cons of photoionization detection in buffer gases are as follows:

(1) It is necessary to use a very pure buffer gas in order to prevent the atoms being detected from participating in the chemical reactions with the residual impurities for a long time of detection. For example, in the case of metal atoms, even negligible impurities of oxygen are dangerous. This disadvantage is very essential in analytical applications when the background impurities of the buffer gas restrict the detection sensitivity of trace atoms.
(2) A considerable pressure of the buffer gas brings about a collisional broadening of spectral lines, thus screening such important details of the spectrum as the hyperfine and isotopic structures. So the detection of atoms in the buffer gas may be Z-selective (selective with respect to the type of atom) and not A-selective (nonselective with respect to the number of neutrons, that is, the mass number A).

In summary, it may be said that the pros of detection in buffer gases are methodic by character while the cons are principal. Therefore the photoionization detection in a vacuum is the most promising trend, and its shortcomings must be overcome in a certain way. One of the possibilities based on photoionizing the atoms in a oven cavity, as they cross the irradiated region many times, was considered in Section 4.2.4. Another approach based on the combination of the pulsed release of atoms from the sample surface and their subsequent pulsed photoionization. This interesting approach is worthy of more detailed consideration.

4.4. Detection of Atoms on the Surface

The detection of the atoms on the sample surface solves two difficult problems on the way to attaining the ultimate aim, that is, the detection of single atoms in a sample. First, it is easy to perform pulsed release of the atoms on the surface into a vacuum, and then to photoionize them. This allows irradiating almost all the atoms, that is, realizing $\eta_{\text{irr}} \simeq 1$ in (4.1). Second, this makes it possible to collect atoms on the surface from a big volume for a long time for their subsequent pulsed release. This operation permits realizing at the same time the ultimate values of η_{del} and η_{irr} in relation (4.1).

4.4.1. Pulsed Release of Atoms

Pulsed release of the atoms from the sample surface can be realized by several methods: pulsed heating of the sample surface with subsequent evaporation or desorption of the neutral particles from the surface or sputtering of the particles from the surface by electron or ion beams. Both of these methods were demonstrated with success in experiments with atoms and molecules.

Laser desorption and subsequent photoionization were experimentally investigated in many experiments with molecules on the surface. The formation of molecular ions was observed as UV laser pulses of moderate-intensity (10^4–10^6-W/cm^2) irradiated molecular crystals (Antonov *et al.*, 1980c, 1981b). A detailed study of this process showed that the process had two stages. The laser radiation pulse slightly heated the surface, which resulted in increasing the probability of molecular desorption from the surface during a laser pulse. The molecules that had escaped the surface were ionized then by the same radiation (Egorov *et al.*, 1983). This mechanism made it possible to realize the conditions when these two functions of laser pulse are divided between two different laser pulses—IR desorbing and UV photoionizing pulses, as is shown in Fig. 4.10a (Antonov *et al.*, 1983; Egorov *et al.*, 1984a,b). This method of resonance photoionization mass spectroscopy of molecules is considered in more detail in Chapter 6.

Pulsed sputtering of atoms from the surface by an ion beam, with subsequent photoionization and analysis of the photoions with a mass spectrometer, was demonstrated with success (Winograd *et al.*, 1982; Kimock, *et al.*,

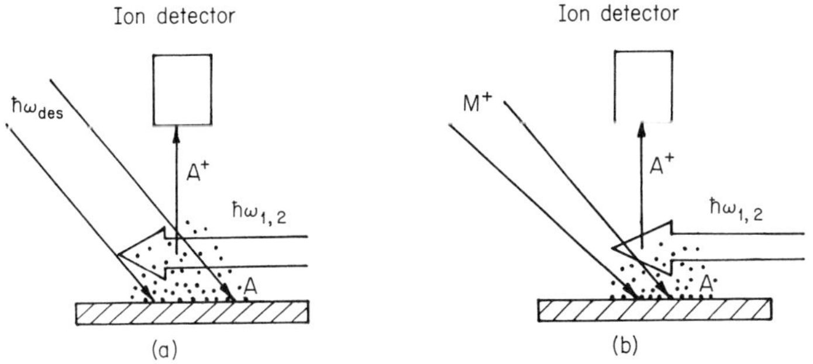

FIG. 4.10. Laser photoionization detection of atoms or molecules adsorbed on the surface (a) through their thermal desorption when the surface is under pulsed laser heating or (b) through sputtering of the surface layer with a pulsed ion beam.

1983; Parks *et al.*, 1983), as is shown in Fig. 4.10.b. The sputtering of the sample surface in the case of bombardment by ions with an energy of several keV is widely used (Behrisch, 1981). Its efficiency is described by the sputtering yield Y_{sp}, which is the average number of atoms escaping the sample surface per one incident particles:

$$Y_{sp} = \frac{\text{number of escaping atoms}}{\text{number of incident ions}}.$$

This quantity varies within wide limits (from 0 to 100), but its typical value ranges from 1 to 5. Above the threshold energy lying in the range 20–40 eV, the value of Y_{sp} increases and reaches its maximum when the ion energy is 5–50 keV. The neutral atoms escape the surface both in the ground state and in low-lying excited states. The fraction of excited atoms depends on the conditions of the experiment, particularly the state of the surface, and may become 50% (Winograd, 1984). The secondary ions, the fraction of which varies within wide limits (from 10^{-3} to 10^{-1}) also escape the surface. They can be easily removed with electrostatic protection. It is quite possible to realize sputtering using ion sources (usually Ar^+) with an energy of 20–30 keV, a current pulse of 10 mA, and a duration of 10^{-5} s. Such an ion beam can be focused onto a spot that is several millimeters in size.

The method of pulsed ion sputtering is quite suitable for detecting atomic traces when the laser system has a low pulse-repetition rate. Indeed, pulsed release of atoms and their synchronous irradiation in a vacuum do not call for a high pulse-repetition rate. Thus photoionization can be carried out using the methods of both resonant and nonresonant photoionization of excited atoms, since it is not difficult in this case to satisfy condition (4.6) because of a low laser pulse repetition rate. This essentially simplifies the requirements on the lasers used to realize step photoionization of atoms.

4.4.2. Accumulation of Atoms on the Surface

In the case of photoionization detection of atoms adsorbed on the surface, it is possible to accumulate atoms for a long time with their subsequent pulsed desorption and photoionization in a vacuum. Such a concept was elaborated for molecules at the Institute of Spectroscopy (Egorov *et al.*, 1983, 1984a,b) and for inert gas atoms in Oak Ridge (Lehmann *et al.*, 1983; Chen *et al.*, 1984; Hurst *et al.*, 1984b).

Particles can be accumulated on a cooled surface due to their adsorption. Because of this their concentration on the surface N_{surf} may be much higher than above the surface in a vacuum N_{vap}. In a simplest approximation, the variation of particle concentration on the surface is defined by the difference of particle flows towards the surface and from the surface,

$$\frac{dN_{\text{surf}}}{dt} = N_{\text{vap}} \frac{v_0}{4} - \frac{1}{\tau_{\text{ad}}} N_{\text{surf}}, \tag{4.13}$$

where v_0 is the average velocity of particles, and τ_{ad} is the average time of the particle's stay on the surface.

The time τ_{ad} is related to the adsorption energy E_{ad} and the surface temperature T by the Arrhenius formula,

$$\tau_{\text{ad}} = t_0 \cdot \exp(E_{\text{ad}}/kT), \tag{4.14}$$

where $t_0 = 10^{-12}$–10^{-14} s is the characteristic period of vibrations of an adsorbed particle at the surface.

When $t \gg \tau_{\text{ad}}$, the particle concentration on the surface is in equilibrium,

$$N_{\text{surf}} = N_{\text{vap}}(v_0 \tau_{\text{ad}}/4). \tag{4.15}$$

From this the enhancement of the detection sensitivity G can be evaluated from comparison of two cases: (1) during desorption of all the molecules from the S_{des} area and their subsequent photoionization in the V_{irr} volume; and (2) during photoionization of the equilibrium molecules in the same V_{irr} volume of gas:

$$G_{s/v} = \frac{N_{\text{surf}} S_{\text{des}}}{N_{\text{vap}} V_{\text{irr}}} = \frac{V_0 \tau_{\text{ad}}}{4} \frac{S_{\text{des}}}{V_{\text{irr}}}. \tag{4.16}$$

If the irradiation geometry has the same form as in Fig. 4.10a, when $V_{\text{irr}} \simeq aS_{\text{des}} \simeq a^3$ (a is the laser beam diameter), the maximum gain in sensitivity is

$$G_{s/v} = v_0 \tau_{\text{ad}}/4a, \tag{4.17}$$

where $v_0 \tau_{\text{ad}}/a$ is the ratio of the distance covered by a molecule in the time τ_{ad} to the diameter of irradiated region a. In other words, due to the accumulation effect it is possible to collect on the surface the particles from an equivalent volume $v_0 \tau_{\text{ad}} a^2$ with their subsequent photoionization in a smaller volume $\simeq a^3$. For example, with $v_0 = 2 \times 10^4$ cm/s, $\tau_{\text{ad}} \simeq 1$ s, $a \simeq 0.5$ cm the gain factor will be $G_{s/v} \simeq 10^4$.

This method of sensitivity increase was checked for the case of napthalene molecules adsorbed on the surface of a molecular crystal (rhodamine 6 G) to 200 K. In this case $E_{\text{ad}} = 0.5$ eV, $\tau_{\text{ad}} = 0.18$ and $G_{s/v} = 2 \times 10^3$ (Egorov *et al.*, 1983, 1984a,b). This method was used to detect only 10^2 molecules of naphthalene and anthracene on the surface of graphite (Antonov *et al.*, 1983).

A simlar concept was developed for photoionization detection of inert gas atoms distributed initially over a big volume (10^3 atoms of Kr in several litres of vacuum space) (Lehmann *et al.*, 1983; Chen *et al.*, 1984; Hurst *et al.*,

1984b). The authors of these works called the accumulation of atoms on a cold surface the "bunching of atoms". For Xe and Kr atoms on the surface of nickel and copper, $t_0 \simeq 10^{-15}$–10^{-20} s, $E_{ad} = 0.22$–0.28 eV. The temperature at which these atoms stay on the surface for 10^{-6} s is equal to 110 K (for Kr) and 130 K (for Xe). When the surface is heated by a laser to a higher temperature for a time longer than 10^{-6} s, full desorption of atoms from a metal surface can be attained. This method allows detection of 10^3 atoms of Kr through photoionization. The scheme of step photoionization of ^{81}Kr atoms with a VUV laser source ($\lambda_1 = 116.5$ nm) and two other lasers ($\lambda_2 = 558.1$ nm, $\lambda_3 = 1.06$ μm) is given in Chapter 3 (Fig. 3.17b). Isotopic selectivity was attained not by isotopic-selective laser excitation but through photoion mass separation. The problem of laser photoionization detection of rare isotopes is discussed comprehensively in Chapter 9.

5. Multistep and Multiphoton Ionization of Molecules

That it is possible to create a laser-selective molecular detector based on a combination of multistep molecular photoionization and mass-spectrometric registration of the photoions produced was realized fairly long ago (Ambartzumian and Letokhov, 1972)[1]. Our laboratory started developing this direction of laser photoionization spectroscopy in parallel with the work on laser detection of atoms. The first successful experiments on the photoionization of molecules by VUV laser radiation were performed in 1975 (Andreyev *et al.*, 1975), and later the two-step photoionization of H_2CO molecules by two-color laser radiation was carried into effect (Letokhov, 1976a,b; Andreyev *et al.*, 1977).

The situation with molecules, however, is much more complicated than that with atoms. Actually, in spite of the fact that selective laser photoionization of atoms has been extensively developed, the method of laser photoionization spectroscopy of atoms cannot be directly transferred to molecules for a number of reasons:

(1) The excited states of polyatomic molecules are of a much more complicated nature, and molecular absorption spectra often have little structure. This makes it much more difficult to realize photoionization of polyatomic molecules with high selectivity.

(2) In contrast to atoms, molecules in intermediate excited states can undergo different photophysical and photochemical transformation (for example, photodissociation), which can compete with photoabsorption and photoionization of processes.

(3) The products of photoionization of polyatomic molecules, especially with the use of intense laser radiation, can be very diverse, and this essentially requires the use of the mass-spectrometric technique of analysis and detection of photoions.

[1] We note that two-step photoionization of molecules was proposed for laser separation of isotopes (Robieux and Auclair, 1963), but this method has not been developed.

(4) Molecular photoabsorption cross sections, as a rule, are several orders of magnitude smaller than the atomic photoionization cross sections, which makes it necessary to use more powerful sources of radiation. At the same time, the molecular absorption and photoionization spectra usually lie in the UV and VUV regions, which are relatively inaccessible to laser sources of radiation.

Nevertheless, in recent years, considerable progress has been made in the development of photoionization laser spectroscopy of molecules (Johnson, 1980; Antonov an Letokhov, 1981; Schlag and Neusser, 1983; Antonov *et al.*, 1984). This is explained, first of all, by progress in laser technology, especially the development of lasers in the UV and VUV regions. Second, the combination of the method of resonance photoionization of molecules with mass-spectrometric identification of the photoions formed in the form of the so-called "two-dimensional mass–optical spectrometer", proposed previously (Ambartzumian and Letokhov, 1972; Letokhov, 1976a), has turned out to be very effective. This method was used to realize, for the first time, laser detection of single molecules in a volume irradiated by a single laser pulse (Antonov *et al.*, 1981a).

5.1. Methods and Regimes of Multiphoton Ionization of Molecules

Several basic schemes of multistep excitation, shown in a simplified manner in Fig. 5.1, can be used for resonance photoionization of molecules. Two-step photoionization via intermediate excited electronic–vibrational states (Fig. 5.1a) is the simplest type of selective stepwise photoionization of molecules. Since with excitation of electronic states the energy required to ionize molecules decreases by several election volts, it is possible to photoionize only excited molecules by appropriately selecting the quantum energy of the ionizing radiation $\hbar\omega_2$ (see Fig. 5.1a). The selectivity of the ionization in this case is determined by the selectivity of photoionization of molecules in the intermediate state. The first experiments in this direction were performed using independent laser pulses in the first and second steps (Andreyev *et al.*, 1977; Antonov *et al.*, 1977b). This made it possible not only to optimize photoabsorption and photoionization of excited molecules but also to investigate the kinetics of the electronically excited states.

Since for most molecules the bands of intense electronic excitation are situated in the ultraviolet region of the spectrum ($\lambda < 300$ nm), which is not easily accessible by tunable lasers, the methods of multiphoton ionization of molecules have been extensively developed. The method is based on the use

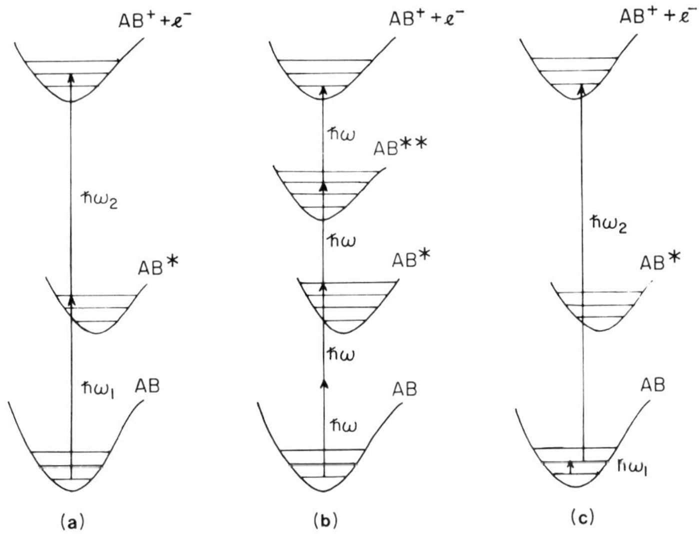

FIG. 5.1. Scheme of resonance laser photoionization of molecules: (a) via an intermediate electronically excited state (two-color resonant photoionization); (b) via an intermediate two-photon resonance (multiphoton ionization); and (c) via a vibrational excited state (IR–UV photoionization).

of a frequency-tunable dye laser with sharp focusing of the radiation down to the diffraction limit (Johnson *et al.*, 1975; Petty *et al.*, 1975). In this case, a radiation intensity exceeding 10^{10} W/cm^2, with which the multiphoton absorption and ionization processes of molecules proceed efficiently, is achieved at the focal point. As the wavelength of the radiation is tuned, resonances arise at the doubled or tripled frequency of the laser with subsequent single- or two-photon ionization of the molecule (Fig. 5.1b). The presence of intermediate resonances greatly increases the yield of ions, and, as a result, a multiphoton ionization spectrum forms. This method turns out to be very effective for investigating the spectra of excited electronic states of molecules, especially those that are forbidden for single-photon transitions.

Vibrational spectra are most characteristic of polyatomic molecules, so that the scheme of two-step photoionization via intermediate vibrationally excited states (Fig. 5.1c) must in principle have higher selectivity than the scheme via the electronic states. In this scheme, however, due to the small change in the ionization energy of excited molecules (0.1–0.3 eV), ionization of unexcited molecules occurs as well, which greatly complicates the experimental realization of this scheme. For this reason, only a small number of experiments have been performed along this line (Sudbo *et al.*, 1979a; Esherick and Anderson, 1980).

Several characteristic regimes of photoionization of molecules can be singled out depending on the intensity of the laser radiation in the UV or visible region of the spectrum:

(1) Resonance two-step (or more) photoionization in weak fields, when the saturation parameter of optical transitions $G \ll 1$, where G is determined by the relation (2.38). In this case, we have a pure n-step process, which terminates when the molecule reaches the ionization limit. The power-law dependence of the yield of photoions on the radiation intensity is determined by the number of absorbed photons, and the yield of ions itself greatly depends on the properties of the intermediate electronically excited states.

(2) Resonance stepwise photoionization in strong fields (saturation parameter $G \simeq 1$). This case is of greatest interest for detection of molecules since the efficiency of photoionization of a molecule approaches 100%. However, on reaching the first ionization potential, the molecules as a rule continue to absorb energy, and this leads to their subsequent fragmentation. Fragmentation of molecules under the action of intense laser radiation, first observed in experiments on multiphoton ionization (Zandee and Bernstein, 1979), have become the object of a large number of investigations from the points of view of studying the physics of the processes that occur and of finding the optimum conditions for photoionization detection of molecules.

(3) Multiphoton ionization. The intensity of the radiation is high enough for ionization of molecules via intermediate multiphoton resonances. Subsequent resonant optical transitions fall in the region of strong saturation ($G \gg 1$). Strong fragmentation of molecules, up to their atomization, occurs.

5.2. Resonance Stepwise Photoionization of Molecules

In multistep photoionization, the quantum transition of a molecule at each step is a resonant single-photon transition. For example, in the case of two-step photoionization, the energy of the first photon corresponds to the transition of the molecule into a real excited state. Since a transition into the ionization continuum occurs at the second step, the condition for a resonance is automatically satisfied. Due to the complicated quasi-degenerate structure of the levels of polyatomic molecules in an intermediate electronic state, as well as due to the presence of different intramolecular relaxation mechanisms, the phase memory of the system in the intermediate state is lost very rapidly. This permits examining the two-step process noncoherently and describing it by simple equations for the occupancy of the levels (see Section 2.1.2.) With respect to selectivity of ionization, this method has the important advantage of permitting independent selection of the wavelength of the exciting and ionizating radiation using two time-synchronized lasers. Variation of the delay time of the ionizing pulse relative to the exciting pulse gives the additional possibility of investigating the kinetics of photophysical processes in excited intermediate states of the molecule.

The investigation of stepwise photoionization in a mass spectrometer, when fragment ions are observed together with molecular ions, is of special value. In this case, we are dealing with a new method, in which it is not simply the occupancy of the excited state of the molecule that is being probed, but the capability of the molecule to give rise to different fragments in a two-step process is investigated. The appearance of fragment ions accompanying photoionization from an electronically excited state can depend on such factors as the presence of excitation in particular types of vibrations and the equilibrium nuclear configuration of the molecule in the excited state.

In strong laser fields, when the saturation parameter $G \simeq 1$, the nature of the interaction of the molecule with the laser radiation becomes more complicated: molecular ions formed by two-step photoionization as well as neutral fragments in competing photodissociation processes can absorb additional photons, and this in general leads to the appearance of a large number of fragment ions. For this reason, a problem that is important for the detection of single molecules is finding the conditions under which the efficiency of stepwise photoionization is close to unity without loss of information on the starting molecules due to the disappearance from the mass spectra of peaks of molecular and other characteristic ions. On the other hand, the investigation of the mechanisms of interaction of polyatomic molecules with intense laser radiation is of interest in itself and permits obtaining new information on the excited states of a molecule and its neutral and ionized fragments.

5.2.1. Experimental Technique

An important factor in all experiments on resonance photoionization of molecules is the use of intense laser radiation, which permits attaining quite high ionization efficiency in multistep photoionization via an intermediate electronically excited state; in order to achieve saturation of the optical transition in the first step, the energy fluence of the first radiation pulse $\Phi_1(\omega_1)$ must satisfy condition (2.94):

$$\Phi_1(\omega_1) \geqslant \Phi_{\text{sat}}^{(1)} = 1/(2\sigma_{\text{exc}}), \tag{5.1}$$

where σ_{exc} is the cross section for resonant excitation. This relation is valid in the optimum case, when the duration of the exciting pulse τ_{p} is shorter than the relaxation time of the excited state, T_1. Appreciable depletion of the excited electronic state due to stimulated transitions into the ionized state is achieved with an energy fluence Φ_2 in the second laser pulse with frequency ω_2:

$$\Phi_2(\omega_2) \geqslant \Phi_{\text{sat}}^{(2)} = 1/\sigma_{\text{ion}}, \tag{5.2}$$

where σ_{ion} is the cross section for the transition from the excited state into the ionization continuum. For characteristic cross sections of excitation and ionization of molecules of 10^{-17}–10^{-18} cm^2 and lifetimes of intermediate excited states $T_1 \simeq 10^{-8}$–10^{-10} s, the energy densities and intensities of saturation of optical transitions constitute, respectively, 0.1–1 J/cm^2 and 10^7–10^9 W/cm^2. In the case of multiphoton ionization, the characteristic cross section of two- and three-photon transitions constitute, respectively, 10^{-30}–10^{-34} cm^4/W and 10^{-43}–10^{-49} cm^6/W^2, so that in order to achieve an acceptable ionization efficiency it is necessary to use radiation intensities $I > 10^{10}$–10^{11} W/cm^2.

Other important requirements imposed on laser radiation sources in experiments on multistep photoionization are related to the spectral characteristics of the molecules. The electronic absorption bands of most molecules are situated in the ultraviolet region of the spectrum, so that photon energies $\hbar\omega \gtrsim 3$–5 eV are required in order to excite electronic singlet states. The ionization potentials of polyatomic molecules usually lie in the range 8–12 eV, so that efficient photoionization of molecules out of electronically excited states in a single step requires photons energies $\hbar\omega = 4$–8 eV.

Considerable progress in quantum electronics in the development of UV and VUV lasers has led to the fact that experimentalists now have at their disposal, in addition to the harmonics of solid state ruby and neodymium lasers, a number of powerful laser radiation sources based on excimer molecules XeCl ($\hbar\omega = 4$ eV), KrF ($\hbar\omega = 5$ eV), and ArF ($\hbar\omega = 6.4$ eV) (Rhodes, 1979). Use is also being made of the VUV laser based on the F_2 molecule ($\hbar\omega = 8$ eV) (Woodworth and Rice, 1978), which has replaced the VUV H_2 laser used in the early experiments on the photoionization of molecules (Knyazev *et al.*, 1975). In experiments with tunable radiation, the first and second harmonics of dye lasers ($\lambda = 260$–460 nm) pumped by nitrogen or Nd:YAG lasers are used.

Radiation pulses from gas-discharge lasers are usually synchronized with the help of hydrogen thyratrons, which serve to commutate the discharge circuits of lasers with nanosecond triggering precision (Antonov, 1978).

To record photoions without separation by mass, a simple ionization chamber in the form of a cylindrical or flat capacitor is used (Andreyev *et al.*, 1975; Freuholz *et al.*, 1980). To analyze photoions by mass in the presence of fragmentation, a mass spectrometer of any type that records ions with the help of a secondary electronic multiplier, which provides a detection efficiency of 30–100%, can be used. However, a time-of-flight mass spectrometer, which permits recording the entire mass spectrum of photoions for a single laser pulse, is the most suitable. Definite measures must be adopted to eliminate possible nonselective photoionization of molecules in the ion source by photoelectrons and desorption of ions from electrode surfaces under the action of scattered light. The presence of traces of oil and lubricants in the vacuum system can result in a large background due to ionization of impurities by intense UV radiation. The vapor pressure of the specimen in the ion source of the mass spectrometer usually does not exceed 10^{-5} Torr, which guarantees that the molecular photofragmentation precesses will be collisionless.

5.2.2. Two-step Photoionization Spectra of Molecules

The method of stepwise laser photoionization permits recording molecular absorption spectra at very low concentrations, because the act of absorption of a photon by a molecule is recorded by the appearance of an ion with maximum sensitivity. We note that, in contrast to the fluorescence method, it is not necessary for the excited state to emit.

This important advantage of the photoionization technique was already used in the experiments on the one-photon ionization of molecules in a beam by VUV radiation from a common source (an H_2 lamp). The technique was employed to observe, for example, the narrow autoionization resonances of NO in a cooled molecular beam to have a distinct vibrational structure in the photoionization spectrum in the range 1350–1185 Å (Ng *et al.*, 1976).

The spectrum of stepwise photoionization of molecules via intermediate electronically excited states is determined by the dependence of the photoionization yield on the frequency of the laser radiation at each step. Far from saturation, the yield of ions from two-step photoionization in the absence of relaxation of the intermediate state is determined by the simple expression:

$$N_i = N_{exc}\sigma_{ion}(\omega_2)\Phi_2 = N_0 q \sigma_{exc}(\omega_1)\Phi_1\sigma_{ion}(\omega_2)\Phi_2, \tag{5.3}$$

where N_i is the number of ions formed per pulse, N_{exc} the number of excited molecules, N_0 the number of molecules in the irradiated volume, σ_{exc} the cross section for excitation of molecules, q the fraction of molecules interacting with the exciting radiation at frequency ω_1, σ_{ion} the cross section of photoionization of molecules from the excited state, and Φ_1 and Φ_2 are the energy fluences of the radiation pulses at the first and second steps in photons per square centimeter. The function $\sigma_{exc}(\omega_1)$ determines the excitation spectrum of the molecules. The cross section for ionization of excited molecules $\sigma_{ion}(\omega_2)$ depends on both the specific excited electronic–vibrational level and the excess energy of two photons, $\hbar(\omega_1 + \omega_2)$, above the ionization limit of the molecule.

For ordinary single-photon ionization in the case of direct ionization, the molecule goes from the ground state into the ground state of the ion, and the dependence of the photoionization cross section on the radiation wavelength is approximated by a step function, which reflects the Franck–

Condon factors for transitions to the vibrational levels of the ion. Due to the superposition of transitions from different vibrational sublevels of the ground electronic state of polyatomic molecules, under ordinary conditions the stepwise nature of the dependence is smoothed out and the ionization efficiency reaches a peak smoothly in the region of vertical Franck–Condon transitions (Akopian, 1977). This also remains in force for two-step photoionization. However, in this case, the transition of the molecule into the state of the ion occurs from the excited electronic–vibrational levels and must therefore be included in the Franck–Condon factor between the two states.

The first experimental investigations of molecular two-step photoionization spectra were performed for the nitrogen dioxide molecule (Antonov *et al.*, 1977b). The molecule NO_2 is a convenient object for such experiments, because it absorbs in the visible region of the spectrum. The two-step photoionization spectrum of this molecule was investigated in the region of dye-laser wavelengths 447–457 nm. The ionization potential of the molecule is $I = 9.78$ eV, so that with the use of a hydrogen laser with photon energy $\hbar\omega_2 = 7.7$ eV at the second step, the excess of energy of two photons above the ionization potential constituted $I - \hbar(\omega_1 + \omega_2) = 0.8$ eV. Figure 5.2 shows the dependence of the yield of two-step photoionization of nitrogen dioxide on the radiation wavelength λ_1 at the first step. The maximum magnitude of the photosignal corresponds to the formation of 3×10^4 ions/pulse.

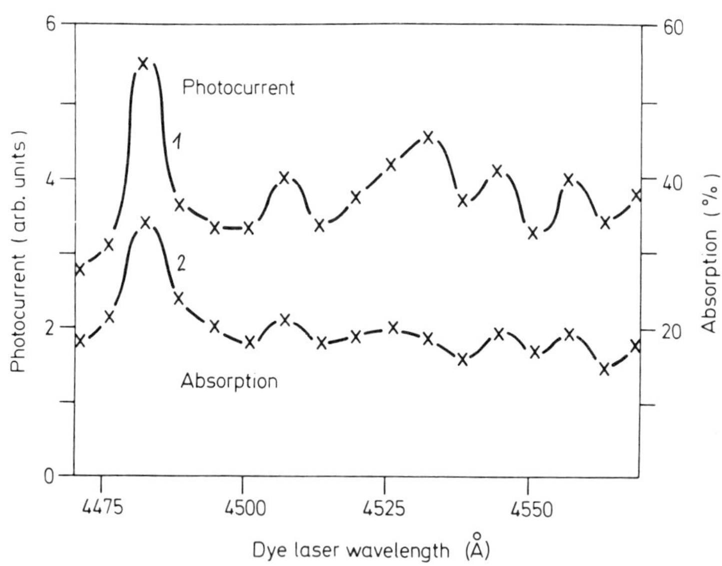

FIG. 5.2. Two-photon photoionization (curve 1) and electronic absorption spectrum (curve 2) of nitrogen dioxide molecules. [From Antonov *et al.* (1977b).]

The same figure shows the usual measured spectrum of electronic absorption of the molecule. It follows from Fig. 5.2 that the two-step photoionization spectrum reproduces, to within the measurement error, the absorption spectrum of the molecule. This indicated that transitions into the ionization continuum far from the ionization limit from different electronic–vibrational states have a much weaker spectral dependence. This characteristic was confirmed in a number of experiments with other molecules (Brophy and Rettner, 1979a; Antonov and Shibanov, 1982).

A characteristic of the two-step photoionization of nitrogen dioxide under the conditions indicated was the unexpectedly low magnitude of the cross section for photoionization from the excited state $\sigma_{\text{ion}} = 10^{-21}$–$10^{-22}$ cm^2 (Antonov, 1982). At the same time, for a number of molecules, the measured cross section of photoionization out of excited electronic states falls in the range 10^{-17}–10^{-19} cm^2 (Boesl *et al.*, 1981). This can be explained by the considerable change in the configuration of the molecule accompanying a transition into the ionization state (Nakayama *et al.*, 1959). Indeed (Fig. 5.3), in the ground state the molecule is bent, forming an angle $\theta_A = 134°$, while the ion has a linear configuration. This leads to the fact that with single-photonionization the yield of photoions near threshold is quite low due to the small Franck–Condon factors (Donnelly and Kaufman, 1978). The

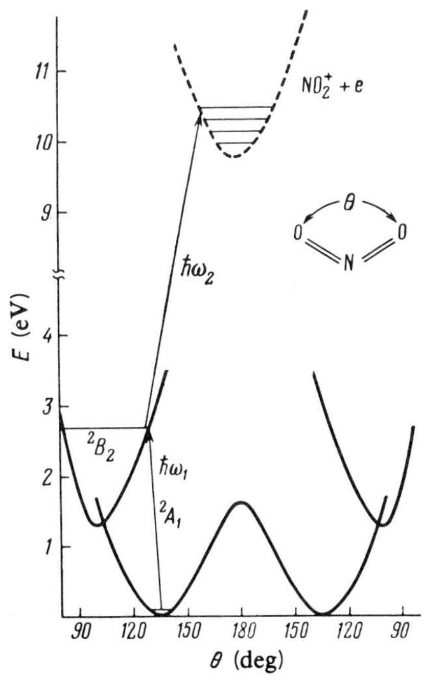

FIG. 5.3. Potential energy curves for the ground and electronically excited states of the NO_2 molecule and ion. [From Antonov (1982).]

situation is even less favorable for two-step photoionization. Absorption by the molecule in the first step for radiation wavelength in the range 440–460 nm is due primarily to the transition into the electronically excited state $^2B_2$, in which the molecule is even more strongly bent than in the ground state: $\theta_B = 111°$ (see Fig. 5.3). For this reason, the Franck–Condon factors for the transition from the $^3B_2$ state into the ground state of the ion must be very small.

A more complicated picture of two-step photoionization can be observed when the energy of the two photons is near the ionization limit of the molecule. As for single-photon ionization, in the two-step process discrete states of the neutral molecule can be excited above the ionization limit. These states can then decay via nonradiative processes into the ionization continuum (autoionization). Single-photon ionization states are well known for many molecules, but in two-step photoionization, other states can be excited. The relation between direct ionization and autoionization processes in two-step photoionization of polyatomic molecules was studied for diatomic molecules of alkaline dimers (Leutwyler *et al.*, 1980, 1982) and for polyatomic molecules (Duncan *et al.*, 1981). In the experiments, the wavelength of the radiation at the first step was fixed and the spectrum of the photocurrent was recorded by tuning the wavelength of the ionizing radiation. The use of supersonic cooled molecular beams permitted removing the hot vibrational bands and Doppler broadening from the optical transitions of the molecules.

In the case of diatomic molecules of alkali-metal dimers, sharp autoionization resonances were observed, which featured even isotopic effects, e.g., for isotopic K_2 molecules (Leutwyler *et al.*, 1980). To illustrate, Fig. 5.4 shows the two-step photoionization efficiency curve for the $^{39,39}K_2$ molecule in the case of excitation of its ($B^1\Pi_u$; $v'' = 1$) intermediate state. The character of the spectral dependence is governed more critically by the vibrational quantum number of the intermediate state than by the total energy of the two photons. Measuring such dependences enables the photoionization potential of the molecule (indicated by the asterisk in Fig. 5.4) to be accurately measured.

In the case of naphthalene molecules, a stepwise dependence of the ion yield on the wavelength of ionizing radiation was observed, which indicates the direct ionization mechanism for this molecule. The spectral dependence of the yield of benzene photoions near the ionization threshold exhibited a complicated structure, which is explained by the superposition of direct ionization and autoionization of Rydberg states. The characteristic width of the peaks (about 100 cm^{-1}) gives a total decay rate of states relative to all processes (dissociation, autoionization) equal to 2×10^{13} s^{-1}. The observation of these ionization peaks indicates that the fraction of the decay into the ionization continuum is quite high (Duncan *et al.*, 1981).

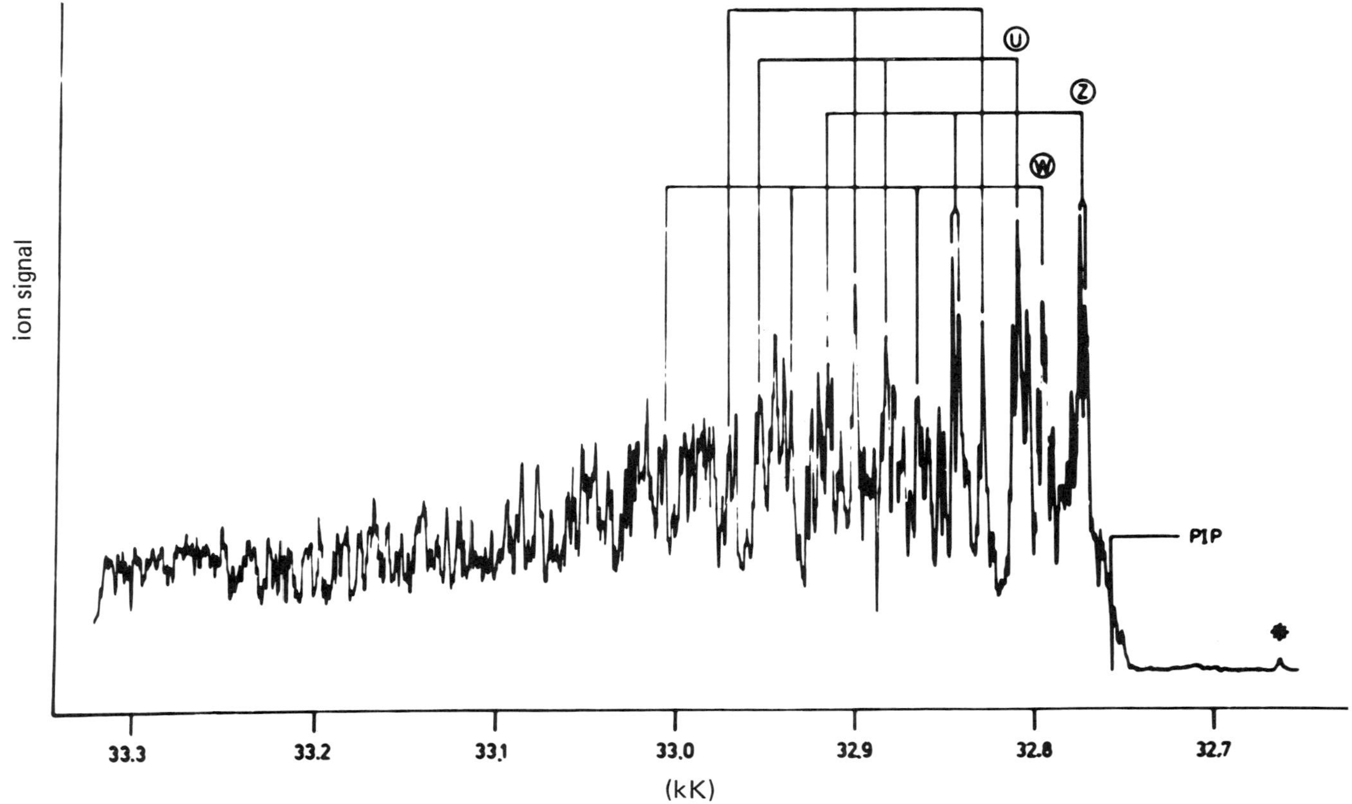

FIG. 5.4. Photoionization spectrum of the $^{39,39}K_2$ molecule excited to the $(B^1\Pi_u;\ v'' = 1)$ vibronic state. Obtained by varying the radiation wavelength λ_2 at the second state. [From Leutwyler *et al.* (1980).]

5.2.3. Fragmentation of Molecules in the Case of Stepwise Photoionization

The formation of fragment ions with single-photon ionization from the ground electronic state with thermal excitation has been studied in detail for many molecules (Reid, 1971). The primary process is usually detachment of the electron, and then, in the presence of sufficiently high vibrational or electronic excitation of the ion, cascade fragmentation of the ion occurs. The existing theory of the formation of mass spectra of polyatomic molecules is based on the assumption that the vibrational energy of the excited ion is distributed equally over the degrees of freedom (statistical approximation). Within the framework of this model it makes no difference how the molecular ion with a given store of vibration energy was formed.

In contrast to single-photon ionization, the method of two-step photoionization permits investigating the formation of fragment ions with transitions from high-lying vibrational levels of the intermediate electronic state. In addition, absorption of two photons by the molecule can lead to excitation of other states of the ion, which can also change the picture of fragmentation of the ion.

Fragmentation of the benzaldehyde molecule accompanying two-step photoionization ($\hbar\omega_1 + \hbar\omega_2 = 3.7 + 7.7 = 11.4$ eV) was investigated by Antonov *et al.* (1977a, 1978). The energy of the two photons exceeded not only the ionization potential of the molecule, $I = 9.5$ eV, but also the appearance potential of the fragment ion with detachment of the hydrogen atom, so that a peak corresponding to the fragment ions $C_6H_5CO^+$ was observed in the mass spectrum of the photoions (see Fig. 6.2a,b). An interesting feature is the much larger (compared with single-photon ionization at the same energy) yield of fragment ions (see Fig. 6.2a). This could be due primarily to the fact that the form of the distribution of the vibrational energy of ions over different modes can differ considerably for single- and two-step photoionization (Antonov, 1982). Absorption of a photon with energy 3.7 eV by a benzaldehyde molecule corresponds to an electronic transition of the type $n \rightarrow \pi^*$, i.e., a transition of an electron from a nonbinding atomic orbital of oxygen into an antibonding molecular π^* orbital. The restructuring of the electronic configuration occurs primarily in the aldehyde group CHO. Since the antibonding orbital of the molecule is occupied, the parameters of the bonds in the aldehyde group, especially the C=O bond, change considerably. This leads to the fact that in an electronic transition, vibrations are excited primarily in the region of the aldehyde group. After absorption of light with $\hbar\omega_1 = 3.7$ eV, the vibrational degrees of freedom of the molecule, after conversion into the triplet state, contain 0.6 eV of energy. After absorption of the second photon, ionization occurs from the molecular state with a large store of vibrational energy, localized in the aldehyde group, which increases the probability of formation of the fragment ion with detachment of the aldehyde hydrogen atom.

Another factor that leads to a large yield of fragment ions in two-step photoionization is that in the case of single-photon ionization, the nuclear configuration of the neutral molecule and of the ion differ little from one another due to the fact that the *n* electron of the oxygen, which does not participate in the formation of the molecular bond, is removed. For this reason, in spite of the fact that the photon energy of 11.4 eV exceeds the ionization threshold by the quite large amount of 1.9 eV, according to the Franck–Condon principle, ions with small vibration excitation will be formed with highest probability (Fig. 5.5). The excess energy of the photon above the ionization limit will be carried off by the electron. In two-step photoionization, the intermediate electronic state has a considerably changed nuclear configuration in the region of the chromophore, which can lead to a displacement of the peak in the vibrational energy distribution function of

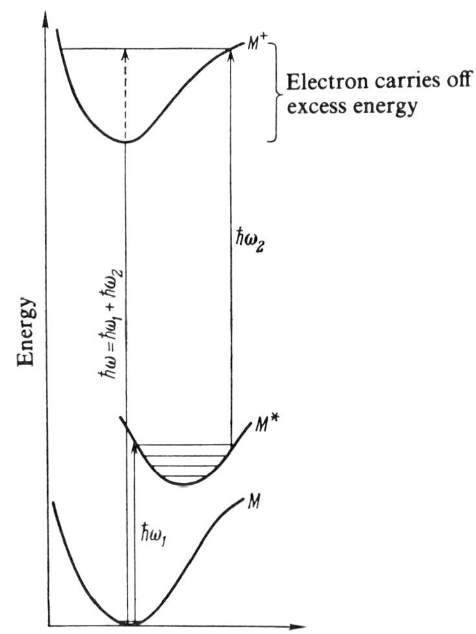

FIG. 5.5. Franck–Condon transitions with single- and two-step photoionization. The energy excess above the ionization potential of the molecule with single-photon ionization is carried away by the electron; with two-step photoionization, it is transformed into the vibrational energy of the ion. [From Antonov (1982).]

the ions formed (see Fig. 5.5). This must likewise facilitate an increase in the fraction of fragment ions.

A different picture of fragmentation is observed in photoionization of polyatomic molecules in a strong laser field, capable of saturating the optical transitions of the molecules. In this case a molecule can acquire an energy greatly exceeding the ionization potential, which gives rise to strong fragmentation of the molecule.

We shall briefly examine the results obtained for the benzene molecule, which is a very convenient object for investigation of photoionization and photofragmentation processes of polyatomic molecules (Rockwood *et al.*, 1979; Antonov *et al.*, 1980a,b; Reilly and Kompa, 1980; Seaver *et al.*, 1980), because the amount of available thermochemical and spectroscopic data is largest for this molecule. The energy of the first excited electronic state S_1 of the benzene molecule is greater than one-half the ionization potential ($I = 9.25$ eV), and this permits photoionizing excited molecules by a second photon in a single-frequency laser field ($2\hbar\omega > I$).

Figure 5.6 shows the mass spectra of photoions of benzene, formed under the action of the radiation of an excimer KrF laser with $\hbar\omega = 5$ eV. For low fluxes of radiation energy, $\Phi < 10^{-4}$ J/cm^2, only molecular ions formed with two-step photoionization (Fig. 5.6a) are observed in the mass spectrum.

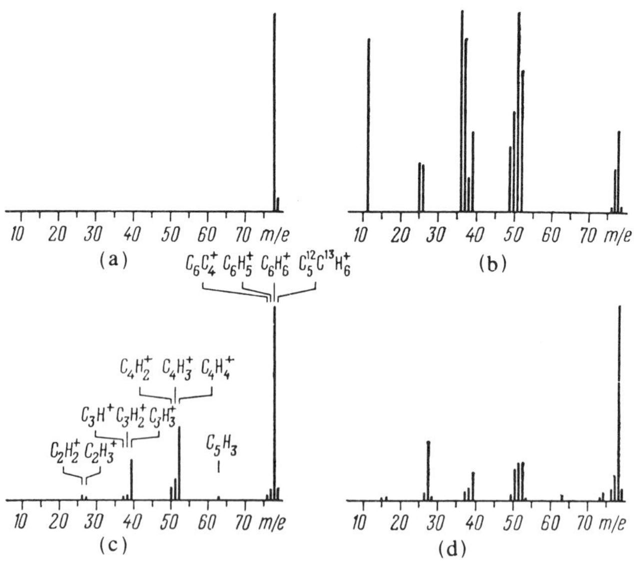

FIG. 5.6. Mass spectra of the benzene molecule. (a) Photoionization mass spectrum with radiation intensity $I = 10^5$ W/cm^2 ($\lambda = 248$ nm, $\tau_p = 20$ ns); (b) with $I = 2 \times 10^6$ W/cm^2 [from Antonov *et al.* (1980a)]; (c) with $I = 3 \times 10^8$ W/cm^2 [from Reilly and Kompa, (1980)]; and (d) electron impact mass spectrum ($E_{el} = 70$ eV) for comparison.

As the energy of the laser pulses increases, increasingly lighter fragment ions appear in the mass spectrum (Fig. 5.6b). For radiation energy fluxes $\Phi > 5\ \mathrm{J/cm^2}$, fragmentation becomes very intense (Fig. 5.6c) and exceeds fragmentation under electron bombardment with electron energies of 70 eV (Fig. 5.6d). Under these conditions, one of the strongest peaks in the mass spectrum is the ionic component C^+ (see Fig. 5.6c). The appearance potential of atomic carbon ions is about 27 eV, and therefore, in order for this ion to form, the molecule (or its fragments) had to absorb not less than six photons. The total yield of ions is very large and approaches 10% with a radiation energy flux of about $10\ \mathrm{J/cm^2}$ (Antonov *et al.*, 1980b). Aside from the high efficiency, the photoionization of benzene by excimer laser radiation is selective. For example, even with maximum radiation intensities, such atmospheric components as H_2 and H_2O are practically completely absent in the mass spectra of the photoions.

The lifetime of the benzene molecule in the intermediate electronically excited state is higher than the duration of the laser pulse. This leads to high efficiency of formation of molecular ions with absorption of two UV photons. Further formation of fragment ions will occur during the laser pulse in the process of successive photodissociation of ions, beginning with the molecular ions. The appearance of fragment ions with intensities of 10^7–$10^9\ \mathrm{W/cm^2}$ (see Fig. 5.6) cannot be explained by absorption of photons by the neutral molecule with a transition into highly excited autoionized states (Zandee and Bernstein, 1979; Lubman *et al.*, 1980), because the rate of decay of autoionized states, as a rule, exceeds 10^{12}–$10^{13}\ \mathrm{s^{-1}}$, i.e., it greatly exceeds the rate of laser-induced transitions above the ionization limit. Indeed, for benzene, the absorption cross section is around $10^{-17}\ \mathrm{cm^2}$, and the stimulated transition rate $\sigma I/\hbar\omega$ at a radiation intensity of $10^8\ \mathrm{W/cm^2}$ focused pulsed-dye-laser radiation) is around $10^{10}\ \mathrm{s^{-1}}$, i.e., it is a few orders of magnitude lower than the decay rate of the autoionization states located above the ionization potential.

The mechanism of fragmentation of polyatomic ions upon absorption of photons (Boesl *et al.*, 1980) was directly proved correct when measuring the kinetic energy of photoelectrons (Meek *et al.*, 1980). The modification of photoionization spectroscopy, associated with the registration of photoelectron spectra, is considered elsewhere in this text (Chapter 11). However, here we should like to note that such measurements enable us to estimate the store of energy in a molecule at the moment of electron detachment. In the case of the two-step photoionization of the benzene molecule by KrF laser radiation at $\lambda = 248$ nm, it was shown that the maximum photoelectron energies did not exceed the difference of the energy of two photons ($2\hbar\omega =$ 10 eV) and the ionization potential of the molecule ($I = 9.25$ eV). Hence, it follows for sure that ionization occurs once the molecule is excited above

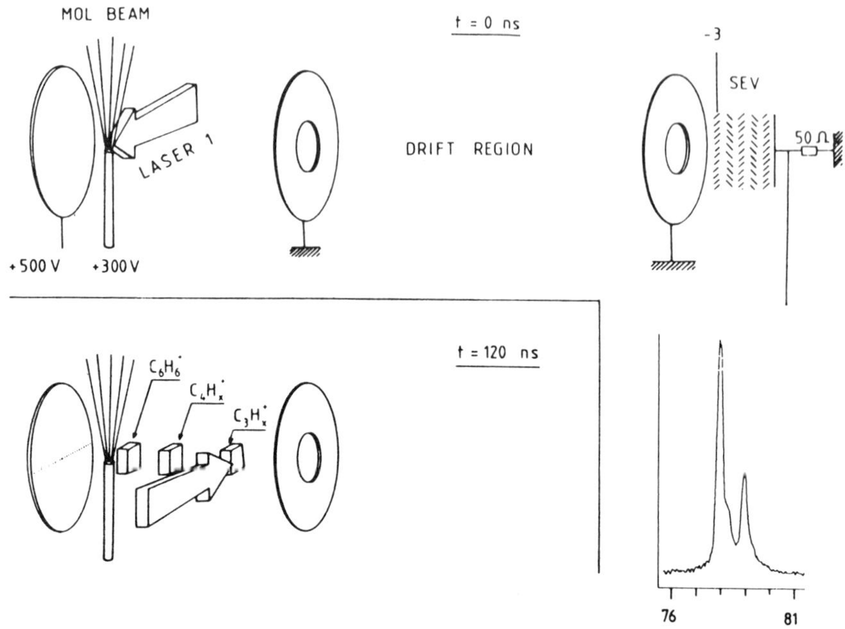

FIG. 5.7. Study of the photofragmentation of the benzene molecule under the effect of two temporally and spatially separated laser pulses in a time-of-flight mass spectrometer. The first laser pulse causes the photoionization of the molecule, and the second pulse excites various molecular ions. [From Schlag and Neusser (1983).]

the ionization potential. This conclusion was also confirmed by the experiments on the photoionization of benzene in a two-frequency field (Boesl *et al.*, 1980).

Other direct evidence of photon absorption by the molecular ions produced was obtained in experiments with two laser pulses of different wavelengths that were separated not only temporally but also spatially (Boesl *et al.*, 1982a). The idea of this experiment is illustrated in Fig. 5.7. The first laser pulse generates various photoions which are then accelerated in an electric field. As a result, the ions that differ in mass become spatially separated. Varying the focus position of the second laser pulse enables us to exert influence on different molecular ions and observe depletion of different mass peaks produced by the first pulse. The experiment confirmed the fact that the dominant part in photofragmentation is played by the absorption of a few photons by the molecular ions produced.

To understand the processes involved in the accumulation of energy and fragmentation in the ionic subsystem, it is of great interest to investigate the spectral dependence of the yield of different fragment ions. Figure 5.8 shows

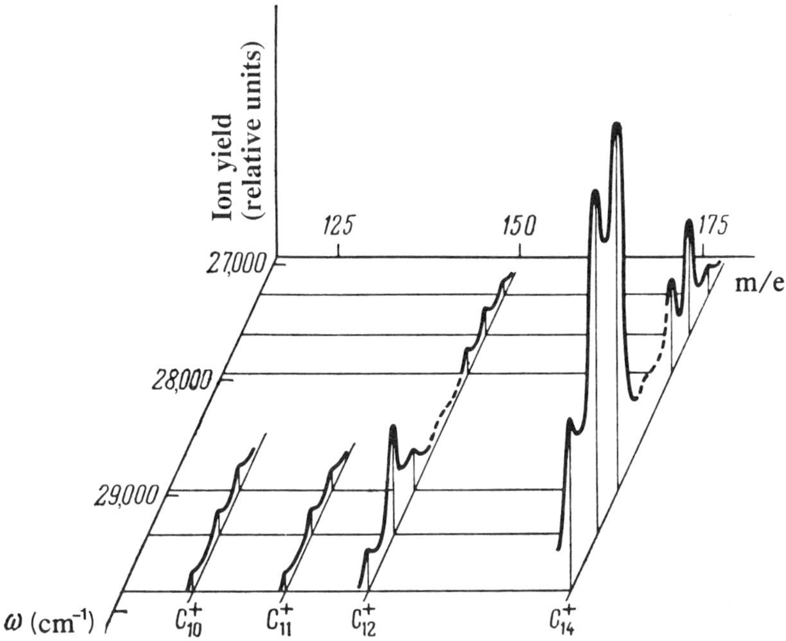

FIG. 5.8. Mass–optical spectrum of the anthracene molecule with laser pulse energy density of 0.06 J/cm^2. [From Antonov and Shibanov (1982).]

the dependence of the yield of molecular and fragment ions of anthracene on the frequency of laser radiation (two-dimensional mass–optical spectrum) with an energy density of 0.06 J/cm^2. In spite of the fact that all isotopic and fragment ions were actually resolved in the mass spectrum, for simplicity, for each group of ions with a constant number of carbon atoms C_n, the total magnitude of all peaks with different numbers of hydrogen atoms is plotted in the figure. For the C_{14}^+, molecular ions, the peaks in the photoionization spectrum correspond precisely to the vibrational structure of the electronic transition in the first step (Antonov and Shibanov, 1982). Among the fragmentations, ions of the type C_{12}^+ have the highest yield (see Fig. 5.8). The spectrum of the yield of these ions is on the whole similar to the spectrum of the C_{14}^+ molecular ions. However, the relative magnitude of the signal of C_{12}^+ fragment ions increases more rapidly with increasing photon energy. In addition, an appreciable increase in the yield of C_{12}^+ ions is observed at a laser radiation frequency of 29.4×10^3 cm^{-1}. For the lighter C_{10}^+ and C_{11}^+ fragment ions, the dependence of their yield on the radiation frequency is even stronger. A yield of these ions, sufficient for reliable recording, was observed only in the shortest wavelength region of the spectrum. The increase in the yield of fragment ions with a decrease of the wavelength of the laser radiation is qualitatively explained by the fact that with a constant number of absorbed quanta, the degree of excitation of the photoions formed increases. Thus, within the spectral interval in which the mass–optical spectrum of anthracene was measured, the energy of a single photon varied by 0.37 eV. With three-step photoionization, the energy absorbed by the molecule increases with decreasing wavelength already by 3×0.37 eV = 1.1 eV, and this leads to more efficient fragmentation of the molecule.

Thus, on the whole, the spectral dependence of the yield of fragment ions is analogous to the dependence for molecular ions, while the increase of the fraction of fragment ions with a decrease in the radiation wavelength is related to the increase in the absorbed energy.

5.2.4. Statistical Model of Fragmentation of Molecules

Comparison of the picture of fragmentation with laser photoionization in a strong resonant field and by electron bombardment (see Fig. 5.6) shows that in both cases the qualitative composition of photoproducts is practically identical, but the degree of fragmentation in the first case can be much higher. With ionization by electron bombardment, the molecule immediately acquires an energy of about 20 eV, exceeding the ionization potential, after which ionization of the molecule occurs with subsequent cascade decay of the overexcited ion into fragments. In spite of the fact that the energy of the ionizing electrons is quite high (usually 70–100 eV), due to the Franck–Condon principle, in a vertical transition only a small fraction of this energy in the form of vibration and electronic excitation is transferred to the ion. With repeated absorption of photons under the action of quite intense UV radiation, this restriction is absent, and the energy absorbed per molecule can greatly exceed 30 eV. This is facilitated by the fact that the electronic structure of the ions absorbs radiation resonantly in the visible and UV region of the spectrum.

The relative yield of different fragment ions in laser photoionization corresponds qualitatively to their appearance potentials: peaks of fragment ions with low appearance potentials have the highest intensity in the mass spectrum (Antonov *et al.*, 1980a).

All this indicates that, at least for benzene, the direction of fragmentation of the molecule is determined primarily by the energetics of the molecular bonds with any method of ionization, and the difference of the fragmentation pattern with ionization by intense laser radiation is related only to the high magnitude of the absorbed energy.

In this connection, an attempt was made to calculate the fragmentation of molecular ions of benzene in the field of intense UV laser radiation based on the statistical theory of unimolecular reactions (Rebentrost *et al.*, 1981;

Rebentrost and Ben-Shaul, 1981). For the case of benzene, it is well known that the basic channels of fragmentation of the initial $C_6H_5^+$ ion with excitation energies of 5–10 eV are the reactions:

$$\begin{aligned} C_6H_6^+ &\rightarrow C_6H_5^+ \rightarrow C_6H_4^+ \\ &\rightarrow C_4H_4^+ \rightarrow C_4H_3^+ \rightarrow C_4H_2^+ \\ &\rightarrow C_3H_3^+ \rightarrow C_3H_2^+ \end{aligned} \tag{5.4}$$

The threshold of formation of the primary fragments $C_6H_5^+$, $C_4H_4^+$, and $C_3H_3^+$ lies near 4 eV. Taking (5.4) as the first step of fragmentation, all decays of the following type are also included in the analysis:

$$C_mH_n^+ \rightarrow C_kH_l^+ + C_{m-k}H_{n-l}. \tag{5.5}$$

Here it is assumed that the molecular ions of benzene, formed with the absorption of two UV photons by the molecule, continue to absorb radiation, and it is also assumed that all the absorbed energy is uniformly distributed by means of rapid nonradiative relaxation of electronic states over vibrational degrees of freedom of the ion. Then, by examining the probability of decay and the energy content of all fragments statistically (according to the RRKM theory), the total pattern of fragmentation can be calculated (Andlauder and Ottinger, 1971).

Figure 5.9a shows the results of such a calculation of fragmentation of benzene by KrF laser radiation. The cross section for absorption of radiation by $C_6H_6^+$ ions was chosen to be equal to 2×10^{-18} cm^2. Comparison of the results of measurements (Fig. 5.9b) shows satisfactory qualitative agreement between the model calculation and experiment.

Further refinement of the theoretical model of fragmentation of molecules in an intense UV laser radiation field must include a more realistic examination of the process of absorption of radiation energy by the ionic system.

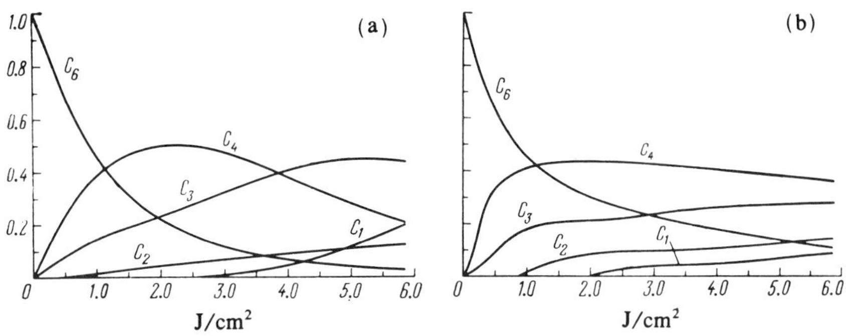

FIG. 5.9. Dependence of the relative intensities of different benzene C_n^+ ion groups on the energy flux of laser radiation. (a) Theoretical calculation and (b) experiment. [From Rebentrost *et al.* (1981).]

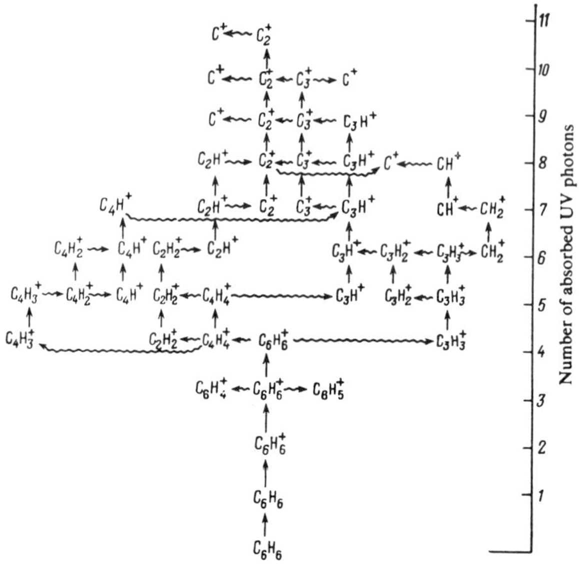

FIG. 5.10. Fragmentation channels for benzene under the action of intense laser radiation with $\hbar\omega = 4.78$ eV. [From Dietz *et al.* (1982b).]

Actually, already with an excitation energy of molecular benzene ions of 7–8 eV, the rate of their dissociation, calculated according to RRKM, is about $10^{10}\ s^{-1}$ (Dietz *et al.*, 1982b). For this reason, during a nanosecond pulse, different fragments are formed, which in their turn can likewise absorb radiation and thus form parallel fragmentation channels. Figure 5.10 shows a diagram of the basic channels of formation of benzene fragments under the action of radiation with a photon energy of 4.78 eV. A calculation using this more complete scheme leads to better agreement with experiment and permits explaining a number of peculiarities in the intensity ratios of the ionic components in the mass spectrum within groups with the same number of carbon atoms C_n (Dietz et al., 1982b). Although detailed calculations were made for the benzene molecule only, this model must in general be valid for other polyatomic molecules.

5.2.5. Role of Neutral Fragments

Experiments performed with a large number of molecules (Seaver *et al.*, 1980; Antonov *et al.*, 1980b; Dietz *et al.*, 1980; Reilly and Kompa, 1979) have shown that in many cases the picture of the interaction of molecules with intense radiation can differ considerably from the one examined above. The degree of ion fragmentation can be much higher, and the relative yield of molecular ions can be very low even with low laser-radiation intensities.

This is related to the dissociation of molecules accompanying excitation into intermediate electronic states. In this case, the main ionization channel could be photoionization of neutral molecular fragments. Thus, in one of the first works (Reilly and Kompa, 1979), in the photoionization of acetaldehyde the energy of two UV photons was much lower than the ionization potential of the molecule, which led to complete disappearance of the molecular ion due to the high rate of the competing process of dissociation from the highly excited state. Dissociation also plays a large role in photoionization of the benzene molecule by ArF laser radiation ($\lambda = 193$ nm) via the intermediate electronically excited state $^1B_{1u}$ (Reilly and Kompa, 1980). Analysis of data on the photoion yield in this case permitted estimating (with some assumptions about the cross section of photoionization out of the excited state) the lifetime of this state, which turned out to be on the order of 10^{-11} s (Reilly and Kompa, 1980).

In the experiments examined, the dissociation time of the molecules investigated in the intermediate excited states was much shorter than the duration of the laser pulse, which results in a high quantum yield of dissociation and, as a result, effective suppression of direct stepwise photoionization of molecules. The neutral molecular fragments formed, which absorb at the frequency of the laser radiation, can then ionize during the laser pulse. However, ionization of neutral molecular fragments can play a large role in the formation of the photoionization mass spectrum of polyatomic molecules, even in the presence of stable intermediate states. The transition of a molecule into a state above the ionization limit from the electronically excited state with a large reserve of vibrational energy or with the nuclear configuration of this state differing considerably from the configuration of the ion, can result in a sharp increase in the fraction of dissociating molecules and, correspondingly, a drop in the yield of molecular ions (Antonov *et al.*, 1980b). For this reason, in order to achieve maximum efficiency of stepwise photoionization of molecules with a minimum degree of fragmentation, the energy of the ionizing photon must exceed the vertical ionization potential of the molecule from the excited state. This condition is apparently satisfied in the photoionization of benzene by radiation at the fourth harmonic of the Nd:YAG laser (Pandolfi *et al.*, 1981).

5.3. Multiphoton Ionization of Molecules

Multiphoton ionization represents the case of ionization of molecules in an extremely intense laser field, $I \gtrsim 10^{10}$ W/cm^2. Multiphoton processes are usually described by the first nonvanishing term in the nonstationary perturbation series expansion (see Section 2.4). The general expression for the probability of n-photon ionization has the form $W^{(n)} = \sigma^{(n)} I^n$, where I is the radiation intensity, and $\sigma^{(n)}$ is the cross section of the multiphoton process. The expression for $\sigma^{(n)}$ contains a summation over all molecular states, including the continuum. For example, for the case $n = 4$, which is typical for multiphoton ionization of ions, the expression for $\sigma^{(n)}$ has the form

$$\sigma^{(4)}(\omega) \sim \left| \sum_1 \sum_2 \sum_3 \frac{\langle f|V|3\rangle\langle 3|V|2\rangle\langle 2|V|1\rangle\langle 1|V|i\rangle}{(E_3 - E_i - 3\hbar\omega)(E_2 - E_i - 2\hbar\omega)(E_1 - E_i - \hbar\omega)} \right|^2 . \quad (5.6)$$

In this expression i and f correspond to the initial bound state and the final state in the ionization continuum; the indexes 1, 2, and 3 indicate the intermediate first, second, and third quantum states, respectively; ω is the frequency of the laser radiation; and V is the molecule–field interaction Hamiltonian. The numbers in the matrix elements indicate virtual states accompanying absorption of one, two, or three photons. When the radiation frequency ω is tuned, depending on the position of the intermediate molecular levels, resonances can arise at the fundamental, doubled, and tripled frequencies. The ionization yield in this case increases considerably if the selection rules for multiphoton transitions are satisfied, and thus the multiphoton ionization spectrum forms. Since the electronic absorption bands of most molecules begin in the UV region of the spectrum, when we use radiation from tunable dye lasers, two- and three-photon resonances are typically recorded. The characteristic cross sections for the two- and three-photon transitions constitute, respectively, 10^{-48}–10^{-54} cm^4 s and 10^{-80}–10^{-86} cm^6 s^2, with the radiation intensity I expressed in photons per square centimeter seconds (Rothberg *et al.*, 1981).

The multiphoton ionization method is now widely used for spectroscopy of excited states of polyatomic molecules, especially states forbidden to single-photon transitions (for reviews, see Johnson, 1980; Robin, 1980). An important advantage of the method is the possibility of investigating Rydberg states lying in the VUV region of the spectrum, as well as ionizing practically any molecules with the help of tunable lasers in the visible region. The presence of two- or three-photon absorption from the ground state in the first step fundamentally requires high radiation intensities, and for this reason the mass spectrum of the photoproducts is very complicated. Fragmentation of molecules accompanying multiphoton ionization, as a rule, exceeds fragmentation under electron impact.

5.3.1. Spectra of Multiphoton Resonances

Information on multiphoton ionization is obtained primarily in the form of spectra. It is significant that the selection rules in multiphoton transitions differ from the single-photon rules. For example, a transition between levels with identical parity is forbidden for single-photon transitions but is allowed for two-photon transitions. Thus spectroscopic data obtained with multiphoton ionization supplement the usual single-photon spectroscopy. In many cases, it is possible to investigate electronic states of molecules that previously were not observed at all.

Figure 5.11 shows the spectrum of multiphoton ionization of the benzene molecule with wavelength tuning of the laser radiation over the range 365–400 nm (Johnson, 1976). The first question that arises in the analysis of the spectra is identification of the multiphoton resonances in the observed ionization spectrum. As indicated by the arrow in Fig. 5.11, the three-photon ionization potential of benzene corresponds to a wavelength of 402 nm, so that the signal at shorter wavelengths corresponds to three-photon ionization and to four-photon ionization in the region of 400–450 nm. The observed structure of the spectrum in the region of 365–400 nm can in principle arise as a result both of resonances with the bound molecular states at the doubled frequency of the radiation and of autoionization states above the ionization limit in the case of a three-photon transition. However, from an analysis of the vibrational structure and its comparison with the well-known single-photon spectrum in the region of the autoionization states, it follows that the observed structure arises due to two-photon resonances. The only state, known up to the present time, in this region is the $^1B_{1u}$ state. The two-photon transition into this state can be vibrationally allowed and can give rise to

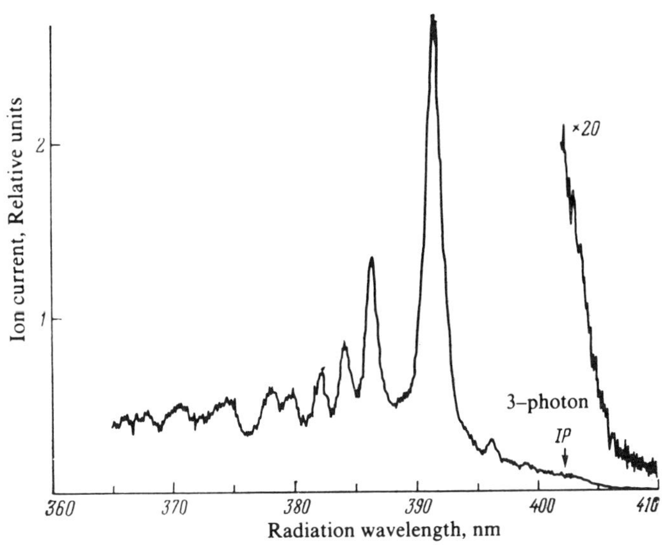

FIG. 5.11. Photoionization spectrum of two-photon resonances ($^1E_{1g} \rightarrow {}^1A_{1g}$) of benzene. [From Johnson (1976).]

the multiphoton ionization spectrum. However, the position of the maximum peak in the spectrum, as well as the observed shift accompanying deuteration, contradict this proposition. Four different types of states are allowed in the two-photon transition from the ground state into a state with D_{6h} symmetry. These are ${}^1A_{1g}$, ${}^1E_{1g}$, ${}^1E_{2g}$, and ${}^1A_{2g}$. The transition into the ${}^1A_{2g}$ state is forbidden if the two photons are identical. The presence of E_{2g} vibrations in the transition indicate a Jahn–Teller effect in the excited state, which indicates either a ${}^1E_{1g}$ or ${}^1E_{2g}$ state. The symmetry of the ${}^1E_{1g}$ state was established from measurements of the ionization yield using linear and circular polarization. A new ${}^1E_{ig}$ electronic state, previously predicted theoretically, has thus been observed and identified based on multiphoton ionization spectra. A more detailed study of two-photon resonances in benzene by MPI techniques has been done by Whetten *et al.* (1983).

The spectrum of three-photon resonances in benzene in the region of 400–500 nm corresponds practically completely to transitions into Rydberg states, and, in addition, the three-photon Rydberg structure is practically identical to the single-photon absorption spectrum, with the exception of the difference in the intensities, indicating the strong effect of the ${}^1A_{1g}$ components in the two-photon virtual state.

The multiphoton ionization spectra of a large number of polyatomic molecules have been investigated in an analogous manner (for reviews see Johnson, 1980; Antonov and Letokhov, 1981). In spite of the fact that the method of multiphoton ionization of molecules appeared comparatively recently, it is now already the standard method and is widely used to investigate the electronic spectra of molecules.

5.3.2. Fragmentation Accompanying Multiphoton Ionization

Historically, the fact that photoionization of polyatomic molecules by powerful laser radiation is usually accompanied by intense fragmentation was first discovered precisely in experiments on multiphoton ionization (Zandee and Bernstein, 1979). It was previously assumed that the number of photons absorbed by a molecule corresponds to the minimum number required by the energy to attain the lowest ionization potential. The formation of fragments with high appearance potentials was neglected. However, fragmentation of the molecule down to atomic ions C^+ was recorded already in the first experiments on multiphoton ionization of benzene (Zandee and Bernstein, 1979). It follows from the appearance potentials of the fragment ions that at least nine photons, each with an energy of 3.17 eV, are required for the formation of C^+ ions. The dependence of the yield of light fragments on the radiation intensity in this case was stronger than for heavy ions, but nevertheless, the power does not exceed 3.5 even for C^+ ions, which indicates the strong saturation of multiphoton absorption. The power dependence is actually determined by the first step of the two- or three-photon absorption; further transitions, leading to ionization and fragmentation of molecules, are usually resonant transitions and are close to being saturated.

In addition to the mechanisms discussed in Section 5.2 for the formation of photoionization mass spectra of molecules, such as the photodissociation of molecular ions and photoionization of neutral fragments in strong fileds, which are typical for experiments on multiphoton ionization with $I \gtrsim 10^9$–10^{10} W/cm^2, other mechanisms of the interaction of molecules with radiation can also play an important role. These primarily include the process of accumulation of energy by molecules in a laser field. A model of multiphoton excitation according to which polyatomic molecules, having absorbed an energy exceeding the ionization potential, do not have time to lose an electron and continue to be excited along the network of high-lying autoionization levels with subsequent autoionization accompanied by fragmentation of molecules, was proposed (Zandee and Bernstein, 1979; Boesl *et al.*, 1980). In the case of the benzene molecule, this possibility was investigated experimentally using synchronized UV and visible laser pulses with wavelengths of 241 and 483 nm, respectively (Boesl *et al.*, 1982a). With an intensity of 3 MW/cm^2, the action of a single UV radiation pulse led to the formation of only a molecular ion. On the contrary, with the visible radiation intensity being equal to 0.4 GW/cm^2 (which is sufficient for the multiphoton ionization of benzene, provided that selection rules allow for the two-photon transition from the ground state), no ionization signal was observed when the doubled visible-radiation frequency was tuned to the S_1 state vibrational sublevel forbidden for the two-photon transition. Under these conditions, the combined action of UV and visible radiation led to intense fragmentation in the mass spectrum, typical for multiphoton ionization. In this case, however, it remains unclear whether the molecule is excited by the visible radiation along the network of autoionization states or the visible radiation gives rise to fragmentation of the ions. To clarify this, the visible radiation was delayed relative to the UV radiation by 17 ns, an amount sufficient for autoionization of the intermediate state. In this case, the photoionization mass spectrum turned out to be completely analogous to the preceding case. This leads to the conclusion that the process of excitations of neutral molecules along the network of autoionization levels cannot compete with autoionization even with intense radiation 10^9 W/cm^2, and therefore the main input of energy to the medium with multiphoton ionization occurs in the ionic subsystem.

A difference occurs, however, in the fragmentation of ions by high-intensity radiation. Figure 5.12 shows the photoionization mass spectra of benzene, measured with radiation intensities of 3×10^8 W/cm^2 and 2.7×10^{11} W/cm^2 (Hering *et al.*, 1981). Comparison of these mass spectra shows that under

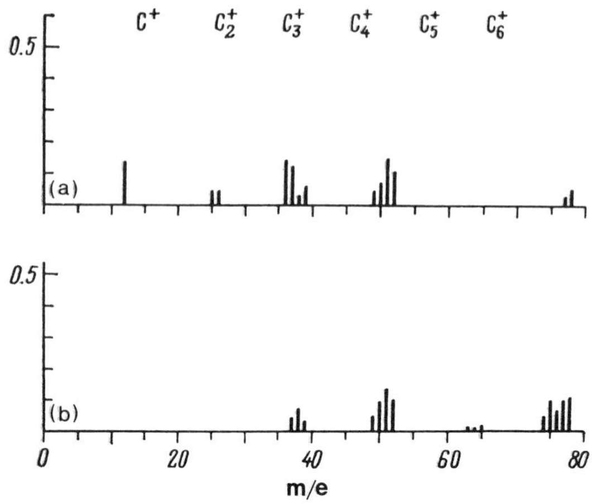

FIG. 5.12. Photoionization mass spectra of benzene with different intensities and durations of laser radiation. (a) $I = 3 \times 10^8$ W/cm^2 (0.6 J/cm^2), $\tau_p = 20$ ns [from Reilly and Kompa (1980)] and (b) $I = 3.7 \times 10^{11}$ W/cm^2 (7.5 J/cm^2), $\tau_p = 20$ ps [from Hering *et al.* (1981)].

the action of a picosecond pulse (Fig. 5.12b) whose intensity is three orders of magnitude greater than in the case of the nanosecond pulse (Fig. 5.12a), fragmentation in the mass spectrum is considerably lower even with an energy flux in the radiation pulses an order of magnitude greater than in the first case (see Fig. 5.12). This difference in the action of nanosecond and picosecond laser pulses on a benzene molecule is related to the fact that in the weaker field ions fragment in successive processes of absorption of photons and dissociation of ions, beginning with the molecular ion (see Section 5.2), while in an extremely strong laser field ($I > 3 \times 10^{11}$ W/cm^2), the rate of radiation-induced transitions is so high that the ions have time to acquire a large amount of energy before they dissociate. Deactivation of reexcited molecular benzene ions apparently occurs predominantly with detachment of hydrogen atoms (see Fig. 5.12b).

5.3.3. "Entropy" Model of Fragmentation

A detailed theoretical calculation of the fragmentation of polyatomic molecules in an intense laser radiation field in an exceedingly difficult problem and requires knowledge of an enormous amount of spectroscopic and photochemical data for all intermediate photoproducts. For this reason, in many investigations (Rebentrost *et al.*, 1981; Rebentrost and Ben-Shaul, 1981; Silberstein and Levine, 1980, 1981) a great deal of attention was devoted to developing approximate computational methods. As already discussed, fragmentation of polyatomic molecules with multiphoton ionization depends strongly both on the intensity and on the wavelength of the laser radiation. Under identical conditions, the photoionization mass spectra of isomers likewise differ considerably (Lubman *et al.*, 1980). While such a dependence of fragmentation on the radiation parameters and the molecular structure is of great interest for different applications, the degree to which this dependence reflects the specific (dynamic) properties of the processes of photoexcitation and photofragmentation of individual molecules in a strong laser field remains an important question. To investigate this question, a method of calculating the statistically most favorable pattern of fragmentation of molecules with multiphoton ionization was developed by Silberstein and Levine (1980, 1981). According to their work, the statistically most favored fragmentation pattern is one in which the maximum entropy of the medium consisting of noninteracting particles—the initial molecules and their neutral and ionized fragments—is attained. The only parameter determining fragmentation here is the average magnitude of the absorbed energy per initial molecule.

Let χ_j be the number of molecules or radicals of type j, and let X_{ij} be the fraction of these particles in the quantum state i. Then,

$$\sum_i X_{ij} = 1, \tag{5.7}$$

$$\sum_j \chi_j = \chi, \tag{5.8}$$

where χ is the total number of fragments. The entropy of the system S is expressed as follows:

$$S = -R\sum_j \chi_j \ln \chi_j - R\sum_j \chi_j \sum_i X_{ij} \ln X_{ij} + R\chi \ln \chi, \tag{5.9}$$

where R is the gas constant. The problem thus reduces to finding the values of χ_j and X_{ij} that maximize expression (5.9) under the following restrictions: (a) conservation of the average absorbed energy $\langle E \rangle$:

$$\langle E \rangle = \sum_j \chi_j \sum_i E_{ij} X_{ij}, \tag{5.10}$$

where E_{ij} is the energy of the quantum state i for a particle of type j, and (b) conservation of matter and charge:

$$\chi_k = \sum_j a_{kj} \chi_j, \tag{5.11}$$

where χ_k is total number of atoms of the kth element or total charge, a_{kj} is number of atoms of the kth element (or charge) of the j-type particle.

The results of the calculation of fragmentation of the benzene molecule together with the measured mass spectra are shown in Fig. 5.13. The good agreement between the experimentally obtained mass spectra and the computed spectra for $\langle E \rangle = 37$ eV is evident. This indicates the considerably statistical character of the fragmentation of a molecule in multiphoton ionization ($\lambda = 504$ nm, $E_{\text{las}} = 7.5$ mJ). In the case of the triethylenediamine molecule, a disagreement with experiment is observed in the intensity ratios of the ion peaks in the region $M/e = 54$–58. Of the five molecules investigated by Silberstein and Levine (1981); this is the largest disagreement with experiment, and it is apparently related to the nonstatistical nature of the processes involved. Nevertheless, in most cases the "entropy" model of molecular fragmentation by intense laser radiation agrees well with experiment.

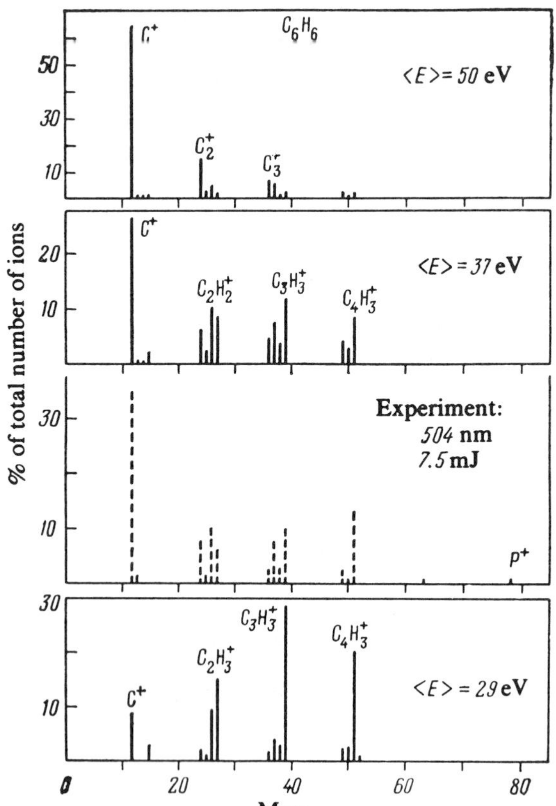

FIG. 5.13. Comparison of photoionization mass spectra, calculated in the statistical limit, with experimental results for the benzene molecule. [From Silberstein and Levine (1981).]

5.4. Molecular Photoionization via Highly Excited Vibrational States

For purposes of highly selective and sensitive detection of molecules, it was suggested that use should be made of laser resonant excitation of molecular vibrations with subsequent photoionization of the vibrationally excited molecules, as illustrated in Fig. 5.1c (Letokhov, 1976a,b). As a matter of fact, to realize this idea proved very difficult because of too small a shift of the VUV absorption band of the molecule, caused by its vibrational excitation. For simple and polyatomic molecules, this difficulty is being overcome to a variable degree and by various methods. Therefore, let us consider these cases individually.

5.4.1. Simple Molecules. Successive Excitation of Vibrational and Electronic States

The IR absorption spectrum of simple molecules features a well-defined vibrational–rotational structure, which makes it possible to excite certain (v, J) states by means of tunable laser radiation. The comparatively high-energy vibrational excitation, especially where vibrational overtones are being excited, provides a marked shift of the electronic absorption band. The difficulties associated with the need to employ a tunable VUV laser radiation can be overcome in two ways: (1) by using a UV laser radiation to excite an intermediate electronic state, which then becomes ionized following the absorption of one more photon: and (2) by using an intense visible laser radiation to effect two-photon excitation of an intermediate electronic state.

These possibilities were demonstrated in experiments with the NO molecule (Esherick and Anderson, 1980; Sudbo and Loy, 1981). Figure 5.14 shows the schemes of multistep ionization through intermediate vibrational and electronic states, which were realized in these works. The NO molecule has its first excited electronic state ($A^2\Sigma^+$ in the region of 4.4×10^4 cm^{-1} and a vibrational frequency of around 1900 cm^{-1} in the electronic ground state $X^2\Pi$, which is split into two states, $X^2\Pi_{3/2}$ and $X^2\Pi_{1/2}$, 123 cm^{-1} apart because of the orbital angular momentum of the molecule (Herzberg, 1950). This enables us to implement the most diverse multistep photoionization schemes. First, a radiation in the region of 5.3 μm can be used to excite the level $v = 1$ and one in the region of 2.7 μm, the level $v = 2$. Second, use can be made of a radiation at 470 nm for two-photon excitation of the $A^2\Sigma^+$ electronic state and one at 227 nm for direct one-photon excitation of this state. Subsequent absorption of two visible or one UV photons leads to the ionization of the NO molecule. With both these multistep photoionization

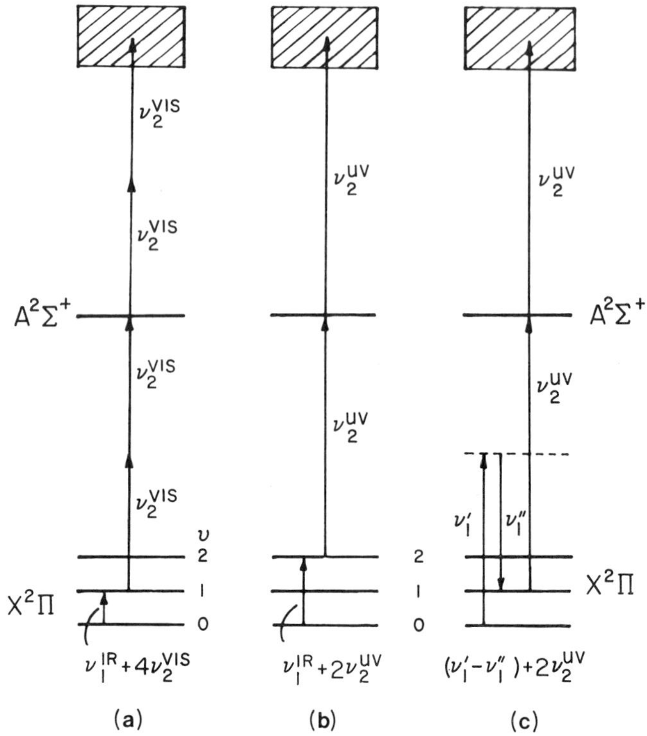

FIG. 5.14. Various schemes of multiphoton ionization of the NO molecule by way of vibrational excitation of the ground electronic state $X^2\Pi$, followed by excitation of the intermediate electronic state $A^2\Sigma^+$ and its ionization upon absorption of an additional photon(s). (a) Excitation of the ground state and two-photon excitation of the electronic state by visible radiation. [From Esherick and Anderson (1980).] (b) Excitation of a vibrational overtone and one-photon excitation of the electronic state by UV radiation. [From Sudbo and Loy (1981).] (c) Raman excitation of the ground state and one-photon excitation of the electronic $A^2\Sigma^+$ state. [From Esherick and Owyoung (1983).]

schemes, it proved possible to observe an ionization signal when exciting a specified (v, J) level of the $X^2\Pi_{3/2}$ or $X^2\Pi_{1/2}$ state.

The vibrational excitation of the molecule can also be effected as a result of the Raman process in a biharmonic light field. The subsequent photoionization of the vibrationally excited molecules provides one more highly sensitive Raman spectroscopy technique (Beterov *et al.*, 1982; Esherick and Owyoung, 1983). For molecules, the capabilities of this method were demonstrated also in the experiments with NO (Fig. 5.14c) (Esherick and Owyoung, 1983). The given vibrational–rotational sublevel ($v = 1, J$) of the molecule was first excited by means of two pulsed lasers operating at wavelengths of $\lambda_1' =$

577 nm and $\lambda_1'' = 647$ nm and having a narrow emission spectrum (0.002 cm^{-1}). Then, after a delay of 20–30 ns to exclude multiphoton processes in the multiple-frequency field, a UV radiation pulse at $\lambda_2 = 236$ nm was switched on, which was capable of ionizing the vibrationally excited molecules. By varying the difference between the wavelengths λ_1' and λ_1'' while keeping λ_2 fixed, it was possible to measure the Raman excitation spectrum of the NO molecule at sufficiently low pressures. Figure 5.15 illustrates the experimental results of measuring the Q branch of the NO molecule by the sensitive gain ionization technique.

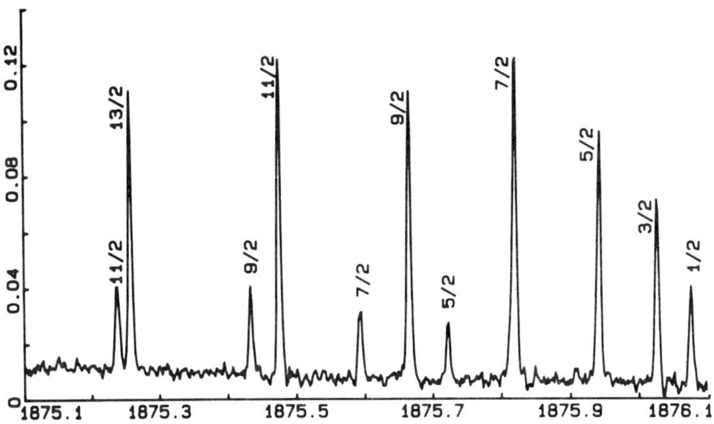

FIG. 5.15. Raman vibrational excitation spectrum of the NO molecule, measured by the ionization detection technique at a partial pressure of 0.01 Torr. One Torr of argon serves to redistribute population among rotational sublevels (8 μJ UV at 236.2 nm). Two series of the Q branch of both the ground states $^2\Pi_{3/2}$ and $^2\Pi_{1/2}$ are well resolved. [From Esherick and Owyoung (1983).]

The Raman excitation of molecular vibrations, followed by ionization of the vibrationally excited molecules, appreciably extends the scope for experimentation, as the tunable IR laser technology is still inadequately developed.

5.4.2. Polyatomic Molecules: Multiphoton Excitation of Vibrations by Infrared Radiation

The photoionization method seems to be the most interesting for the polyatomic molecules for which the standard methods of spectroscopy (absorption, fluorescence, Raman, etc.) are not sensitive enough. As concerning the polyatomic molecules, the selective channel of IR vibration excitation can be used. In this case the resonant multiphoton IR excitation of high-lying vibrational states with subsequent excitation from them of the electronic states leading to the ionization of the molecule seems more feasible. Figure

5.16 shows some possible schemes of molecular photoionization via high-lying vibrational states.

A powerful IR laser pulse whose frequency is tuned to resonance with the vibrational transition can be used to provide for resonant excitation of vibrational levels of a polyatomic molecule up to the dissociation limit (Ambartzumian *et al.*, 1974). A discussion of the problem of multiphoton excitation of molecular vibrations by IR radiation can be found in a number of monographs (Letokhov, 1983a, Chapter 5; Bagratashvili *et al.*, 1985c). The only experiment on the molecular photoionization according to the scheme shown in Fig. 5.16a has been performed. A powerful CO_2 laser pulse excited high-lying vibrational levels of the SF_6 molecule in a molecular beam. Owing to the presence of highly excited molecules, the molecular photoionization spectrum suffered a slight deformation in the long-wavelength wing of the VUV region (Sudbo *et al.*, 1979a). That the magnitude of the spectrum deformation was small was explained by the small fraction of the highly excited SF_6 molecules.

Large polyatomic molecules can be ionized by powerful IR laser radiation alone (Bagratashvili *et al.*, 1983). It is customary to assume that the maximum vibrational energy E_{vib}^{max} of a molecule is limited by its dissociation energy D_0. This is true only for small molecules or for molecules subject to collisions. However, where the IR field is strong enough to provide for a sufficiently high multiphoton excitation rate (see Letokhov, 1983a, Chapter 5),

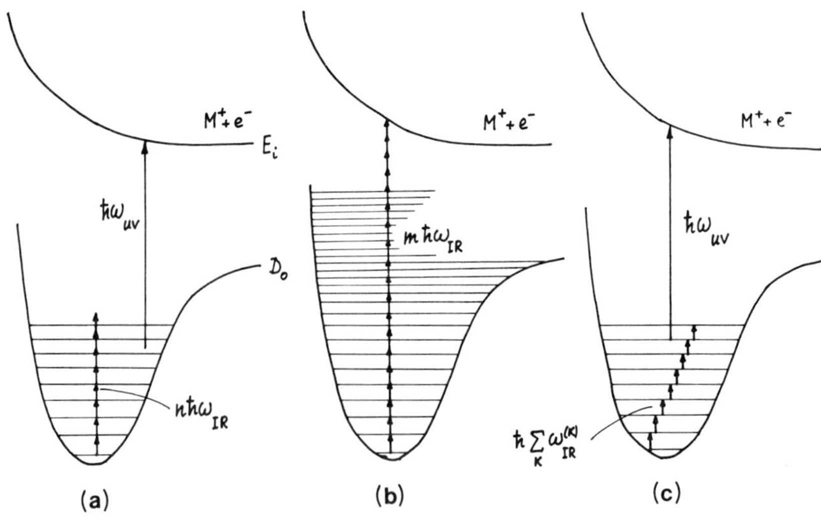

FIG. 5.16. Possible schemes for resonant molecular photoionization via high-lying vibrational states by way of multiphoton resonant vibrational excitation with an IR laser radiation. (a) IR + VUV excitation, (b) IR multiphoton excitation, and (c) IR multistep + VUV excitation.

a large polyatomic molecule with a limited monomolecular decay rate can be vibrationally overexcited well above the dissociation threshold, so that $E_{vib}^{max} - D_0 > D_0$. Such a competition between the two processes is quantitatively described by the statistical (RRKM) theory of monomolecular decay (Robinson and Holbrook, 1972). In the semiclassical approximation of this theory ($E_{vib}^{max} \gg E_0$, where E_0 is the energy of the zero-order molecular vibrations), the following estimate can be obtained for E_{vib}^{max} at a given multiphoton excitation rate W:

$$(E_{vib}^{max} - D_0)/D_0 = [1 - (W/K)^{1/(s-1)}]^{-1} - E_0/D_0 - 1, \qquad (5.12)$$

where s is the number of vibrational degrees of freedom, and K the frequency factor appearing in the expression for the monomolecular decay rate in the semiclassical approximation, whose magnitude is of the order of 10^{15} s^{-1}. With IR multiphoton excitation, an excitation rate of $W = \sigma_{vib} I \simeq 10^9$ s^{-1} can easily be ensured. Then, according to (5.12), a strong overexcitation is possible for a molecule with a number of degrees of freedom of $s \gg \ln(K/W)$. Because of the nonadiabatic coupling of the vibrational and electronic motions, the molecule might be expected to transfer from very high vibrational levels within the electronic ground state (but at an energy of $E_{vib} > D_0$) onto the excited electronic terms leading to the ionization decomposition of the molecule.

An experiment of this kind was performed with the anthracene ($C_{14}H_{10}$) molecule ($s = 66$, $D_0 = 4.8$ eV) for which the estimate (5.12) gives $E_{vib}^{max} - D_0 \simeq 3D_0$. With so strong an overexcitation, the ionization of the molecule ($I = 7.4$ eV) is quite possible. In fact, the formation of anthracene molecular ions was observed when the molecules were irradiated by a sufficiently powerful (about 10^9-W/cm^2) short (70-ns) CO_2 laser pulse under collisionless conditions (pressure $= 4 \times 10^{-6}$ Torr) in a time-of-flight mass spectrometer. These ions can be believed to appear as a result of the IR multiphoton ionization of the molecules in accordance with the scheme of Fig. 5.16b. Of course, such a multiphoton ionization technique is applicable only to large polyatomic molecules, since for polyatomic molecules with a small number of atoms, the maximum possible degree of overexcitation above the dissociation threshold is comparatively low (Sudbo *et al.*, 1979b).

It is expected that methods for a multistep soft vibrational excitation of molecules by means of multifrequency IR laser pulses (as shown in Fig. 5.16c) may be developed. This will enable us to deposit substantial amounts of energy (1–3 eV) in the molecules. In this case, it will probably be possible to realize the idea of highly selective ionization of molecules by way of their strong vibrational and subsequent electronic excitation.

6. Detection of Molecules by Multistep and Multiphoton Resonant Ionization

It is much more difficult to detect single molecules than single atoms. An even simpler task, such as the selective detection of traces of a polyatomic molecule (around 10^6 molecules) has not yet been solved by physical methods. The basic method of identifying trace quantities of molecules is mass-spectrometric analysis.

The analysis of matter by the mass-spectroscopic method is based on the transformation of the molecules of a sample into ions that can be further separated and analyzed due to their different mass-to-charge ratios. Mass spectrometrists have always been trying to create ion sources with high ionization efficiencies and moderate molecular fragmentation. The progress that has been made in this direction has made the mass-spectrometric method one of the most sensitive methods of analysis. Under favorable conditions it is possible to analyze substances with no more than 2.5×10^{-14} g (Hunt and Crow, 1978). The characteristic efficiency of the transformation of molecules into ions (with their subsequent detection in mass spectrometers) is one in one hundred thousand (Ligon, 1979) and, hence, is far from being optimal. The situation is somewhat worse for selectivity. As a rule, molecular mass spectra are the result of fragment ions. Often no molecular ions are observed in the mass spectra at all. Despite the fact that an analysis of the fragments yields valuable information on the molecular structure, it is rather difficult to analyze multicomponent mixtures. Due to an overlap of the mass spectra it is especially difficult to analyze molecular components that are only present in small amounts. A preliminary separation of the components in a mixture by column chromatography can partially overcome this problem (Ligon, 1979).

Therefore, we are tempted to try to combine the existing methods for detection and identification of complex molecules (mass-spectrometric analysis, in particular) with selective molecular photoionization techniques (Ambartzumian and Letokhov, 1972; Letokhov, 1976a,b). The first experiments in this direction were performed in 1977 (Antonov *et al.*, 1977a, 1978).

Molecules can also be detected without resort to mass-spectrometric analysis, by photoionizing them in a buffer gas and then registering the electron–ion pairs by monitoring the electron avalanche developing in a proportional counter (Freuholz *et al.*, 1980), as described in Section 5.3. Unfortunately, the sensitivity of that kind of molecular detection technique is not very high, and so it is desirable that the molecules should undergo preliminary separation, for example, by the method of chromatography (Klimcak and Wessel, 1980), or that the molecular ions produced should be subsequently separated by way of plasma chromatography (Lubman and Kronick, 1982). Finally, especially effective appears the method whereby trace molecules are first accumulated on a surface by absorption and then pulse-desorbed and selectively photoionized with a laser radiation in the chamber of a time-of-flight mass spectrometer (Egorov *et al.*, 1983, 1984a,b). Let us consider these three possible approaches to the photoionization detection of single molecules individually.

6.1. Detection of Molecules by Multiphoton Resonant Ionization Mass Spectrometry

The use of selective two-step IR–VUV molecular photoionization in mass spectrometers was proposed in 1972. The possibility of creating a universal detector for trace amounts of complex molecules, a so-called two-dimensional optical mass spectrometer, has been discussed many times (Letokhov, 1976a,b). This is illustrated in Fig. 1.4. A laser with a tunable frequency ω_1 selectively excites the vibrational states of the molecules (the electronic states of some molecules). Due to such excitation the edge of the photoionization band, which usually lies in the VUV region, is shifted by a small amount. The second laser, which operates in the VUV range, performs the photoionization of the molecules. Its frequency ω_2 lies in the region of the maximum slope of the edge of the photoionization band. In this case the selective pre-excitation of the molecules by the tunable laser (a rather small value of $E_{exc} = 0.1–0.5$ eV) brings about a detectable photoionization cross section (10^{-1}–10^{-2}% with molecular distribution over rotational states taken into account), i.e., a change in the photocurrent. The photoions enter a standard mass spectrometer and the mass spectrum, i.e., the photocurrent $i = f(M/e)$, is recorded. In addition, this modified mass spectrometer is capable of measuring the photocurrent for a given value of M/e as a function of the frequency ω_1. In this case the IR spectrum of traces of a complex molecule will be recorded. Since the laser frequency and the frequency of molecular absorption coincide, the molecules will be excited, and, hence, the amplitude of the ion photocurrent will vary. Instead of the usual nonselective ionization by an

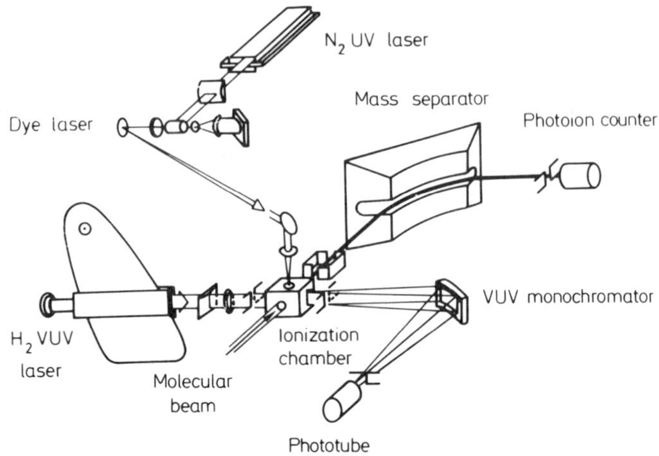

FIG. 6.1. Experimental setup for the study of the two-step photoionization of a molecule in a molecular beam with a mass spectrometer. [From Antonov *et al.* (1977a, 1978).]

electron beam (or by wide band VUV radiation), the laser mass spectrometer can selectively photoionize molecules and simultaneously produce an optical (IR and visible) absorption spectrum and a mass spectrum.

6.1.1. Two-Dimensional Mass–Optical Spectroscopy of Molecules

The first experiments on the two-step resonant photoionization of molecules, involving the mass analysis of the photoions produced, were carried out in 1977 (Antonov *et al.*, 1977a, 1978).

The experimental setup for studying the stepwise photoionization of polyatomic molecules in the mass spectrometer consists of a static magnetic mass spectrometer and time-synchronized exciting and ionizing lasers (Fig. 6.1). The intermediate electronic states of benzaldehyde, benzophenone, and *p*- and *o*-nitrotoluene were excited using a nitrogen laser ($\lambda_1 = 337$ nm) with a pulse energy of 0.2 mJ and a pulse duration of 12 ns. The absorption spectrum of NO_2 was measured by the two-step photoionization method using a dye laser tuned in the 440- to 470-nm range in the first step. In all cases the excited molecules were ionized by a hydrogen H_2 VUV laser with a generation energy of 0.5 μJ and a pulse duration of about 1 ns. The laser pulses were directed into the ion source of the mass spectrometer at right angles to each other (Fig. 6.1). The region of laser beam intersection was $0.1 \times 0.1 \times 1\ cm^3$. The generated photoions passed through the cylindrical ion optics and the magnetic separator and were registered by a secondary electron multiplier. The vapor of the molecules was admitted into the ionization chamber through a glass multichannel inlet at an estimated pressure of 10^{-4}–10^{-6} Torr.

The mass spectrum of the two-step photoionization of benzaldehyde (Fig. 6.2) exhibits a peak for the molecular ion and a peak for the fragment $C_6H_5CO^+$, which is the result of a dissociation of the hydrogen atom from the aldehyde group. The characteristic feature of two-step photoionization is that the yield of the fragment ion is almost twice that obtained by single-photon ionization, in which the photon energy is equal to the total energy of the two laser quanta. This suggests that the theory of formation in the mass spectrum based on the assumption that the excess energy of the ion is uniformly distributed over the degrees of freedom (statistical approximation) is of limited use. In the first stage the molecule undergoes electronic excitation of the type $n \to \pi^*$, i.e., a transition of the p electron of the oxygen atom to the antibonding π orbital. In effect, the vibrational energy is localized in the aldehyde group. Apparently this increases the probability of eliminating the hydrogen atom from the aldehyde group in the subsequent ionization of the molecule in the excited state.

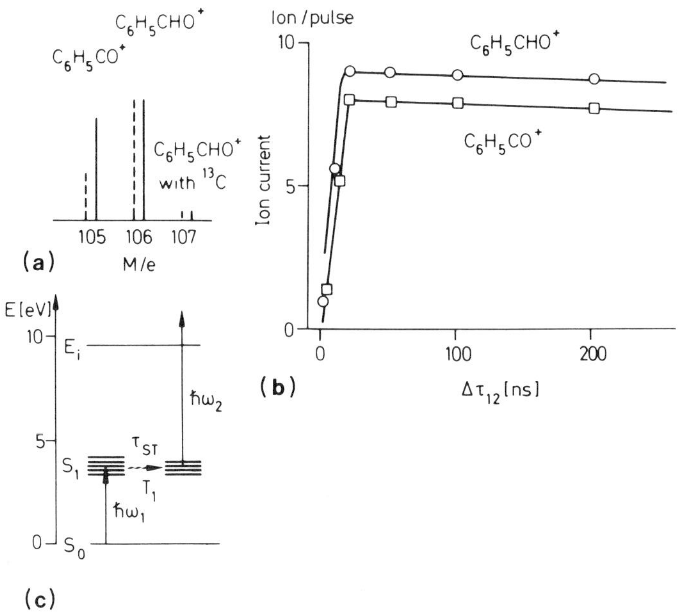

FIG. 6.2. Two-step photoionization of benzaldehyde. (a) Mass spectrum of two-step photoionization (solid lines) and single-photon ionization (dashed lines). (b) Dependence of molecular and fragment photoion yield on the delay time between the two laser pulses. (c) Quantum transitions for two-step photoionization. [From Antonov *et al.* (1978).]

Figure 6.2b illustrates the dependence of the molecular and fragment photoion yields on the delay time of the ionizing pulse. The increase in the signal in the zero-delay range is due to the finite duration of the laser pulses. A small decrease in the 1-μs region is caused by escape of the excited molecules from the interaction region. The absence of a rapid decrease in ion current at delay times less than 1 μs agrees with the available data on the fast ($\tau_{ST} \simeq 10^{-12}$ s) conversion of the molecule to the triplet state (Fig. 6.2c).

A more efficient and simple scheme for the mass analysis of photoions by means of a quadrupole filter was utilized in the experiments on the two-step molecular photoionization (Boesl *et al.*, 1978; Zacharias *et al.*, 1981; Danon *et al.*, 1982). The mass resolution $m/\Delta m$ of a quadrupole mass spectrometer (around 20 cm in length and 1 cm in field radius) is usually limited to 100 to avoid transmission losses (Dylla and Jarell, 1976). It can be increased to 1000 but at the cost of a 10^3- to 10^4-fold reduction in the spectrometer transmission. Therefore, such a spectrometer is practically incapable of resolving isotopic molecules of the same mass, for example, $^{15}N^{16}O$ and $^{14}N^{17}O$. It is, however, quite possible to do this by taking advantage of the isotopic selectivity of the optical channel, i.e., by making use of selective two-step photoionization. Such an experiment was carried out by Zacharias and his co-workers (1981). In the experiment, NO molecules were ionized in two steps: first they were excited to the ($A^2\Sigma^+, v' = 0, 1$) state by means of a tunable dye laser operating at a wavelength of $\lambda_1 = 226$ or 215 nm, and then the excited molecules in this state were excited above the ionization threshold with the fourth-harmonic output of a Nd:YAG laser ($\lambda_2 = 266$ nm). The ion signal NO^+ was observed at NO concentrations down to 10^5 cm^{-3} and was found to increase linearly with pressure. This corresponded to a detection sensitivity of around 2.5×10^4 cm^{-3} for NO molecules in the same quantum state. Similar experiments involving a two-frequency irradiation at the same wavelengths as above were performed with the CH_3 and CD_3 radicals obtained upon the multiphoton collision-free photoionization of CH_3I and CD_3I molecules (Danon *et al.*, 1982). The minimum absolute number of radicals detected in the two-frequency irradiation region (8×10^{-4} cm^{-3}) was about 10^4 for CD_3. Thus, the highly sensitive technique of stepwise photoionization mass spectrometry is of great interest as far as the registration of the products of chemical relaxations is concerned. This aspect is considered in more detail later in the text (Chapter 10).

A special place among simple molecules is occupied by H_2, which plays an important role in many chemical and physical problems. To detect this molecule in specified quantum states, use is being made of various laser spectroscopy techniques, but the sensitivity of all of them is not high enough. Therefore, the development of a photoionization mass spectrometry approach to the detection of H_2 is sufficient demonstration of the great capabilities of this method. Figure 6.3a shows the potential curves of some

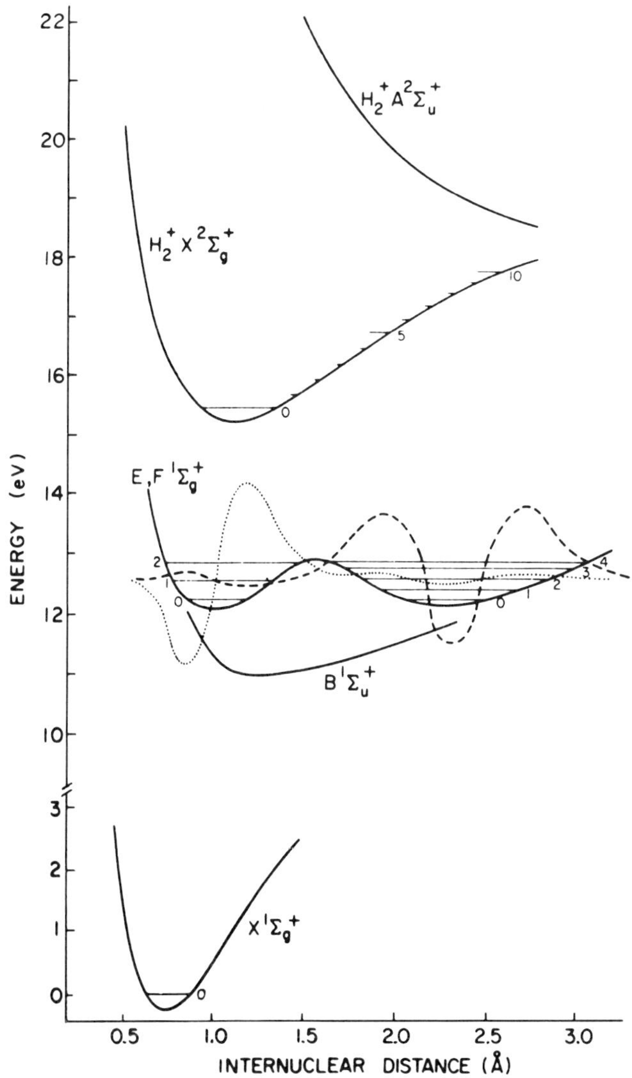

FIG. 6.3. (a) Potential energy curves of some electronic states of H_2. The curve for the $C^1\Pi_u$ state located in the vicinity of the $E^1\Sigma_g^+$ state is not shown. [From Marinero *et al.* (1983).] (b) Schemes of three- and two-photon ionization of H_2 by means of a single-frequency VUV laser radiation.

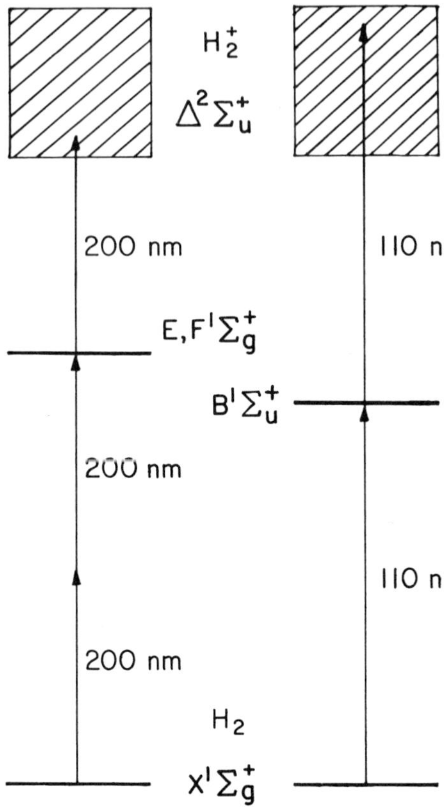

(b)

FIG. 6.3. (*Continued*)

electronic states of the H_2 molecule (Marinero *et al.*, 1983). Shown to the right in the figure are schemes of multiphoton ionization of H_2 by means of a VUV laser radiation.

First, one can effect the two-photon excitation of the E, $F^1\Sigma_g^+$ state with two minima and a lifetime of around 100 ns (Kligler and Rhodes, 1978; Kligler *et al.*, 1980). The corresponding two-photon transition frequencies are as follows: (0, 0) band, 98350–99156 cm^{-1}; (0, 1) band, 84995–95261 cm^{-1}; and (0, 2) band, 90646–91074 cm^{-1}. To excite these transitions, use was made of a laser radiation in the range 196.6–197.2 nm. The absorption of a third photon of the same wavelength leads to the ionization of the H_2 molecule (Marinero *et al.*, 1982). The detection sensitivity reached 10^8 cm^{-3} at a laser energy of 10–20 μJ.

Second, it is possible to carry out direct two-step ionization with one-photon excitation on the $B^2\Sigma_u^+(v' = 0, J') \leftarrow X^1\Sigma_g^+ \ (v'' = 0, J'')$ transition by means of a shorter-wavelength radiation in the range 1108–1116 Å (Rottke and Welge, 1983). Used in this experiment were a special setup to generate such VUV radiation and effect with its aid the photoionization of H_2 and a quadrupole mass spectrometer to detect the H_2^+ ions. The tunable radiation in the above-indicated wavelength range was obtained by mixing ($\nu_{VUV} = 2\nu_{UV} + \nu_{vis}$) in krypton the frequencies of a UV laser (second-harmonic output of a Nd:YAG laser) (Hilbig and Wallenstein, 1982). A detection sensitivity of around 10^{10} H_2 molecules in the mass filter chamber was attained at a VUV radiation pulse energy of 1 nJ (10^9 VUV photons). Based on these data, a conclusion was drawn that to reach a detection sensitivity of about 10^6 H_2 molecules in the same quantum state would be really possible.

Apart from experiments involving soft resonant stepwise photoionization under the effect of VUV laser pulses of low or moderate intensity, successful experiments were also performed on the multiphoton ionization of polyatomic molecules under the action of intense UV laser pulses of tunable frequency (Zandee and Bernstein, 1979; Lichtin *et al.*, 1980). Such experiments commonly use a time-of-flight mass spectrometer, making it possible to register with moderate resolution the entire mass spectrum during a single laser pulse. Of course, the mass spectrum is in this case sufficiently intense because of the strong fragmentation of the molecules. A typical result of measuring the two-dimensional mass–optical spectrum of the 1,4-Diazabicyclo[2.2.2]octane (DABCO) polyatomic molecule is shown in Fig. 6.4.

The mass spectra obtained always show the presence of fragment ions, any of which may have a spectral dependence of the yield of its own, differing from that of other ions. So the laser selective photoionization tenchique enables us to obtain fairly characteristic, two-dimensional molecular "charts". The highly informative, two-dimensional mass–optical spectra can, in principle, form the basis for attaining the selectivity necessary to detect different molecules similar in composition and molecular mass.

The time-of-flight mass spectrometer is ideally suited for use in conjunction with molecular photoionization by a pulsed laser radiation. Both the sweep-type mass spectrometer (Fig. 6.1) and the quadrupole mass spectrometer allow only a single spectrum line to be observed during a single laser pulse. This materially lowers the photoionization detection sensitivity, especially where the fragmentation of molecules is noticeable. The time-of-flight mass spectrometer is capable of registering the entire mass spectrum during a single laser flash. The photoionization of molecules takes place within a strictly limited, small region (a few cubic millimeters) and during a short time interval (5–20 ns), and the energy of the photoions produced can be easily controlled by varying the laser radiation intensity and wavelength.

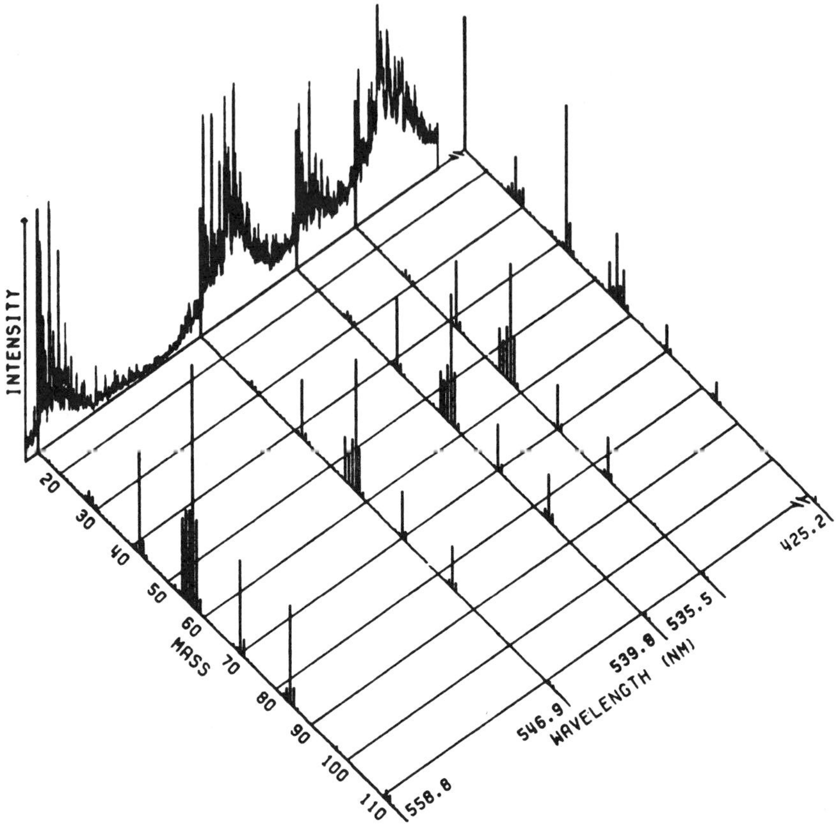

FIG. 6.4. Two-dimensional vibrational–mass spectrum for multiphoton ionization of DABCO in a strong laser field. Shown in the background is the radiation wavelength dependence of the net photoion yield. [From Lichtin *et al.* (1980).]

With the drift tube being a few tens of centimeters long, it usually proves possible to reach a resolution of the order of 150–200 at almost a 100% ion transmission. The resolution can be substantially improved by using the time-of-flight mass spectrometer modification known as the "reflectron" (Mamyrin *et al.*, 1973). In this mass spectrometer (Fig. 6.5), the ions reflect backward from an electrostatic mirror, the ions with higher velocities penetrating deeper into the reflecting system. In this way, the mass-peak broadening due to the ion velocity distribution is compensated for. When this technique was used to register the photoionization mass spectrum of the benzene molecule, the ion packet duration was shortened to 8 ns (Boesl *et al.*, 1982b). This was tantamount to increasing the resolution $m/\Delta m$ to 3900

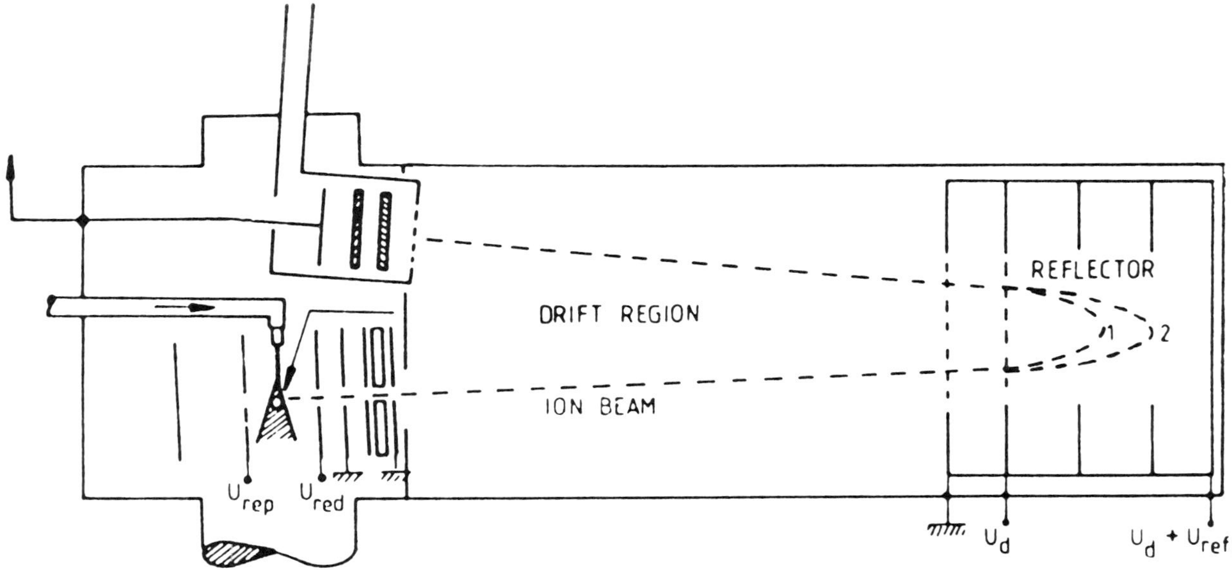

FIG. 6.5. Schematic diagram of a reflecting time-of-flight mass spectrometer. Paths 1 and 2 correspond to ions having equal masses but different kinetic energies. Faster ions (path 2) have longer flight paths in the reflecting system, and so they come back to the detector simultaneously with the slower ions (path 1). [From Schlag and Neusser (1983).]

for masses of $m \simeq 100$. The dramatic improvement of the photoionization mass spectrum resolution provided by the time-of-flight reflectron for the benzene molecule can be achieved. This technique is believed to be capable of increasing the resolution of the time-of-flight photoionization spectrometry still more, up to around 10,000 (Schlag and Neusser, 1983).

6.1.2. Detection of Single Molecules in a Photoionization Mass Spectrometer

During irradiation by intense resonant UV laser radiation there is a possibility that the molecule (or its fragments) absorbs more photons than is necessary for simple ionization. The result is a highly fragmented mass spectrum, and the information of the initial molecules may be lost (see Section 5.2.3). An ultimate ionization yield with moderate molecular fragmentation can be attained in two-step photoionization only when the intermediate electronically excited state is stable throughout the duration of the laser pulse and the laser quantum energy in the second step exceeds the vertical ionization potential of the molecule in the excited state (Antonov *et al.*, 1980a,b). Compared with ionization, the latter corresponds to a lower molecular dissociation yield after the absorption of a second photon.

It was under these optimal conditions that the detection of single polyatomic molecules was demonstrated (Antonov *et al.*, 1981a). The experiment was performed on naphthalene using an excimer KrF laser ($\lambda = 248$ nm) to excite and ionize the molecules in intermediate electronic states. The photoions were recorded by a time-of-flight mass spectrometer especially designed for the laser-photoionization experiments.

The radiation of the KrF laser (pulse duration was 15 ns and maximum energy was 10 mJ) was focused into the ion source of the time-of-flight mass spectrometer by a quartz lens with $f = 1$ m. The cross section of the laser beam in the region of photoionization was $(1 \times 3)\ \text{mm}^2$ and the photoionization volume was $1.2 \times 10^{-2}\ \text{cm}^3$. After being accelerated and separated in the field-free region, the resulting photoions reach the cathode of the electron multiplier. The average single-ion pulse from the multiplier was 3 mV strong and 15 ns long. Taking into account a 75% transmittance of photoions to the multiplier and a 70% detection efficiency, the total detection efficiency was 50%. The mass resolution of the ions was about 200. The required very low pressure of naphthalene was attained by filling the mass spectrometer with naphthalene vapor at a pressure of 10^{-6} Torr and subsequently evacuating the mass spectrometer in the course of several days. The residual naphthalene vapor (10^{-5} Torr) was irradiated in the presence of air.

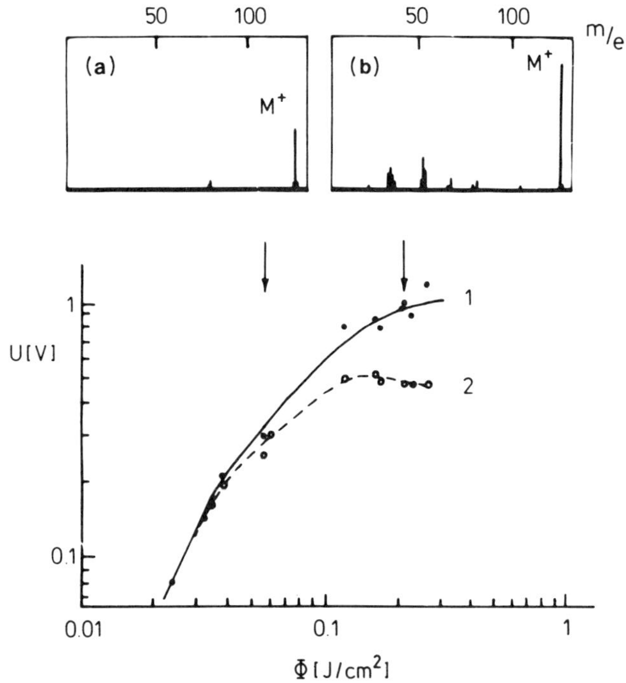

FIG. 6.6. Dependences of the total photoion signal (curve 1) and the molecular ion signal (curve 2) of naphthalene on the energy flux of the laser pulses at $\lambda = 248$ nm. [From Antonov *et al.* (1981a).]

Figure 6.6 illustrates the dependence of the total ion signal (curve 1) and the signal for the molecular ions of naphthalene (curve 2) on the energy flux of the laser pulses as well as the mass spectra of the resulting photoions. In the region of low energy flux the dependence of the ionization yield of naphthalene is quadratic, and only molecular ions can be observed in the mass spectrum (Fig. 6.6a). With laser energy fluxes over 0.05 J/cm^2, both the molecular ion and the total ions signals of naphthalene undergo saturation. In the region between 0.15 and 0.20 J/cm^2 the total ion signal reaches a plateau, which is followed by a slight increase in the fraction of fragments in the mass spectrum (Fig. 6.6b). According to Antonov *et al.* (1980a), a slight decrease in molecular ion yield may be observed due to their dissociation in the laser field.

The quadratic behavior of the photoion yield in the low energy flux region is due to a two-step molecular photoionization. This is possible because the ionization potential of naphthalene, $I = 8.12$ eV, is smaller than the energy of two laser quanta, $2\hbar\omega = 10$ eV. The saturation of the ionization yield of naphthalene in the 0.05 to 0.10-J/cm^2 region is qualitatively consistent with a saturation energy of the electronic absorption in the first step, estimated from the known molecular absorption cross section in the gas phase. The fact that at pulse energies between 0.15 and 0.20 J/cm^2 the total ion signal reaches a plateau suggests that the photon absorption in the second step also reaches saturation. Since the energy of two laser photons is greater than the molecular ionization potential, and photoionization is equivalent to the elimination of a nonlocalized π electron, the possibility of molecular dissociation in addition to ionization after absorption of a second photon may be neglected. Thus, the photoionization efficiency with such energy fluxes is almost 100%.

Under these conditions (a laser-pulse energy flux of 0.15 J/cm^2, with which the photoionization efficiency is close to 100%), the mass spectrum of naphthalene was recorded with high reliability. To attain a very low partial pressure of napthalene, the mass spectrometer was evacuated for a week. The ion peaks in the mass spectrometer (Fig. 6.6a) correspond to single-ion pulses. In addition to the peaks of the molecular ions of naphthalene, several peaks are observed that are the result of the photoionization of organic impurities in the air.

In this experiment the average value of the naphthalene molecular ion signal was about 1 ion/pulse. Taking into account the fact that the yield of molecular ions at 100% ionization efficiency is about 50% (Fig. 6.6, curve 2) and that the recording efficiency of the photoions is equal to 50%, we may conclude that the observed signal corresponds to 4 molecules in the photoionization volume. This value corresponds to a partial pressure of naphthalene equal to 10^{-14} Torr and a relative concentration in the air of 10^{-9}.

The results obtained are practically limiting with respect to the efficiency of detection of molecules in the photoionization volume and correspond to the detection of single molecules over the course of a single pulse. We note that achievement of high sensitivity of detection of naphthalene molecules contributed greatly to the selectivity of laser stepwise photoionization. The basic components of air were completely absent from the mass spectrum of the photoions because simple molecules do not absorb in the 250-nm region. Even such selectivity of ionization with respect to classes of molecules greatly increases the analytical possibilities of mass spectrometry. In standard magnetic mass spectrometers, for example, there is a constant background at all masses due to scattered ions, which restricts the detectable relative concentrations to the level of 10^{-6}–10^{-9}.

We note that naphthalene molecules are not unique from the point of view of the possibilities of their detection via photoionization. Table 6.1 presents experimental data on the efficiencies of two-step photoionization of a number of molecules in a single-frequency laser field with intensity 10^7 W/cm^2 (Boesl

TABLE 6.1. Experimental Values of the Absorption Cross Sections for Radiation in the First and Second Steps and the Efficiency of Two-Step Photoionization of Some Polyatomic Molecules in a Single-Frequency Laser Field[a]

	$\langle\sigma_1\rangle$ (cm^2)	$\langle\sigma_2\rangle$ (cm^2)	I_{sat} (W/cm^2)	Ionization yield, $I = 10^7$ W/cm^2(%)
Benzene	2.7×10^{-17}	3.4×10^{-18}	1.5×10^8	3.7
Thiophene	1.8×10^{-17}	—	—	0.01
Toluene	6.0×10^{-18}	2.0×10^{-17}	3.3×10^7	6.2
Naphthalene	2.8×10^{-18}	1.6×10^{-19}	2.8×10^9	0.05
Aniline	4.5×10^{-17}	2.0×10^{-17}	2.3×10^7	25

[a] From Boesl et al. (1981). The radiation frequency corresponded to the onset of electronic absorption of the molecule.

et al., 1981). It is significant that fragmentation of molecules is practically absent at this radiation intensity. It is evident from the table that when the laser radiation intensity is close to the saturation intensity, the efficiency of photoionization of molecules is very high.

6.2. Photoionization Detection of Molecules in a Buffer Gas

Separating ions by their mass-to-charge ratio in the course of mass-spectrometric registration improves the selectivity of the laser photoionization technique of detecting molecules. However, such separation in any type of mass spectrometer requires that the ions move without collisions in the electric and magnetic fields used for the purpose. To ensure this, it is, as a rule, necessary that the gas pressure not exceed 10^{-5} Torr. Therefore, when analyzing a sample, for example, at atmospheric pressure, we must reduce the molecular concentration at the entrance to the mass spectrometer by a factor of 10^8 or 10^5–10^6 when an effective differential pumping is used in the ion source. When the relative concentration of the molecules of interest is low (less than 10^{-9}), such a drastic reduction of pressure may result in there not being a single molecule to be detected in the laser photoionization region. For this reason, it is of great interest to develop laser resonant photoionization techniques that allow molecules to be detected at atmospheric pressure directly in air or in a buffer gas.

6.2.1. Detection in an Ionization Chamber

The selectivity of molecular detection in ionization chambers designed to register the net photocurrent depends solely on the selectivity of the laser resonant photoionization of the molecules. A typical experimental setup

(Freuholz *et al.*, 1980) consists of an ionization chamber comprising a coaxial-cylinder (or sometimes parallel-plate) capacitor, a registration system, and a variable-frequency laser source. The ionization chamber can be operated in the following two modes: (1) complete photocharge collection mode and (2) gas amplification mode. In the first mode, the completeness of photoion collection is verified by the presence of a plateau in the dependence of the ion signal on the voltage applied to the chamber electrodes. The ion registration sensitivity is in this case governed by the amplifier noise and typically has a value of about 10^3 pulses/min (Klimcak and Wessel, 1980b).

In the second mode, the electric field strength is set high enough for an electron to acquire between two collisions an energy sufficient to ionize the buffer gas molecules. Under such conditions, every single photoelectron initiates an avalanche with a duration of 10^{-6} s and an amplitude of 10^{-14} C (Freuholz *et al.*, 1980). When the laser fluences used are low, the registration of ions is carried out by counting pulses (avalanches), and when the fluences are high, the ionization rate becomes so high that the avalanches overlap in time, and the registration is in this case effected by measuring the mean photocurrent. Such an ion registration technique allows us to detect each electron–ion pair formed (Williams and Sara, 1962).

The detection of polyatomic molecules present in air in low concentrations by means of a simple ionization chamber was investigated in experiments with the aniline molecule (Brophy and Rettner, 1979b). The energy level of the first electronically excited singlet states S_1 of aniline lies above one-half its ionization potential, and, therefore, an effective two-step photoionization of the molecule can be carried out in a single-frequency laser field. With the aniline concentration in air being as low as 10^{-5}, the two-step photoionization spectrum of the molecule was registered with a signal-to-noise ratio in excess of 10 without accumulation. Estimates show that increasing the power of the laser pulses and reducing their duration should make it possible to detect aniline present in air in concentrations as low as 10^{-11}. Of course, in conditions of an actual experiment, the detection sensitivity may be limited by the background noise signal due to the ionization of other organic impurities present in the air and the multiphoton ionization of the air constituents.

The naphthalene molecule, a typical representative of the class of aromatics, is also well suited for laser photoionization experiments. This molecule was used to investigate the ultimate potentialities of detecting molecules present in a buffer gas (nitrogen) in low absolute concentrations (Klimcak and Wessel, 1980a,b). Measurements were taken in an ionization chamber operated in the complete photoion collection mode, the photoionization region occupying a volume of 0.1 cm^3. The photoionization of the naphthalene molecules was effected according to a two-step scheme by means of laser pulses (second-harmonic output of a dye laser) 1 mJ in energy, 5 ns in duration, and 1 mm in beam diameter. The naphthalene molecule has an absorption maximum on the $S_2 \leftarrow S_0$ transition at a wavelength of 268.5 nm and a minimum at 277.5 nm. The photoionization signal was found to depend linearly on the naphthalene concentration over a wide dynamic range occupying eight orders of magnitude of the naphthalene partial pressure variation, and the maximum detection sensitivity amounted to 5×10^4 mol/cm^3 at a ratio of $S/N = 2$. The sensitivity can, in principle, be further increased 10–100 times by operating the ionization chamber in the avalanche mode.

When detecting molecules by the laser photoionization technique in an ionization chamber, it is especially critical to ensure the selectivity of photoionization, for it is in fact the principal factor limiting the sensitivity of the method. For instance, in the above-mentioned experiment with naphthalene, the photoionization signal due to the ionization of naphthalene was taken as the difference of two signals at the wavelengths corresponding to the absorption maximum and minimum of the molecule. In a number of instances, it may prove useful to utilize a three-step photoionization. On the one hand, this allows employing a longer-wavelength radiation and thus reducing the background ionization, and on the other, the presence of a second resonant transition between discrete levels improves the photoionization selectivity. Such an approach was realized for the iodine molecule and made it possible to attain a sensitivity of detecting iodine in air at a level of 10^{-9} (Darznek *et al.*, 1983).

6.2.2. Photoion Mobility Spectrometer

There is a possibility of enhancing the selectivity of detecting ions by way of their chromatographic separation on the basis of their mobility in a buffer gas. The plasma chromatography technique is widely used for analytical purposes (Karasek *et al.*, 1976; Hagen, 1979). A standard plasma chromatograph uses a β-source (^{63}Ni) to produce buffer gas ions, which then enter into a series of ion–molecular reactions leading to the formation of ions of the impurities present in the gas. The impurity ions enter the drift region of the chromatograph where they move in the buffer gas in an externally applied electric field and become separated in space and time in accordance with their mobilities.

The use of laser resonant photoionization instead of the standard ion production technique allows us, in principle, to avoid many problems associated with the fact that ion–molecular reactions frequently yield a great variety of ion–molecular combinations that complicate the analysis (Lubman

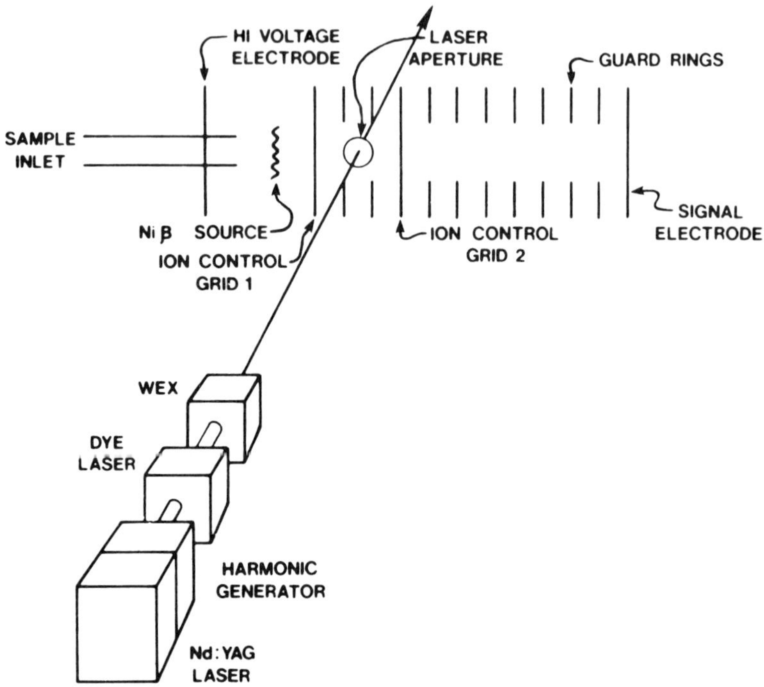

FIG. 6.7. Plasma chromatograph modified to produce ions by the laser resonant photoionization technique. [From Lubman and Kronick (1982).]

and Kronick, 1982). What is more, the separation of the produced photoions by their mobilities considerably improves the selectivity and hence the sensitivity of detecting impurity molecules in air or a buffer gas and thus makes it possible to surmount the main difficulties involved in the nonselective registration of photoions in the ionization chamber.

Figure 6.7 shows a schematic of the ion mobility spectrometer using laser photoionization (Lubman and Kronick, 1982), built around a standard plasma chromatograph. Ions are produced in the region between grids 1 and 2 under the effect of resonant laser radiation. When using laser photoionization, grid 1 is kept positively charged to block the ions coming from the region of the radioactive source. A pulsed voltage is applied to grid 2, which expels photoions into the drift region 8 cm in length. The expulsion pulse duration can be varied over the range 0.05–0.5 ms. The shorter the ion injection duration, the higher the resolution, but the lower the sensitivity. The entire arrangement is operated at a temperature of around 200°C to protect the spectrometer against contamination with the substances being analyzed. The sample is bled in through a quartz leak at a temperature of 275°C. The buffer gas used in the experiments is pure nitrogen, and photoionization is effected according to a two-step scheme by means of a single-frequency UV-laser radiation (second-harmonic output of a dye laser or fourth-harmonic output of a Nd:YAG laser).

This work presents the ion mobility spectrograms for naphthalene, benzene, and toluene, obtained upon the two-step photoionization of these molecules at wavelengths of 280 and 266 nm, the relative naphthalene concentration in the sample being fixed at a level of 2×10^{-8}. All the peaks are well resolved relative to one another and to the residual background noise spikes in the region of long ion-injection times. When using the longer-wavelength radiation ($\lambda = 280$ nm), the spectrum shows mainly the naphthalene peak, since the benzene and toluene molecules do not absorb at this wavelength and therefore cannot be ionized by the radiation. The maximum detection sensitivity for benzene in nitrogen amounted to 10^{-9} and was limited by the presence of residual benzene in the spectrometer and not by the background noise. This experiment demonstrates the essence and capabilities of the technique based on combining the optical and ion-chromatographic selectivities of trace molecule detection.

6.3. Photoionization Detection of Molecules on a Surface

All cases examined above concerned photoprocesses with formation of photoions accompanying irradiation of molecules in a gas, when molecules can be viewed, during the course of the interaction with the laser pulse, as being isolated from collisions with the walls. Meanwhile, there is great interest in obtaining ions by irradiating molecules on the surface of a metal or dielectric. Moreover, this process is actually related to photoionization of molecules in the gas phase.

6.3.1. Production of Molecular Ions upon Irradiation of a Surface

Processes leading to the formation of free ions accompanying the action of laser radiation on the surface of a solid have attracted a great deal of attention for a long time. In investigations conducted during the last few years in different laboratories, laser radiation with parameters varying over a very wide range was used: pulse durations varied from continuous radiation to 10^{-11} s, wavelengths varied from the IR to the UV region, and intensities varied from a few watts to 10^{11} W/cm^2. The action of radiation with such different parameters on a surface must give rise to very diverse processes, leading to the formation of ions. The investigation of these processes is of great practical interest. For example, laser-stimulated formation of ions is already used now in mass spectrometric analysis of a wide class of materials ranging from refractory materials to biological molecules.

The mechanism of ionization accompanying the action of intense laser radiation on a surface with intensity $I = 10^9$–10^{11} W/cm^2, leading to strong heating ($T \gtrsim 1000°C$) of the surface of the solid, intense vaporization of the material, and formation of a plasma, has been studied in the greatest detail (Ready, 1971). In this case the ionization process has a thermal character and is described by the Langmuir–Saha equation. This method for obtaining ions is used in mass spectrometry of inorganic specimens (Knox, 1972). This method is of little use in analyzing specimens containing complicated organic molecules, because it leads to complete decomposition of these molecules due to the strong heating of the material.

Since 1976 a number of laboratories have begun intense investigations of the mechanisms of laser-stimulated formation of ions of organic molecules on the surface of solid specimens (Unsold *et al.*, 1976; Posthumus *et al.*, 1978; Stoll and Röllgen, 1979; Antonov *et al.*, 1980c; Conzemius and Capellen, 1980). Both pulsed radiation with an intensity of 10^4–10^9 W/cm^2 (Unsold *et al.*, 1976; Antonov *et al.*, 1980c) and continuous CO_2 laser radiation (Stoll and Röllgen, 1979) were used in the investigations. One of the mechanisms of formation of ions of organic molecules consists of the following: the pulse of laser radiation heats the surface, intense vaporization of the material leads to the formation of a dense cloud of vapor in which chemical reactions occur (proton transfer, addition of the cation of an alkali metal, etc.), which leads to the formation of ions (Schade *et al.*, 1981; Van der Pegl *et al.*, 1981). Both positive and negative ions—$(M + H)^+$, $(M - H)^-$, and (M + alkali)—and fragment ions are observed in the mass spectrum. To explain the results obtained on the laser microprobe mass analyzer (LAMMA) apparatus, it was proposed that "collective, nonequilibrium processes in the condensed phase" are responsible for the formation of ions (Hillenkamp, 1982).

Antonov *et al.* (1980c, 1981b, 1982a) observed the molecular ion formation upon irradiation of crystalline specimens of the bases of nucleic acids and anthracene with UV laser pulses with moderate intensity ($I = 10^4$–10^6 W/cm^2). An important feature of the observed process is the absence of fragment ions. Only molecular ions M^+ were observed in the mass spectra. It was observed that the ion yield increases with the photon energy. For the case of adenine, estimates showed that with the minimum recorded ionic signal, the heating of the surface did not exceed 100°C, which is much lower than the melting temperature of adenine ($T_m = 360°C$) and cannot lead to intense vaporization of the specimen. These facts indicate that the process differs considerably from the processes described previously. A number of hypotheses were proposed (Antonov *et al.*, 1981b, 1982a) concerning the mechanism of formation of such ions, but the following simple model was proposed as a result of special experiments: the laser radiation pulse slightly heats the surface, thereby increasing the probability of desorption of molecules from the surface during the laser pulse. Molecules leaving the surface are resonantly ionized by the same radiation, as described in Chapter 5. Estimates performed within the framework of the proposed model agree well with the experimental results. This mechanism and its application to the detection of molecules are discussed later in the text.

6.3.2. Formation of Molecular Ions from Crystals

In experiments with a surface, the specimen is usually placed directly into the ion source of the mass spectrometer (Fig. 6.8). This permits analyzing the masses of the ions forming under the action of the laser radiation on the surface. When a time-of-flight mass spectrometer is used, the electrical pulse can be applied to the ion gun with different delay τ_d relative to the laser pulse, and this facilitates the bunching of ions with different masses with respect to initial velocities (Wiley and McLaren, 1955). The regime in which

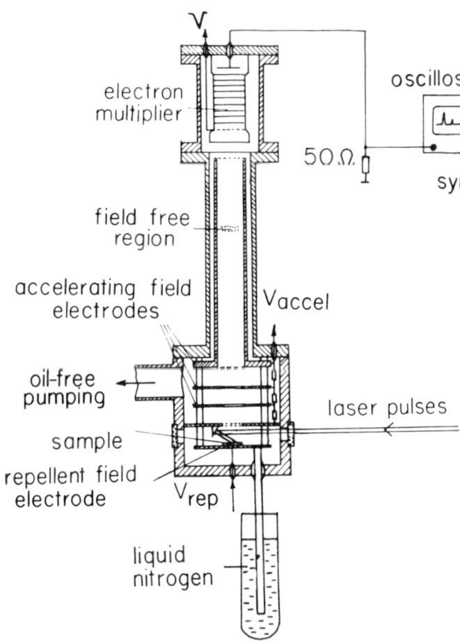

FIG. 6.8. Experimental setup for the laser photoionization detection of molecules on surface in a time-of-flight mass spectrometer. [Egorov *et al.*, 1983.]

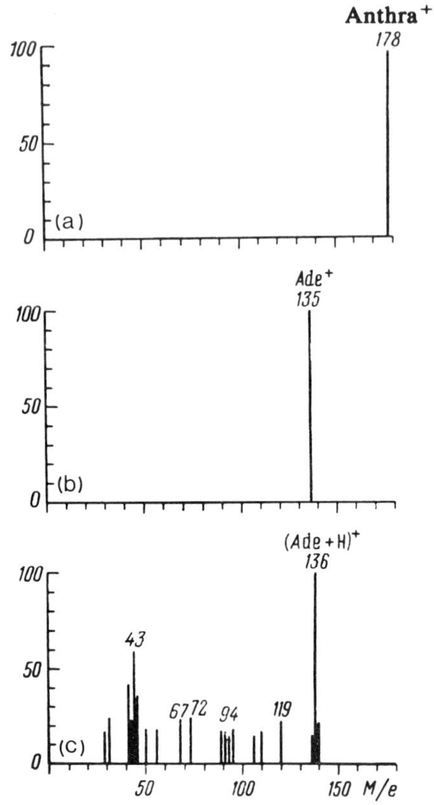

FIG. 6.9. Characteristic mass spectra of ions forming upon irradiation of crystalline powders of (a) anthracene and (b) adenine by KrF laser radiation. [From Antonov *et al.* (1980c, 1981b).] The mass spectrum of adenine (c) obtained on the LAMMA apparatus is presented for comparison. [From Schueler and Krueger (1980).]

the ion pulse is recorded with velocity bunching permits working with maximum mass resolution. In the absence of such bunching ($\tau_d = 0$), it is possible to measure the distribution function of ions over the initial velocity $f(v_0)$.

Different types of specimens were used in the experiments: fine-crystalline adenine powder; single-crystalline anthracene films; polycrystalline rhodamine 6G films, obtained with rapid vacuum evaporation of a solution of rhodamine 6G in alcohol, and fine-crystalline peptide powder. The electrode on which the specimen was placed could be cooled to $T = 200$ K.

The characteristic mass spectra of ions, formed upon irradiation of crystalline powders of adenine and anthracene, are shown in Fig. 6.9. The mass spectrum of adenine obtained on the LAMMA apparatus is presented for comparison (Fig. 6.9c) (Schueler and Krueger, 1980). As is evident from the figure, the first two mass spectra consist only of molecular ions, in contrast to the third mass spectrum (see Fig. 6.9c) in which the protonated molecular ion $(M + H)^+$ dominates, and there are a number of fragment ions.

The ion yield depends sharply on the energy flux Φ of laser pulses. For the case of adenine crystals, this dependence is approximated by $N_i \sim \Phi^7$.

A comparison of the yield of molecular ions obtained upon irradiation of adenine by nanosecond ($\tau = 20$ ns, $\lambda = 249$ nm) and picosecond ($\tau = 30$ ps, $\lambda = 266$ nm) pulses (Antonov *et al.*, 1982b) shows that the formation of ions is determined by the density of the absorbed energy $\Phi\varkappa$, where $\varkappa$ is the coefficient of absorption of the crystal per unit length, and not by the radiation intensity. If it is assumed that the energy absorbed in the crystal is completely thermalized, then, neglecting the thermal conductivity, the maximum possible heating of the surface with a characteristic energy flux $\Phi = 4$ mJ/cm^2 constitutes $\Delta T = \Phi\varkappa/c\rho = 200°C$; here c and ρ are the specific heat capacity and specific weight of adenine. Under such heating, the rate of desorption of molecules from the surface can be expected to increase considerably.

It was observed that together with the formation of molecular ions, upon irradiation of adenine neutral molecules also fly off from the surface (Antonov *et al.*, 1982b). The surface of the specimen was irradiated with picosecond pulses, and the neutral molecules were recorded with the help of a photo-ionizing KrF laser pulse, which irradiated a small region above the surface of the specimen. This observation suggested the following elementary mechanism of ion formation: the pulsed heating of the surface of the specimen with irradiation → desorption of neutral molecules → photoionization of free molecules above the surface.

This hypothesis was checked as follows (Egorov *et al.*, 1983, 1984b). The probability of desorption of a molecule per unit time W_{des} or the inverse lifetime of an adsorbed molecule on the surface are determined by the Arrhenius formula:

$$W_{des} = \tau_{ads}^{-1} = \Omega \exp(-E_{ads}/kT), \tag{6.1}$$

where $\Omega = 10^{12}$–10^{14} s^{-1} is the rate factor, and E_{ads} is the energy of adsorption of a molecule on the surface with temperature T. For example, for adenine the probability of desorption decreases by a factor of ten due to cooling of the surface heated by the laser pulse approximately $\tau_{cool} = 80$ ns after termination of the laser pulse ($\tau_p = 15$ ns). Therefore a much larger number of neutral molecules accumulate in the space above the surface after termination of the laser pulse than during it. The number of molecules desorbed after the laser pulse can be measured by irradiating the surface with a second pulse of laser radiation after a time τ_d ($\tau_f > \tau_d > \tau_{cool}$), where τ_f is the time for molecules to fly away from the irradiated region. The flux of energy in the second laser pulse Φ_2, of course, must be less than the energy flux Φ_{th}, thus producing the appearance of an ionic signal upon irradiation of the surface by a single laser pulse.

To check this model, an experiment was performed with two KrF laser pulses ($\lambda = 249$ nm) (Egorov *et al.*, 1983). The pulse of the probing KrF laser with a delay of $\tau_d = 2\ \mu s$ was oriented along the surface of the specimen, touching it so that all molecules leaving the surface after the first KrF laser pulse entered the region of ionization of the probing pulse. Measurements were performed with an energy flux $\Phi_1 = 4\ mJ/cm^2$ in the first pulse and $\Phi_2 = 2.3\ mJ/cm^2$ in the second. The ionic signal was absent when the specimen was illuminated only by the second pulse. The measurements showed that the number of ions formed from neutral molecules under the action of the second pulse is 11 times greater than the number of ions forming under the action of the first pulse on the adenine surface. If the fact that the ion yield for the second pulse depends on $U_2 \sim \Phi_2^2$ (two-step ionization) is taken into account, then for equal fluxes of energy the ratio of the number of ions will equal $(U_2/U_1)(\Phi_1/\Phi_2)^2 = 36$.

These experimental results suggest that ions forming under the action of the first pulse appear as a result of photoionization of molecules that flew off from the surface during the laser pulse. Desorption of neutral molecules, which are subsequently ionized by the second laser pulse in the gas phase, continues for a longer period of time after termination of the first pulse.

Inasmuch as ion formation is preceded by thermal evaporation of molecules from the surface, the ionic signal N_i is proportional to the product of the desorption probability P_{des} and the multiphoton ionization yield $Y_{ion}(\lambda)$:

$$N_i \sim N_0 P_{des}(\Phi, \lambda) Y_{ion}(\lambda), \tag{6.2}$$

where N_0 is the number of irradiated molecules on the surface.

The probability of desorption of a molecule P_{des} during the laser pulse τ_p evidently equals

$$\mathrm{P}_{des} = 1 - \exp\left[-\int_0^{\tau_p} W_{des}(t)\,dt\right], \tag{6.3}$$

where W_{des} is the rate of desorption (6.1).

The magnitude of the heating ΔT has a maximum in the region of maximum absorption $\kappa(\lambda)$. Thus, the desorption probability P_{des} depends sharply on the density of the energy absorbed in the specimen $\Phi\kappa(\lambda)$ and together with the spectral dependence of photoionization $Y_{ion}(\lambda)$ determines the resonant character of ion yield. This explains the results obtained with irradiation of adenine by pulses with $\lambda = 249$, 308, and 337 nm (Fig. 6.10). The yield of adenine ions is maximum for radiation with $\lambda = 249$ nm, which falls in the electronic absorption band of adenine. Radiation with $\lambda = 308$ nm falls on the edge of the absorption band, when $\kappa(\lambda)$ and $Y_{ion}(\lambda)$ decrease. For this reason, ions are observed with a considerably higher energy density of laser radiation. Finally, radiation with $\lambda = 337$ nm lies outside the resonant

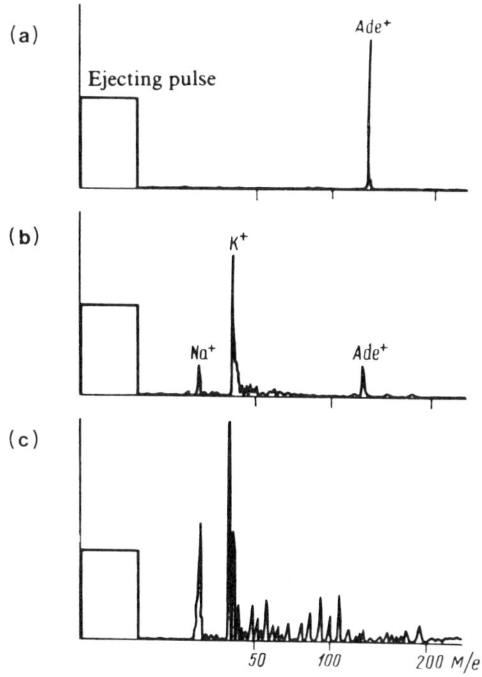

FIG. 6.10. Mass spectra obtained upon irradiation of adenine by pulses with wavelengths of (a) 249 nm (2 mJ/cm^2), (b) 308 nm (12 mJ/cm^2), and (c) 337 nm (42 mJ/cm^2), gradually leaving the electronic absorption band of adenine. [From Antonov *et al.*, 1981b.]

absorption band of adenine, which sharply decrease $\kappa(\lambda)$ and $Y_{ion}(\lambda)$. To observe the mass spectrum it is necessary in this case to use radiation with much higher intensity, when strong heating of the material and formation of plasma due to nonresonant absorption are possible. In this case, strong fragmentation of molecules occurs, and the molecular ion $(M)^+$ is not observed.

6.3.3. Detection of Molecules Adsorbed on a Surface

With the thermal mechanism of desorption of molecules form the surface and their subsequent photoionization in the gas, it is in principle possible to observe ions of any molecules adsorbed on the surface, it they are susceptible to multiphoton ionization. Such effects have been observed in experiments with adenine crystals, if anthracene vapor was present in the volume of the mass spectrometer. When an adenine crystal is irradiated by UV pulses, the mass spectrum contains both adenine and anthracene ions. When the volume above the surface was irradiated, only anthracene molecules were observed, but the signal with $\Phi = 6$ mJ/cm^2 was two to three times weaker. Cooling the ion gun containing the specimen to $T = 200$ K greatly decreased the signal of anthracene ions from the gas phase and increased the signal from the surface. This leads to the conclusion that anthracene ions are formed as a result of photoionization of anthracene molecules adsorbed on the adenine surface. This suggests that ions of both the molecules constituting the crystals and the adsorbed molecules are formed by the same mechanism.

Figure 6.11 presents the dependences of the ionic signals on the energy flux of laser radiation Φ for anthracene adsorbed on the surface of adenine or rhodamine 6G. For anthracene adsorbed on adenine (Fig. 6.11a) at

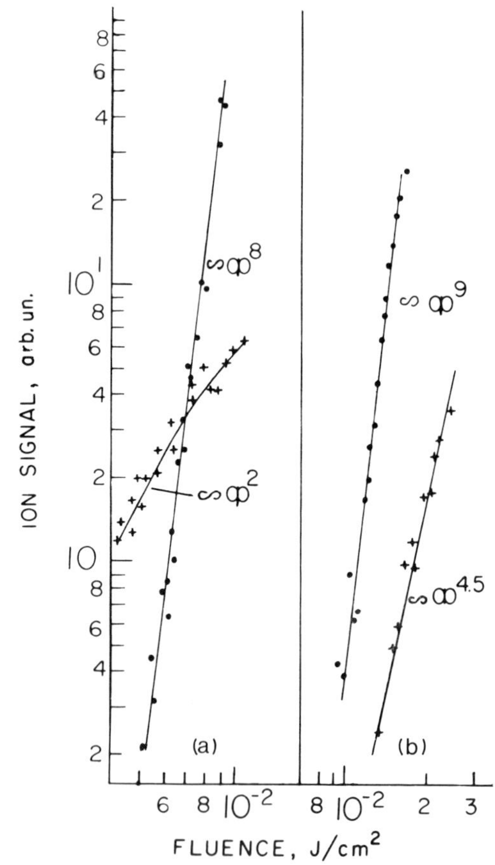

FIG. 6.11. Dependence of the ion signal on the energy flux of the laser pulse Φ upon irradiation of anthracene adsorbed on the surface of (a) adenine and (b) rhodamine 6G at 200 K (●) and 300 K (+). [From Egorov *et al.* (1983).]

$T = 200$ K, the dependence is approximated by $U \sim \Phi^8$, and at $T = 300$ K it is close to the dependence of the ionization yield $Y_{ion}(\Phi)$ of anthracene vapor on the energy flux. These facts can be explained as follows. At low specimen temperature, the number of ions is determined by the number of molecules desorbed during the pulse and by their ionization yield. The rate or desorption depends very sharply on the temperature of the surface and, therefore, on the energy flux of the laser pulse. At $T = 300$ K all molecules are desorbed at the beginning of the laser pulse, so the ionization yield is determinated primarily by the ionization efficiency in the gas phase. This mechanism might be responsible for the results of the other experiments on the observation of ions produced as a result of irradiation of molecules on a surface by UV laser pulses (Opsal and Reilly, 1983).

For anthracene on rhodamine 6G (Fig. 6.11b), the exponent in the dependence $U \sim \Phi^n$ varies from $n = 9$ at $T = 200$ K to $n = 4.5$ at $T = 300$ K. At $T = 200$ K and $\Phi = 16$ mJ/cm^2 the signal due to anthracene ions, when the rhodamine 6G surface is irradiated, is a factor of 10^3 greater than with irradiation of the volume above the surface. Therefore the method of adsorption of molecules on a surface can greatly increase the sensitivity of detection of molecules in the photoionization mass spectrometer.

Comparison of the amplitudes of the ionic signals corresponding to irradiation of a surface and of the volume above the surface permits estimating the lifetime of the adsorbed molecule on the surface τ_{ads} and the energy of adsorption E_{ads}. Indeed, for a "gas–surface" equilibrium the concentration of molecules adsorbed on the surface n_s and in the gas phase n_g are related by the expression

$$n_s = \tfrac{1}{4} n_g v_0 \tau_{ads}, \tag{6.4}$$

where v_0 is the average velocity of molecules in the gas. Assume that all molecules on the surface are desorbed by the first laser pulse ($P_{des} = 1$). Then the number of ions forming under the action of the second laser pulse equals $N_i^s = n_s S_{des} Y_{ion}$, where S_{des} is the irradiated surface area from which desorption occurs. When the volume V_{irr} above the surface is irradiated, the number of ions formed equals $N_i^g = n_g V_{irr} Y_{ion}$. Therefore the ratio of the integrated ionic signals corresponding to irradiation of the surface and of the volume equals

$$N_i^s / N_i^g = \tfrac{1}{4} \tau_{ads} v_0 (S_{des}/V_{irr}). \tag{6.5}$$

The experimental ratio N_i^s/N_i^g in the case of naphthalene molecules adsorbed on rhodamine 6G is 2×10^3 at $T = 200$ K. From here it is possible to determine the lifetime of the molecule on the surface $\tau_{ads} = 0.18$ s and to estimate, from (6.1), the energy of adsorption of naphthalene on rhodamine 6G: $E_{ads} = 0.5$ eV.

These experiments formed the basis for the development of the method for substantial improvement of the sensitivity of the photoionization detection of molecules, which was already discussed in short in Section 5.3 (Egorov *et al.*, 1983, 1984a). The method is based on combining the pulsed desorption of molecules that can accumulate on a surface during a considerable time τ_{ads} with their subsequent photoionization in a vacuum. The improvement of the detection sensitivity is basically due to the fact that the equilibrium number of molecules on the surface being irradiated can under certain conditions substantially exceed the equilibrium number of molecules in the irradiated volume. When the surface molecules are being pulse desorbed, they enter the irradiation region in a concentration exceeding the equilibrium concentration by the amount defined by Eq. (4.16). The nature of desorption being perfectly different from that of photoionization, to optimize these processes it is most advantageous to effect them by means of different laser pulses. The molecules are desorbed under the effect of the first laser pulse irradiating the surface and are then ionized at some distance from the surface under the action of the second laser pulse (or pulses) of a certain wavelength. The use of pulsed desorption allows the thermal decomposition of the molecules to be materially decreased (Meuzelaar *et al.*, 1974), and laser stepwise photoionization is practically the only technique enabling the desorbed molecules to be ionized with a probability close to unity.

One of the main factors governing the sensitivity of detecting surface-adsorbed polyatomics by this method is the maximum proportion of molecules that can be desorbed from the surface without being decomposed. This matter was investigated in the experiments on detecting naphthalene and anthracene molecules adsorbed on the surface of graphite (Antonov *et al.*, 1983). The experimental setup (Fig. 6.12) consisted of a time-of-flight mass spectrometer with a 3×3 mm^2 graphite substrate at $T = 200$ K held to the repulsion electrode of the ion source and temporally synchronized pulsed lasers using CO_2 ($\lambda_1 = 10.6$ μm) and KrF ($\lambda_2 = 248$ nm) molecules. The CO_2 laser emission was made to impinge upon the substrate at a small angle (7°), and the ionizing radiation of the excimer laser was directed parallel with the substrate close to its surface. The delay time τ_d between the desorbing and ionizing laser pulses amounted to 0.7–1 μs.

The naphthalene and anthracene molecules were deposited on the substrate from the gaseous phase and were in equilibrium with their vapors at partial pressures (as estimated from photoionization measurements) of 10^{-12}–10^{-13} Torr.

The photoionization of the molecules desorbed under the effect of the IR radiation was carried out through intermediate electronically excited states by means of the UV radiation pulses from the KrF laser, the energy of two photons ($2\hbar\omega = 10$ eV) being sufficient to ionize both the naphthalene and

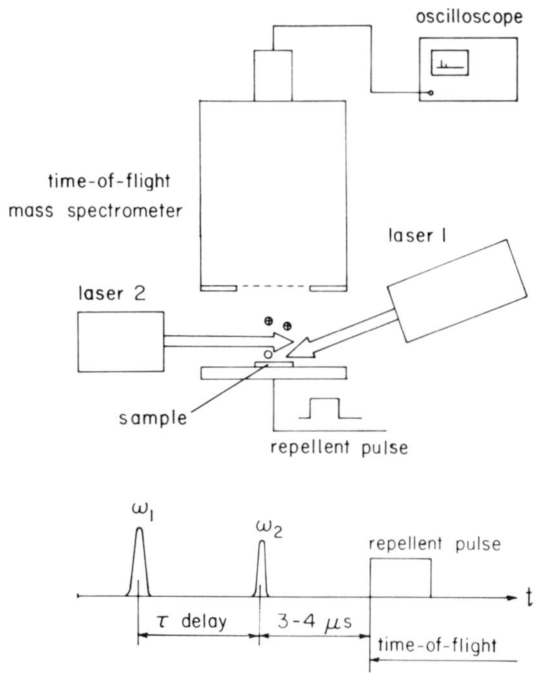

FIG. 6.12. Simplified scheme of an experiment on laser detection of adsorbed molecules by means of their IR laser-induced pulsed desorption (laser 1) and following multistep laser UV resonant ionization (laser 2). [From Antonov *et al.* (1983).]

anthracene molecules. With the KrF laser pulse fluence being as high as 0.2–0.3 J/cm^2, the ionization yield for the naphthalene molecules reached 100% (Antonov *et al.*, 1981a). In the experiment, to reduce the photoionization signal amplitude, the ionizing laser pulse fluence was kept at 0.015 J/cm^2, which provided for a 2% ionization yield.

Figure 6.13 shows the dependences of the naphthalene and anthracene photoionization signals on the desorbing CO_2 laser pulse fluence at a fixed ionizing KrF laser pulse fluence range 0.05–0.3 J/cm^2, which is less distinct in the case of anthracene. The maximum ion signal in the experiment amounted to around 10^3 ions/pulse for naphthalene and 10^4 ions/pulse for anthracene. When the IR laser fluence comes in excess of 0.3 J/cm^2, the yield of molecular ions is observed to decrease drastically, while, at the same time, there appear various fragment ions and the substrate material ions C^+ and C_2^+. The substrate is also observed to incandesce in this fluence region, which points to the fact that the substrate surface is heated to a temperature of $T \lesssim 1000$ K.

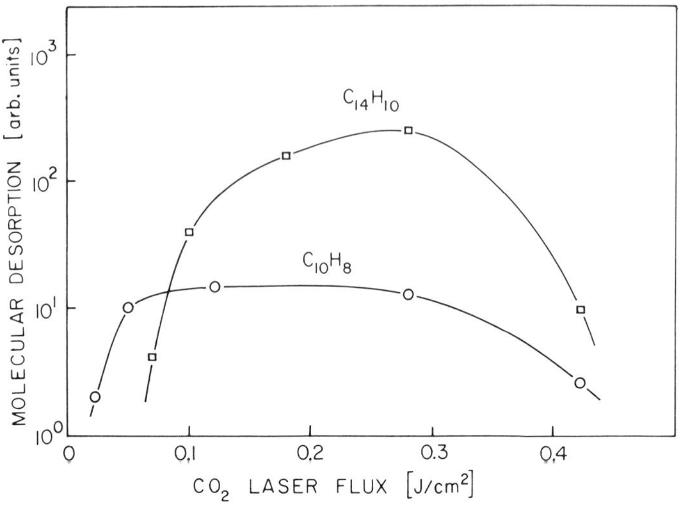

FIG. 6.13. Desorption yield of naphthalene and anthracene molecules from a graphite surface as a function of the energy fluence of a CO_2 laser. Desorbed molecules are detected by means of the resonant photoionization-induced pulsed KrF-laser at $\lambda = 248$ nm and energy fluence 0.015 J/cm^2. [From Antonov *et al.* (1983).]

The rate W_{des} of molecular desorption from the surface is defined by Eq. (6.1). For the pulsed desorption to become saturated, the following condition must be satisfied:

$$W_{des}\tau_p \simeq \tau_p\Omega \exp(-E_{ads}/kT) \simeq 1. \tag{6.6}$$

No further increase in the surface temperature will cause the photoion signal to rise, and it is precisely this fact that is attested to by the appearance of the plateau in the IR laser dependence of the signal (Fig. 6.13). It is not very difficult to estimate the sensitivity of detecting, for example, naphthalene molecules on a surface: in the case of complete desorption of the molecules, the observed photoion signal of 10^3 ions/pulse corresponds to a surface concentration of 10^6 molecules/cm^2 or about 10^{-8} monolayer, provided that the ionization yield is 2% and the photoion registration efficiency is 50%. With an optimum ionizing UV laser pulse fluence, the detection limit for adsorbed molecules will be 50–100 molecules/cm^2 at a photoion signal level of 1 ion/pulse.

The developed photoionization technique for detecting polyatomics proved a success in experiments with tryptophan molecules (Antonov and Letokhov, 1984). The experiments were conducted on the setup shown in Fig. 6.12. Subject to investigation was the detection sensitivity provided by the laser IR desorption–UV photoionization technique in conditions of a conventional analytical sample preparation procedure: a droplet of an aqueous tryptophan solution was dried on a substrate which was then placed in a photoionization mass spectrometer. The multistep resonant photoionization of the tryptophan molecules was effected by means of a XeCl laser pulse ($\lambda = 308$ nm, $\tau_p = 10$ ns). The choice of this type of laser was dictated by the presence of a one-photon resonance with the first electronically excited state of tryptophan. Besides, the dissociation yield of tryptophan in this state is low (McLaren and Shugar, 1964). The absorption of a second photon brings the molecule to a state above the ionization energy ($E_i \lesssim 7.5$ eV) (Akopian and Loginov, 1967). In the experiments, the XeCl laser fluence was taken at 0.1 J/cm^2, based on a trade-off of the need to attain a maximum ionization yield against that to keep to a minimum the fragmentation of the molecules. The main component in the photoion mass spectrum of tryptophan were the R^+ ions (Fig. 6.14a) with a mass of $m = 130$, formed as a result of breakage of the R—CH(NH_2)COOH bond in the tryptophan molecular ion (Fig. 6.14b). It is by measuring exactly this characteristic peak that the tryptophan molecules were detected in the experiments. The minimum concentration of the tryptophan solution at which it proved still possible to detect the R^+

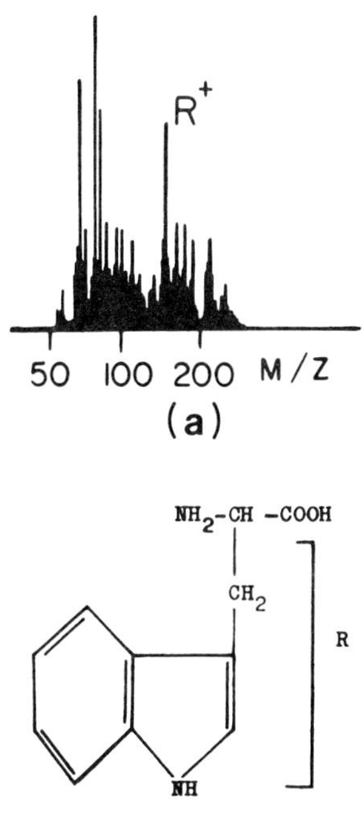

FIG. 6.14. (a) Photoion mass spectra obtained for samples containing 0.01 pg of tryptophan. (b) Molecular structure of tryptophan. [From Antonov and Letokhov, 1984).

ions was equal to 20 pg/ml which corresponded to 10^{-14} g of tryptophan on the substrate (Fig. 6.14a). This level was determined by the background noise signal due to the "memory" of the mass spectrometer.

6.4. Ways of Developing a Universal Laser Photoionization Detector of Molecules

The studies carried out so far support the new approach, put forward ten years ago, for analyzing trace amounts of complex molecules. It is based on a selective multistep photoionization of molecules by laser radiation with a subsequent mass-spectrometric analysis of the ionization products.

From the very start the multistep photoionization of polyatomic molecules in a mass spectrometer was developed to detect trace amounts of complex molecules. Using one or several lasers tuned in resonance to appropriate molecular transitions, including vibrational ones, we can very selectively excite the molecule to any state from which it may be very efficiently ionized.

There are several basic schemes for stepwise molecular photoionization. Below we shall consider them with regard to the maximum selectivity and the ionization efficiency.

6.4.1. Photoionization via Intermediate Electronic States

In this case, use can be made of both the multiphoton ionization of molecules in a strong single-frequency laser field and multi-step photoionization in a single- or multiple-frequency laser field.

6.4.1.1. Multiphoton Ionization of Molecules. The method of multiphoton ionization of polyatomic molecules is currently widely used for systematic investigation of the spectroscopy of excited states of molecules, especially states forbidden for single-photon transitions (Johnson, 1980). An attractive aspect of the method is the relative simplicity of obtaining multiphoton spectra and the higher (compared with fluorescence measurement) sensitivity. Another advantage is the possibility of using visible and near-ultraviolet radiation to investigate electronic states lying in the vacuum ultraviolet region of the spectrum. Although multiphoton ionization in the mass spectrum gives much new information on photoionization and photofragmentation of molecules under the action of intense laser radiation, nevertheless, systematic use of multiphoton ionization of molecules for mass-spectral analysis is hardly suitable due to the high degree of fragmentation and comparatively low efficiency of ionization of molecules.

6.4.1.2. Stepwise Photoionization of Molecules. In this variant, two schemes can be distinguished: (1) stepwise photoionization in a single-frequency laser field and (2) independent steps of excitation and photoionization of molecules using time-synchronized laser pulses.

In the first case, for efficient photoionization of molecules the laser radiation frequency ω must satisfy the condition $2\hbar\omega > I$, where I is the ionization potential of the molecule. In this case, the only restriction on the intensity of the laser radiation is the loss of mass-spectral information due to strong fragmentation of molecules. The efficiency of ionization of molecules in this scheme is very high and, in many cases, approaches 100%. However, for many molecules with this choice of frequency of laser radiation the high-lying electronic states, which can rapidly deexcite via competing channels of photoionization dissociation and electronic conversion, are excited. The absorption spectra of polyatomic molecules in this region often have no structure.

With respect to selectivity the second scheme with independent excitation and ionization frequencies has greater possibilities. When far-UV and VUV radiation are used in the step involving photoionization of molecules from an excited state, two-dimensional spectra of molecules with wavelength tuning of the exciting radiation in the long-wavelength region of electronic transitions, where the molecular spectra have greater structure, can be obtained. An additional channel of information concerning the lifetimes of excited molecules appears in this scheme with ionization of molecules from excited states by a delayed laser pulse. It should be noted, however, that in this scheme extraneous channels of photoionization, whose suppression requires restricting the maximum intensity of radiation both at the first and second steps, can in principle exist. This decreases the efficiency of selective two-step photoionization. Indeed, in the typical case in which $2\hbar\omega_1 < I < 2\hbar\omega_2$, where ω_1 and ω_2 are the frequencies of laser radiation at the first the second steps, respectively, two-step photoionization of molecules through the high-lying electronic states by laser radiation of the second step is possible. In addition, three-step photoionization can exist with resonant absorption by the molecule of a sequence of three photons of the exciting laser with frequency ω_1.

Let us compare the relative contributions of each of the three channels of photoionization of a molecule (see Fig. 6.15). For an estimate, we shall assume that all cross sections of optical transitions indicated in Fig. 6.15 are equal and that the lifetimes of all intermediate levels exceed the laser pulse duration. We shall assume that the statistical weights of the upper levels for all transitions are much greater than those of the lower levels. This permits neglecting the reverse transitions under the action of the radiation. It is clear that for an energy density of the laser radiation close to the saturation energy of the transitions $\Phi \simeq \Phi_{sat} = \hbar\omega/\sigma$, all three channels make approximately the same contribution to ionization of the molecule. Under

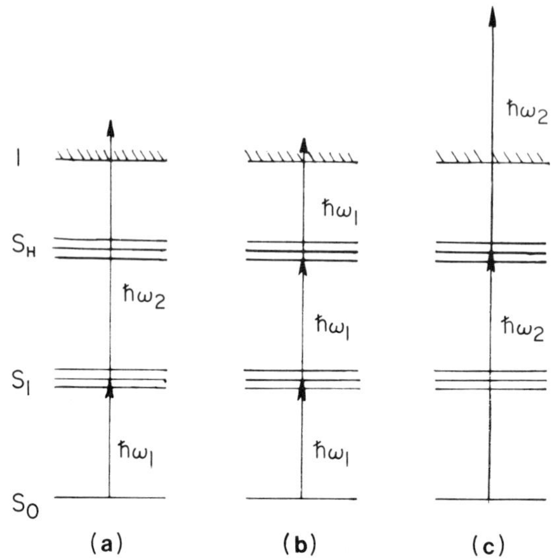

FIG. 6.15. Schemes of the possible additional photoionization channels for molecules in a two-frequency laser field. (a) Selective two-step photoionization, (b) three-step photoionization under the action of laser radiation in the first step, and (c) two-step photoionization of molecules by laser radiation in the second step.

conditions that are far from saturation, the number of ions forming along these channels will be

$$N_i^{(a)} = (\Phi_1/\hbar\omega_1)(\Phi_2/\hbar\omega_2)\sigma^2 N_0, \tag{6.7}$$

$$N_i^{(b)} = \tfrac{1}{6}(\Phi_1/\hbar\omega_1)^3\sigma^3 N_0, \tag{6.8}$$

$$N_i^{(c)} = \tfrac{1}{2}(\Phi_2/\hbar\omega_2)^2\sigma^2 N_0, \tag{6.9}$$

where Φ_1 and Φ_2 are the energy densities of the laser pulses at the first and second steps, and N_0 is the number of molecules in the photoionization volume. It is evident that the numbers of ions in channels (6.8) and (6.9) have higher-order power-law dependences on the energy density of the radiation pulses Φ_1 and Φ_2, so that as the energies of the laser pulses decrease, the ratio of the useful signal of two-step ionization along channel (6.7) to the background (6.8) and (6.9) will increase due to a reduction of ionization efficiency. Choosing, for example, the ratio $N_i^{(a)}/(N_i^{(b)} + N_i^{(c)}) = 10$, we obtain from (6.7)–(6.9) an efficiency of selective two-step photoionization equal to 0.01%.

For real molecules, the situation can be more favorable. The principal difference from the model analysis lies in the fact that in the highly excited electronic states, which participate in transitions along channels (6.8) and (6.9) (see Fig. 6.15), the rates of nonradiative excitation from these states will be strongly suppressed. Thus, for example, in experiments (Andreyev *et al.*, 1977; Antonov *et al.*, 1977b, 1978) on selective two-step photoionization with excitation of a number of molecules (formaldehyde, benzaldehyde, benzophenone, and others) into highly excited states with VUV radiation with photon energy 7.7 eV, the two-stepped ionization signal was not observed.

Under optimal conditions for absence of extraneous channels of photoionization, the selectivity of ionization is determined by the electronic absorption spectra of molecules. It is well known, however, that electronic absorption spectra of polyatomic molecules are much less characteristic in the gas phase than vibrational IR spectra. The selectivity of photoionization through electronic states can be radically increased with the use of cooled supersonic molecular beams. The possibility of using supersonic molecular beams to cool complicated molecules, such as, for example, phthalocyanine ($C_{32}N_8H_{18}$), has been recently demonstrated (Fitch *et al.*, 1978). The absorption spectrum of this molecule under the conditions of deep cooling of vibrational and rotational degrees of freedom consisted of lines with width 0.5–1 cm^{-1}. Separate vibrational–rotational lines were completely resolved in the electronic absorption spectrum of the oxalyl fluoride ($C_2O_2F_2$) molecule, cooled in a pulsed supersonic beam to 0.17 K (Liverman *et al.*, 1979). In this sense, we can talk about the achievement of limiting quantum selectivity in absorption of radiation.

Let us estimate the limiting sensitivity of the method of stepwise laser photoionization through intermediate electronic states in combination with the supersonic cooled molecular beam for detection of impurities in the atmosphere. The expression for the number of selectively forming ions within a single pulse has the form

$$N_{ion} = \eta_{det} Y_{ion} \gamma n_{beam} V_{irr}, \tag{6.10}$$

where η_{det} is the efficiency of detection of photoions, Y_{ion} the yield of two-step photoionization, γ the relative concentration of impurity molecules in air, n_{beam} the density of molecules in the beam, and V_{irr} the photoionization volume. Assuming that the threshold of detection of molecules equals the detection of a single selectivity-formed photon and substituting the values $\eta_{det} = 0.5$, $Y_{ion} = 1$, and $n_{beam} = 2 \times 10^{14}$ cm^{-3}, corresponding to a maximum gas pressure of 5×10^{-3} Torr with which collisionless extraction of ions from the beam is still possible, and $V_{irr} = 0.1$ cm^3, we obtain $\gamma = 10^{-13}$, which corresponds to a detection threshold of 3×10^6 molecules/cm^3 in air.

6.4.2. Photoionization via Intermediate Vibrational States

Vibrational spectra contain the most information on polyatomic molecules, and thus, with respect to the selectivity trace-detection, the stepwise photoionization through intermediate vibrational states is of particular interest. This is rather difficult to achieve, because during vibrational excitation the shift of the photoionization limit is smaller than the width of the photoionization band edge. Thus, a single-photon ionization of unexcited molecules is possible. Besides, after a short-wavelength quantum $\hbar\omega_2$ has been absorbed, photodissociation competes with the photoionization process since the total energy of the laser photons slightly exceeds the molecular ionization potential. Also the Franck–Condon factors for the transitions associated with ionization may be small in the general case. The possibilities of this method can be estimated by considering the simplest model of two-step photoionization, in which the laser radiation of frequency ω_1 selectively excites those molecular vibrations whose excitation is required for the subsequent photoionization brought about by radiation of frequency ω_2 (chosen in the region of maximum change in the photoionization cross section).

Let us assume that as the molecule is being photoionized from the excited vibrational state the photoionization cross section changes by some value $\Delta\sigma$. The situation in which a maximum number of molecules is excited at the frequency ω_1 is optimal for molecular detection. In this case, the energy flux of the first laser pulse should satisfy the condition

$$\Phi(\omega_1) \gtrsim \Phi_{\text{sat}}^{(1)} = \hbar\omega_1/2\sigma_{\text{exc}}, \tag{6.11}$$

where σ_{exc} is the cross section of the transition to the intermediate state. Hence, the number of molecules transferred to the excited state equals

$$N_{\text{exc}} \simeq \tfrac{1}{2}qN_0, \tag{6.12}$$

where N_0 is the number of molecules in the ground state, and q is the relative number of molecules in resonance with the exciting radiation. For a maximum photoion yield, the laser pulse of frequency ω_2 must have an energy flux equal to

$$\Phi(\omega_2) \gtrsim \Phi_{\text{sat}}^{(2)} = \hbar\omega_2/(\sigma_{\text{i}} + \sigma_{\text{d}}). \tag{6.13}$$

In this case each excited molecule dissociates, with a probability close to unity, through photoionization and photodissociation channels. The buildup in the number of ions formed under these conditions is determined by the simple expression

$$\Delta N_{\text{i}}^{(2)} = N_{\text{exc}}\frac{\Delta\sigma}{(\sigma_{\text{i}} + \sigma_{\text{d}})} = N_0\frac{q}{2}\frac{\Delta\sigma}{(\sigma_{\text{i}} + \sigma_{\text{d}})}. \tag{6.14}$$

Thus molecular detection consists of the recording of a small number of selectively formed ions, $\Delta N_i^{(2)}$, against the fluctuation, $(N_i^{(2)})^{1/2}$, of a great number of nonselectively formed ions. This makes it possible to evaluate the minimum number of detectable molecules:

$$N_0^{\min} = \frac{1}{\eta_{\det}}\left(\frac{2}{q}\right)^2\left(\frac{\sigma_i}{\Delta\sigma}\right)\left(\frac{\sigma_i + \sigma_d}{\Delta\sigma}\right). \tag{6.15}$$

Substituting $\eta_{\det} = 0.5$, $q = 0.01$, $\sigma_i/\Delta\sigma = 10$, and $\sigma_d/\sigma_i = 10$, we get $N_0^{\min} = 10^7$. Thus, because of the nonselective photoionization of unexcited molecules, the sensitivity of such a photoionization spectroscopy scheme is worse than that of the scheme involving photoionization through excited electronic states by a few orders of magnitude.

The sensitivity and selectivity of this method can be substantially increased through: (a) multistep vibrational excitation, which will lead to a considerable shift in the photoionization limit of the excited state and, hence, to an increase in the fraction of selective ions; (b) consecutive excitation of vibrational and electronic states (Section 5.4.1), which combines the selectivity of molecular vibrational spectra with such advantages of photoionization through electronically excited states as the absence of background noise due to one-photon ionization and the possibility of obtaining information on the lifetimes of electronically excited molecules; (c) use of a cooled molecular beam, in which case the population factor q of a molecular state in resonance with the laser field may be close to unity. The efficiency of selective photoionization will be limited by the competitive dissociation process during the absorption of short-wavelength photons.

Summarizing the discussion of the possible schemes of resonant stepwise photoionization of polyatomic molecules by laser radiation, we emphasize that it is entirely realistic that a highly selective laser detector of trace quantities of molecules, whose sensitivity will be several orders of magnitude higher than the sensitivity of all existing analytical methods, will be developed in the near future. Success along this path will be achieved as a result of the combination of the most diverse achievements of experimental physics: first, the technique of low-temperature molecular jets; second, the technique of multistep excitation of high-lying vibrational levels by tunable IR laser radiation and subsequent photoionization of vibrationally excited molecules directly or through intermediate electronic states; and third, the technique of time-of-flight mass spectrometric detection of the products of photoionization of molecules.

The main disadvantage of the method of selective laser stepwise photoionization, also of electron impact, is that it is suitable only for molecules that can be put into the gas phase. In this respect, the experiments described in Section 6.3 on the combined formation of molecular ions under irradiation of a surface by pulsed UV laser radiation whose wavelength falls in the absorption band of the material, are of great interest:

$$\begin{pmatrix}\text{moderate pulsed heating of the}\\ \text{surface with resonant absorption}\end{pmatrix} \rightarrow \begin{pmatrix}\text{thermal desorption}\\ \text{of neutral molecules}\end{pmatrix} \rightarrow \begin{pmatrix}\text{multiphoton resonance}\\ \text{photoionization}\end{pmatrix}$$

This process is of interest for photoionization mass spectrometry of molecules on the surface of a material. Moreover, experiments with molecules adsorbed on the surface of a specimen (Egorov *et al.*, 1983, 1984a,b) prove that the sensitivity of photoionization detection of molecules can be radically increased (by a factor of 10^3). The increase of sensitivity is achieved by accumulating molecules on the surface while their density in the gas phase is extremely low. Thus an alternative approach is possible to the development of a laser photoionization detector of molecules based on the following sequence of processes:

$$\begin{pmatrix}\text{accumulation of molecules}\\ \text{on the surface}\\ \text{of the adsorbent}\end{pmatrix} \rightarrow \begin{pmatrix}\text{pulsed thermal}\\ \text{desorption of}\\ \text{accumulated molecules}\end{pmatrix} \rightarrow \begin{pmatrix}\text{multiphoton resonance}\\ \text{photoionization of}\\ \text{desorbed free molecules}\end{pmatrix}$$

In this approach it is possible to use lasers with low repetition frequency. It is not difficult to imagine as well a variation of such a detector in which molecules desorbed in pulses enter a pulsed supersonic jet for cooling and subsequent more selective photoionization detection in the cooled jet.

7. Obtaining Spectroscopic Information on Atoms and Molecules

The resonance photoionization technique has numerous applications in atomic and molecular spectroscopy, especially where high sensitivity is a must. This is actually what this whole book is about, but it seems useful to discuss separately such applications of the technique as substantially facilitate the acquisition of spectroscopic data on atoms and molecules. The possibilities existing here are illustrated by Fig. 7.1 with reference to the simplest case of two-step photoionization. By varying the wavelength λ_1 or λ_2, we can get information on the resonance frequency of the corresponding transition; fine tuning makes it possible to obtain data on the hyperfine and isotope structures of the resonance transitions, their various splittings, etc. Measuring the ionization yield as a function of the delay time τ_d between the ionizing laser pulse at λ_2 and the exciting laser pulse at λ_1 can help acquire data about the lifetime of the atom or molecule in the intermediate excited state.

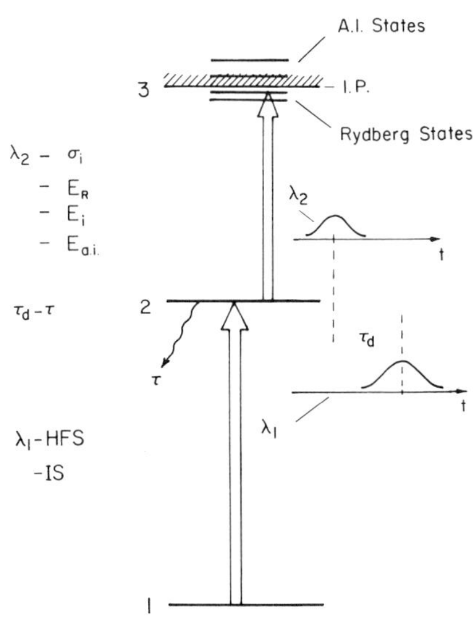

FIG. 7.1. Spectroscopic information obtainable by the stepwise photoionization technique.

Such information can, of course, be obtained with the aid of other laser spectroscopy methods such as the fluorescence technique. In this respect, the use of the photoionization technique is not a matter of principle if atoms or molecules in sufficient quantities are subject to study.

The situation is quite different when it is necessary to measure highly excited states near (below or above) the ionization threshold. To combine the multistep excitation method with the ionization detection technique here is as natural as it is effective. This combined method is capable of providing information about such, as yet, imperfectly studied states, as well as data on ionization potentials. Many of these spectroscopic problems have already been considered to some extent in the preceding chapters, for knowledge of spectral data is absolutely necessary for the realization of the photoionization method. The present chapter specially treats in brief the use of the photoionization technique for obtaining systematic spectral data on atoms and molecules.

7.1. High-Spectral-Resolution Photoionization Spectroscopy

The spectral data whose measurement requires high resolution are the exact frequencies of optical transitions, the isotope and hyperfine structures of energy levels and transitions, and the broadening of spectral lines. Such measurements generally require that the Doppler broadening should be either reduced or altogether eliminated. For this purpose, use is made of the large stock of well-developed laser techniques (Shimoda, 1976). In atomic beams, this can be done by properly selecting the irradiation geometry (transverse irradiation) or by accelerating the atoms (longitudinal irradiation of accelerated atoms). In a low-pressure gas, use is made for this purpose of the well-known Doppler-free nonlinear spectroscopy techniques (Letokhov and Chebotayev, 1977). Let us consider various types of such experiments.

7.1.1. Experiments with Atomic Beams

The transverse atomic beam geometries were employed in the experiments involved in investigations into the isotope shifts of short-lived isotopes of rare earth elements (Alkhazov *et al.*, 1983a, 1984; Fedoseyev *et al.*, 1984; Zherikhin *et al.*, 1984). This version of photoionization spectroscopy of a very small quantity of atoms in a sample has been described in Section 4.2. In the first experiments with Eu atoms, resolution was determined by the radiation linewidth of the narrowband pulsed dye laser used at the excitation step under study, i.e., for all practical purposes, $\Delta\nu_L = 0.03\ \text{cm}^{-1} = 6$ GHz. This is quite sufficient for investigations into isotope shifts where the hyperfine structure of the excitation line is small enough to lie within $\Delta\nu_L$. Where it is necessary to study both isotope and hyperfine structures, use must be made at the transition to be examined of a laser with a narrower spectral bandwidth.

Figure 7.2 shows an optical diagram of a photoionization spectrometer for measuring hyperfine and isotope structures of radioactive atoms (Alkhazov *et al.*, 1985). Employed at one of the excitation steps are a dye laser with a narrow radiation linewidth and a dye amplifier. The duration of the amplified pulse, $\tau_a = 18$ ns, corresponds to a pulse spectral bandwidth of $\Delta\nu_L = 50$ MHz. To use narrowband CW radiation in schemes involving multistep excitation is usually undesirable because of the effects of atomic relaxation from the excited states that result in the optical pumping of atoms at some sublevels and thus reduce sensitivity. A more detailed scheme of irradiation of an atomic beam under "on-line" conditions has been shown in Chapter 4 (Fig. 4.4). Such a photoionization spectrometer allows the examination of isotope and hyperfine structures of a radioactive isotope contained in a sample in an amount as small as 10^6–10^7 atoms. To illustrate, Fig. 7.3 shows optical spectra of the hyperfine structure of the ^{165}Tm radioactive isotope ($T_{1/2} = 30.1$ h) on the $4f^{13}6s^2\ ^2F^0_{7/2} \rightarrow 4f^{12}(^3H_6)5d_{5/2}(6, \frac{5}{2})_{7/2}$ optical transition and the reference stable isotope ^{169}Tm, the isotope shift being clearly manifest (Alkhazov *et al.*, 1985).

The sensitivity of the spectrometer is so high that the instrument allows obtaining high-resolution spectra of isotopes with a half-life as short as 10 s.

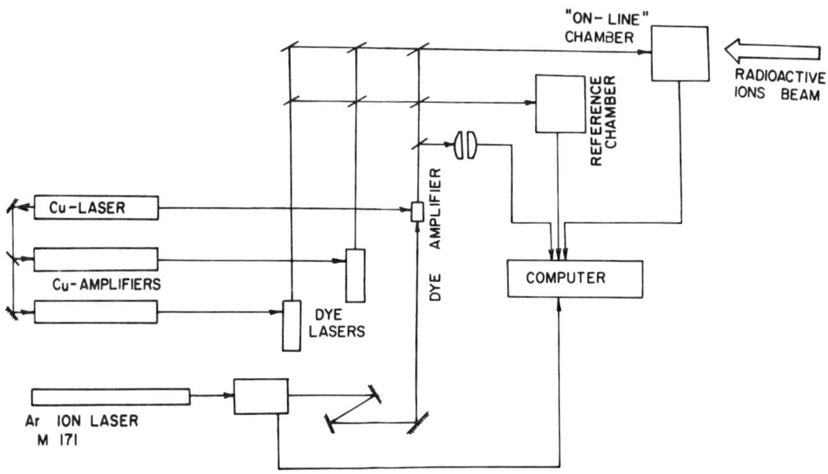

FIG. 7.2. Schematic of a photoionization spectrometer for studying the hyperfine and isotope structures of atoms available in extremely small quantities. Used at the measurement step are a narrowband CW dye laser and a pulsed dye amplifier. [From Alkhazov *et al.* (1985).]

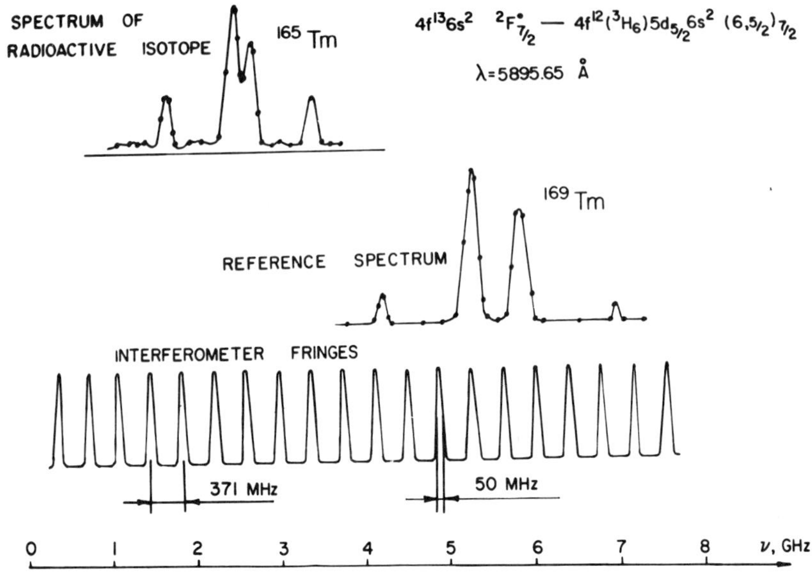

FIG. 7.3. Measurements of the hyperfine and isotope structures of Tm isotopes on the optical transition with the aid of the spectrometer shown in Fig. 7.2. [From Alkhazov *et al.* (1985).]

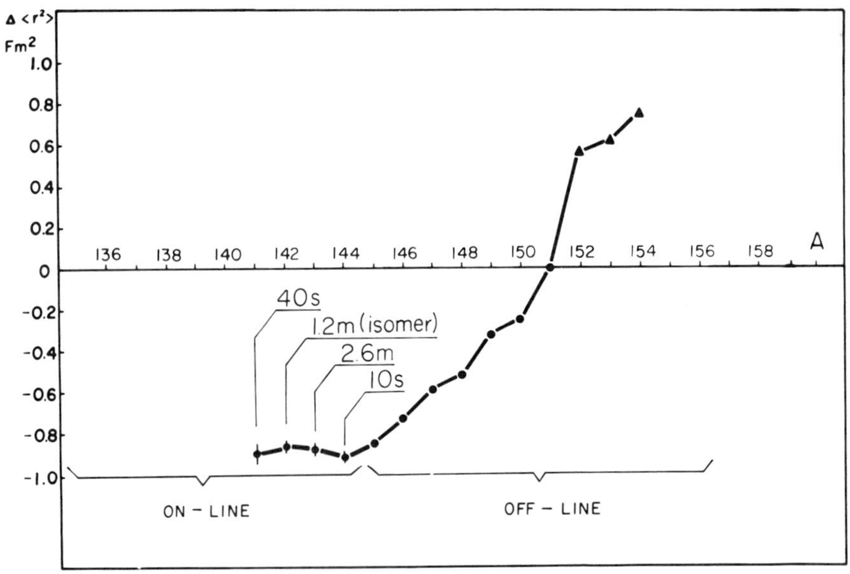

FIG. 7.4. Dependence of the variations of mean-square charge radius of short-lived Eu isotopes on the mass number, obtained from measurements of isotope shifts by the three-step photoionization of atoms in a beam. [From Fedoseyev *et al.* (1984).]

By measuring the isotope shifts for a chain of radioactive isotopes, we can, using standard formulas (see Otten, 1981), derive the dependence of the mean-square charge radius of the nucleus on the neutron number. As an illustration, Fig. 7.4 presents typical results of such measurements for a chain of radioactive Eu isotopes, made with reference to the stable isotope ^{151}Eu. A good idea of the capabilities of the technique can be had from Table 7.1 which lists isotope shifts $\Delta\nu_{is}$ for one of transitions in Eu, measurement accuracies $\delta\nu$ in the case of a pulsed laser with a relatively wide spectral bandwidth, isotope half-lives $T_{1/2}$, and isotope yields n_i from a tantalum target irradiated with a proton beam. If the measurement time $\tau_{meas} > T_{1/2}$, the number of atoms under study in the sample is merely around $N_i T_{1/2}$, i.e., for example, only 10^7 in the case of ^{144}Eu. Thus the high-sensitivity and high-resolution laser photoionization spectroscopy is a fairly useful method for nuclear physics. By the way, nuclear physics requirements governed the choice of rare earth elements for the first experiments.

The choice of the rare earth elements with nuclei in the neighborhood of the closed neutron shell with $N = 82$ is due to the following reasons. First, systematic measurements of the charge radii for chains of isotopes including those with magic nuclei ($N = 82$) helps understand the role of shell effects for neutron-deficient nuclei far from beta-stability of nuclei. Second, of special interest is the change in the nuclear shape from spherical to highly deformed that an atom must suffer as the number of its neutrons is increased from 82 to 90. And third, there may exist a new deformation region at $N < 82$ (Linder and Möller, 1982).

TABLE 7.1. Relative Isotopical Shifts $\Delta\nu_{is}$ for the $4f^7 6s^2\ {}^8S^0_{7/2} \rightarrow 4f^7 6s6p\ {}^6P_{3/2}$ ($\lambda_1 = 5765$ Å). Transition of the Neutron-Deficient Isotopes of Eu, Accuracy of Measurement ($\delta\nu$), Half-Life ($T_{1/2}$), and Yield (N) of Eu Isotopes from a Tantalum Target[a]

A	$\Delta\nu_{is}$ (151 − A) (GHz)	$\pm\delta\nu$ (GHz)	$T_{1/2}$	N (10^5 s^{-1})
150	1.48	0.08	34.2 y	16
149	−1.85	0.08	93 d	10
148	−2.98	0.08	54 d	30
147	−3.39	0.08	22 d	30
146	−4.19	0.1	4.6 d	30
145	−4.9	0.1	5.9 d	20
144	−5.3	0.2	10 s	8
143	−5.05	0.25	2.6 min	7
142	−5.0	0.2	1.2 min	2
141	−5.2	0.3	40 s	1

[a] From Fedoseyev *et al.* (1984).

When conducting experiments involving the measurement of isotope and hyperfine structures of isotopes by the method of multistep excitation and ionization, we should pay special attention to the polarization of the laser light. In particular, the polarization was observed to affect the yield of even and odd Yb isotopes in the selective three-step photoionization through a Rydberg state of the natural isotope mixture (Bekov *et al.*, 1981b). This is explained by the difference in hyperfine structure (HFS) between even and odd isotopes and the fact that different selection rules are effective for transitions between different HFS components for plane and circularly polarized radiation (Fedoseyev and Yusupov, 1984). To avoid the influence of polarization on the relative intensity of HFS components, use should be made of a naturally polarized radiation at all excitation stages.

One of the shortcomings of experiments with transverse atomic beams is the strict anticorrelation between resolution and sensitivity. By reducing the angular divergence of the beam, we can improve resolution but at a sacrifice in the atomic flux and, hence, in sensitivity. The same disadvantage is characteristic of the saturation spectroscopy method with which we isolate from the entire atomic velocity distribution only the small part responsible for the formation of the Lamb dip. Therefore, where the maximum possible spectral resolution is required along with maximum sensitivity, two other methods are more advantageous: two-photon excitation in counter-propagating beams and collinear irradiation of accelerated atoms. The latter technique is considered in Chapter 9 in connection with the problem of the laser photoionization detection of extremely rare isotopes. For this reason, we discuss here only experiments on Doppler-free, two-photon photoionization spectroscopy.

7.1.2. Experiments on Doppler-Free Two-Photon Excitation

The principle of Doppler-free two-photon spectroscopy (Vasilenko *et al.*, 1970) has been briefly described in Section 2.2.4. Its detailed description can be found in a number of monographs (Bloembergen and Levenson, 1976; Letokhov and Chebotayev, 1977). This technique is fairly convenient to combine with photoionization detection for the following two reasons. First, an atom having absorbed two photons is excited into a highly excited state which is comparatively easy to ionize. A two-photon excited atom usually finds itself either in a high-lying Rydberg state which is easy to ionize by collision or electric field (see Section 3.3) or in such a state as becomes photoionized upon absorption of a single photon only. Second, where the energy of two photons, $\hbar(\omega_1 + \omega_2)$, is in exact resonance with a narrow two-photon peak, all atoms move into an excited state, no matter what their velocity.

These advantages of the Doppler-free ionization spectroscopy have been clearly demonstrated in investigations into hyperfine structure and isotope shifts of highly excited atoms and in precision measurement of the optical transition frequency of the positronium atom.

In the experiments with Xe atoms in a gas (Bushaw and Whitaker, 1981), then $6p[\frac{5}{2}], 2$ states of Xe were two-photon excited with a pulsed narrowband radiation at $\lambda = 250.2$ Å. The absorption of a third photon led to the ionization of the atom. Measuring the two-photon absorption spectrum with a resolution of 410 MHz, they measured the hyperfine splitting constants of ^{131}Xe to be A = 407 ± 10 MHz and B = 216 ± 10 MHz.

Experiments with alkali metal and alkaline earth atoms were conducted in a thermonic gas diode to study the isotope shifts for highly excited levels in ^{39,41}K (Niemax and Pendrill, 1980), ^{39,40}K (Pendrill and Niemax, 1982), ^{6,7}Li (Lorenzen and Niemax, 1982), and Ca and Sr (Lorenzen *et al.*, 1983). A specific feature of these experiments is the excitation of Rydberg states in atoms with CW dye laser radiation in the presence of a low-pressure (0.1-Torr) inert buffer gas which provides for the ionization of the atoms by collision (see Section 3.3.4). To illustrate the capabilities of such experiments, Fig. 7.5 shows the two-photon absorption spectrum for the $4s^2\ ^1S_0 \rightarrow 3d5s$

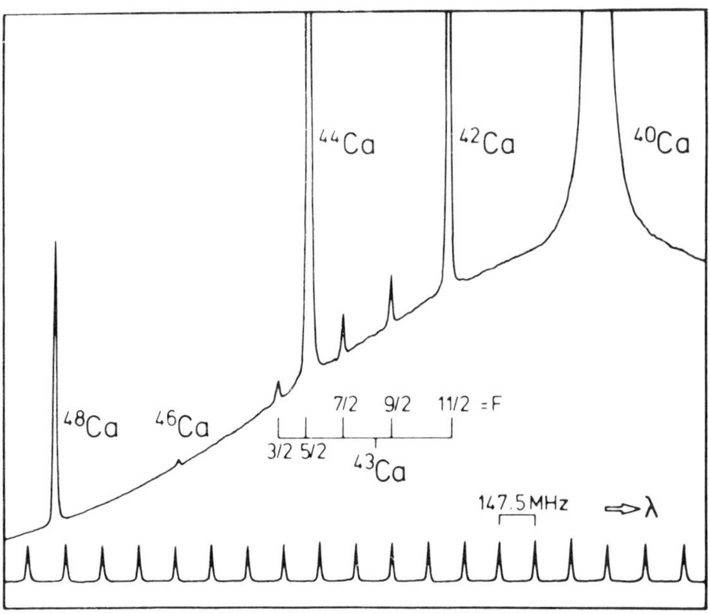

FIG. 7.5. Two-photon absorption spectrum for the $4s^2\ ^1S_0 \rightarrow 3d5s$ ($J = 2$) transition of Ca atoms of natural isotope composition (0.0033% $^{46}_{20}$Ca and 96.947% $^{40}_{20}$Ca). [From Lorenzen *et al.* (1983).]

($J = 2$) transition of Ca atoms of natural isotope composition. The Doppler-free two-photon resonances of both abundant and very rare isotopes are clearly manifest.

Quite a number of works are devoted to studies with this method of fine structural details of highly excited Rydberg states: the hyperfine structure n^2D of the Rb series close to the ionization threshold (Harvey and Stoicheff, 1977), the Rydberg level shifts *msns* due to hyperfine interaction in odd states in alkaline earth elements (Beigang and Timmermann, 1982a,b), isotope and hyperfine structures and singlet–triplet mixing of Rydberg states in Yb (Majewski *et al.*, 1982), *n*-mixing of highly excited ($n > 100$) Rydberg states in Sr due to the contact Fermi interaction of the Rydberg electron with an internal unexcited electron (Beigang and Timmermann, 1983a), singlet–triplet mixing due to the configuration interaction of the 4s*n*d Rydberg states in Ca (Beigang and Timmermann, 1983b), etc. In the experiments with very highly excited Rydberg atoms ($n > 100$), use was made of a modification of the standard thermionic gas diode in order to eliminate the electric field in the atomic excitation region. This was a "ring-thermionic diode" in which 8–10 wires were arranged axially around a tube. Because of the axial symmetry of the device, the first-order electric field at the tube axis disappeared (Beigang *et al.*, 1984).

The combination of the two-photon excitation and ionization detection techniques in the thermionic diode is fairly effective and suitable for taking measurements with a high spectral resolution for any atoms available in sufficient quantities.

An especially vivid example of an experiment in which the combination of a high spectral resolution with a sensitivity adequate for detecting single atoms is absolutely necessary is the experiment on the measurement of the frequency of the $1^3S_1 \rightarrow 2^3S_1$ optical transition of the positronium atom (Chu and Mills, 1982). Existing slow positron sources can yield 20–30 atoms per cm^3 on the average. Not a single laser spectroscopy method except for the photoionization technique can provide the sensitivity necessary to deal with so small a number of particles. For the Ps atom to become photoionized from the 2^3S_1 triplet state, it is sufficient for it to absorb one more (a third) photon of the same frequency. The photoionization event gives rise to a pair $e^+ + e^-$, and to detect it, it is sufficient to register e^+. The small mass of the positronium atom causes a material Doppler broadening of its optical transitions. Therefore, experiments with positronium require that use should be made of the Doppler-free two-photon excitation at $\lambda = 486$ nm (Letokhov and Minogin, 1976). Such an experiment combining the high resolution of two-photon spectroscopy and the high sensitivity of photoionization detection was conducted by Chu and Mills (1982). The high accuracy (10^8) attained by them in measuring the frequency of the 1S → 2S transition corresponds to a 1% accuracy of measurement of the first-order radiation correction of the transition frequency (the Lamb shift) (Chu *et al.*, 1983).

7.1.3. Investigation of Collisional Effects

There are a vast number of techniques specially developed for investigations into collisional effects in atomic and molecular spectra. The photoionization detection technique here proves fairly useful, at least insofar as the following two problems are concerned: studies of (1) highly excited states that easily become ionized upon collisions or in an electric field and (2) far spectral line wings.

A number of works are devoted to investigations of collisional effects for Rydberg atoms with the aid of ionization detection. For example, subject to such investigations were the self-broadening of Rydberg levels in Cs (Weber and Niemax, 1979), frequency shifts and phase-interference effects due to collisions of Rydberg Rb atoms (Stoicheff and Weinberger, 1980), and highly excited Rydberg atoms ($n > 100$) (Stoicheff, 1981) with Rb atoms in the ground state. All these experiments used the ionization detection of the Rydberg atoms in a thermionic diode.

The highly sensitive photoionization detection of excited atoms in the buffer gas of a proportional counter was employed to study the effect of collisions on the far spectral line wings, the experiments requiring a fairly high sensitivity of measurement. Among such experiments are those involved in the research of the collisional broadening on the resonance line wing of Cs in a buffer gas (Ar) (Nayfeh *et al.*, 1977) and observation of satellites on the resonance line wing of Cs in a Cs–Ar system (Nayfeh *et al.*, 1978).

The capabilities of the photoionization technique extend still farther, into the realm of investigations of very weak spectral line wings for excitation even in collision-free conditions. Such lines have been discussed in Section 2.1.3. This matter is important for the determination of the ultimate selectivity of multistep excitation of multilevel quantum systems. Experiments of this type are still to be carried out with atomic beams.

7.2. Kinetics of Intermediate Excited States

Multistep excitation under the effect of laser pulses of varying wavelength provides a natural method for measuring the lifetime of the intermediate excited state of the atom being excited, based on altering the delay of the second laser pulse relative to the first (Fig. 7.1). The time resolution of the method obviously depends on the duration of the pulses. Modern techniques for producing laser pulses whose duration can be varied over a broad range, up to 10^{-13} s, makes the method fairly convenient, especially for such states that relax via nonradiative channels and for highly excited states. The method is being successfully applied both for atoms and for molecules.

7.2.1. Atomic States

The first systematic measurements of the decay rates of atomic states with the aid of the photoionization method were carried out for many levels of the U atom in the course of investigations into the laser isotope separation by the multistep photoionization technique (Carlson *et al.*, 1976; Janes *et al.*, 1976). In this case, advantage is taken of the unique possibility of exciting strictly definite quantum states and probing their population by means of laser radiation. This is especially valuable for identification of the energy levels of complex atoms like the U atom, when other methods prove ineffective. To illustrate, Fig. 7.6a presents the results of measurement of the yield of U ions as a function of the delay time between the exciting laser pulse at $\lambda_1 = 4266.3$ Å and the photoionizing laser pulse at $\lambda_2 = 3371$ Å. A laser pulse duration of $\tau_p = 5$ ns provided for a resolution far in excess of the measured decay time $\tau = 155$ ns. When the sign of the delay is reversed (the λ_2 pulse coming first), the ionization signal disappears. This is a check for the correctness of interpretation of the stepwise ionization scheme. In the above-mentioned work, the lifetimes of tens of energy levels in U were measured by this method. The method has recently been successfully used for studies of the decay of highly excited states in rare-earth elements (Zherikhin *et al.*, 1984a,b). However, the photoionization methods prove most valuable and indispensable in investigations into the excited molecular states.

7.2.2. Molecular States

The method of stepwise laser photoionization of molecules is very convenient for studying excited molecular states, which is difficult or impossible to observe by other methods. For example, gas-phase absorption spectroscopy requires a high concentration of excited molecules or a long absorbing medium, which greatly complicates extension of this method to the case of noninteracting molecules ("isolated molecules") at low pressure. The method of laser-induced luminescence has high sensitivity and can be used for such investigations. However, the region of its applicability is limited to those cases when the excited molecular states have a sufficiently high rate of radiation and, therefore, an acceptable quantum yield of luminescence.

Limitations related to the emissivity of the excited state are completely absent in the method of two-step photoionization, and the potentially high sensitivity is determined by the fact that the absorption of a photon by an excited molecule leads to the formation of an ion, which can then be recorded

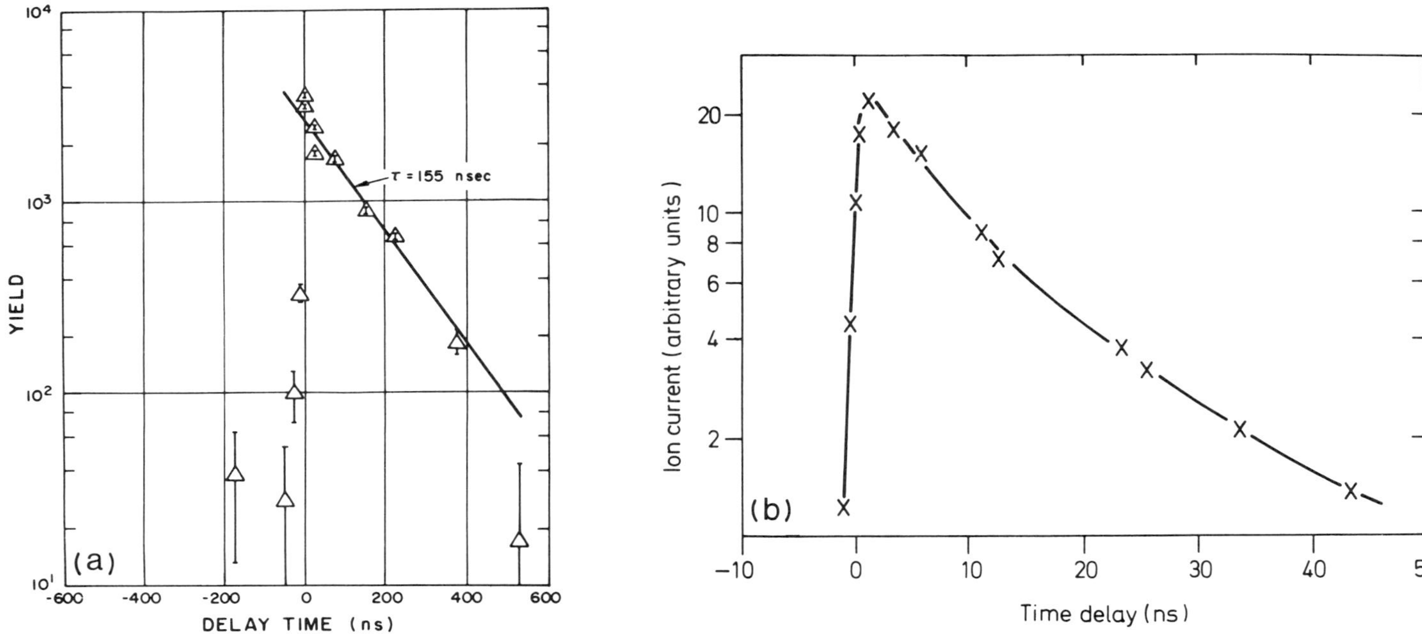

FIG. 7.6. Ion yield in two-step photoionization as a function of the delay time between the exciting and ionizing laser pulses: (a) for the molecule [from Janes *et al.*, (1976)] and (b) for the H_2CO molecule [from Andreyev *et al.* (1977)].

with a probability close to unity. Using time delay of the ionizing laser pulse relative to the exciting pulse, we can investigate the kinetics of electronically excited states by observing the change in the photoion yield. The efficiency of this technique was confirmed by Andreyev *et al.* (1977) for the formaldehyde molecule. This method was later used to investigate the evolution of the T_1 triplet state of the toluene molecule.

The investigation of the decay kinetics of the excited S_1 state of the formaldehyde molecule is of great interest for the study of nonradiative intramolecular transitions. The characteristic magnitude of the molar extinction for the transition $S_1 \rightarrow S_0$ is quite small ($\varepsilon = 10$), since the purely electronic dipole transition is symmetry-forbidden and can be allowed only in vibrations. This results in the fact that the radiative lifetime of the molecule in the S_1 state is about 5 μs (Yeung and Moore, 1973), while the duration of the fluorescence of the molecule in this state lies in the nanosecond range and is due to nonradiative deactivation of the excited state. For this reason, the quantum yield of fluorescence is quite low, they making it difficult to perform an accurate measurement of the lifetime of this state using fluorescence (Sakurai *et al.*, 1971; Aoki *et al.*, 1973). The decay of the S_1 state of formaldehyde was investigated by two-step photoionization with the use of synchronized pulses from nitrogen and hydrogen lasers ($\hbar\omega_1 + \hbar\omega_2 = 3.7 + 7.7 = 11.4$ eV) (Andreyev *et al.*, 1977).

Figure 7.6b shows the measured dependence of the yield of two-step photoionization of formaldehyde on the delay between the exciting and ionizing laser pulses. The growth of the signal near zero delay is due to the finite duration of the laser pulses. The drop in the signal with increasing delay is characteristically not exponential. For a different order of laser pulses, i.e., when the nitrogen laser pulse followed the hydrogen pulse, the two-step photoionization signal was not observed.

It is well-known that with excitation into the S_1 electronic state, the formaldehyde molecule dissociates with a quantum yield close to unity, and in addition, the products of dissociation accompanying absorption of radiation by the molecule in the region of 337 nm are CO and H_2 molecules (Yeung and Moore, 1973). The ionization potentials of the H_2 (15.4 eV) and CO (14 eV) molecules are higher than the energies of the two laser quanta, so the only source of ions is the ionization of molecules from the excited state. In this case, the drop in the output of the two-step photoionization signal accompanying the increase in the delay τ_d between the laser pulses (see Fig. 7.6b) is entirely related to the decay of the excited state of the molecule. Comparison of the average decay time, obtained from this dependence ($\tau_{av} = 14$ ns) with the results of measurements in other works (Sakurai *et al.*, 1971; Aoki *et al.*, 1973), equal to 27 ns and 18 ns, respectively, shows that the latter quantity is close to the value obtained in the experiment for the lifetime of the molecule.

A more complicated situation occurs in the case of the excitation of the toluene molecule into the first singlet state S_1 (Dietz *et al.*, 1982a). It is known that the excited molecule can then undergo different radiative and nonradiative relaxation processes. These include (a) the radiative transition $S_1 \rightarrow S_0$, (b) nonradiative conversion into the ground electronic state $S_1 \rightsquigarrow S_0$, (c) conversion into the triplet state $S_1 \rightsquigarrow T_1$, and (d) the conversion $T_1 \rightsquigarrow S_0$. As far as relaxation of excited singlet states of aromatic molecules is concerned, a great deal of data obtained from fluorescence measurements has been accumulated. Data on the lifetime of nonradiative triplet states in the case of "isolated molecules", however, are practically nonexistent due to the absence of adequate methods for performing measurements (Dietz *et al.*, 1982a). It is therefore of great interest to apply the method of stepwise photoionization to such investigations.

The scheme of the levels and transitions in the toluene molecule (Fig. 7.7) illustrates the idea of the experiment. Toluene molecules are excited into the S_1 state by a laser radiation pulse and are then probed by a sounding laser pulse at different times. Molecules excited first into the S_1 state can either decay into the ground electronic state or make a transition into isoenergetic levels T_1^* of the triplet state T_1. The molecules in the S_1 state ionize with the absorption of a second photon. The T_1^* states likewise can ionize efficiently,

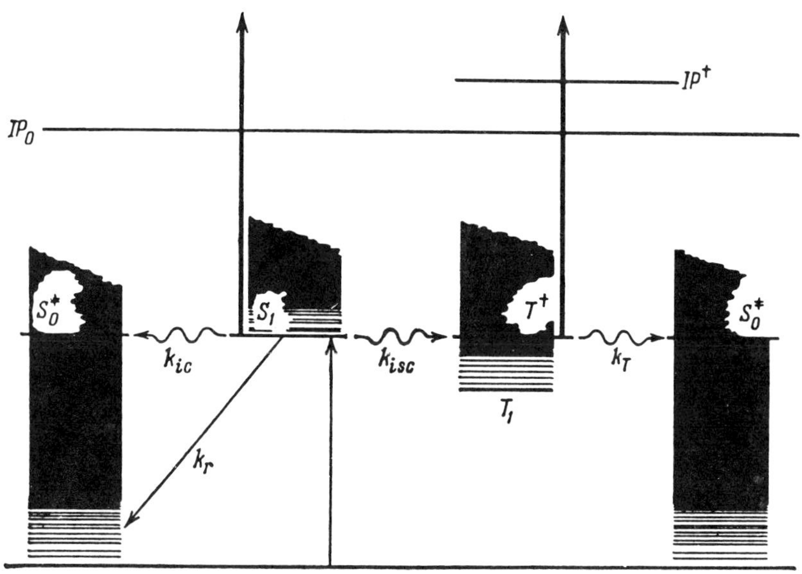

FIG. 7.7. Two-step photoionization scheme for toluene: $S_0^\ddagger$, high-lying vibrational levels of the electronic ground state S_0; S_1, first excited singlet state; T^+, excited vibrational levels of the electronic triplet state T; IP and IP^+, vertical potentials of ionization from the singlet and triplet states, respectively. [From Dietz *et al.* (1982a).]

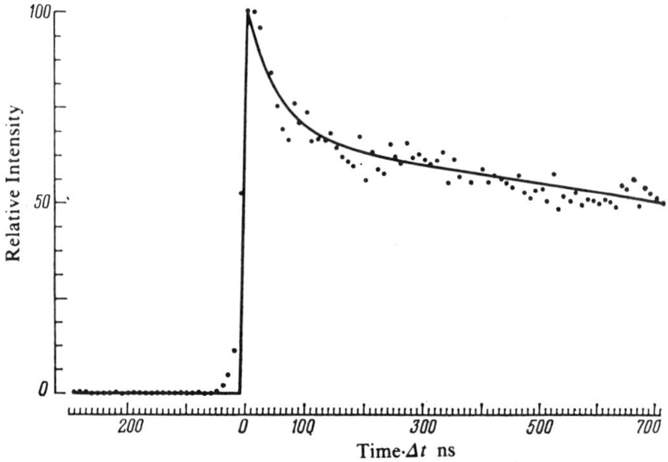

FIG. 7.8. Toluene photoionization signal as a function of the delay time between laser pulses. The fast component corresponds to the decay of the S_1 state and the slow component to the decay of T_1. The laser frequency at the first step is tuned to the σ^1 vibration of the S_1 state (38,608 cm^{-1}). [From Dietz *et al.* 1982a).]

because the energy of the second photon exceeds the vertical ionization potential from the triplet state IP^+.

Figure 7.8 shows the typical dependence of the photoionization signal on the delay time between the exciting and ionizing laser pulses. The double exponential decay is a result of the successive decay of S_1 and T_1^* states:

$$d[S_1]/dt = -(k_r + k_{ic} + k_{isc})[S_1] = -k_s[S_1], \tag{7.1}$$

$$d[T_1^*]/dt = k_{isc}[S_1] - k_T[T_1^*], \tag{7.2}$$

where k_r, k_{ic}, k_{isc}, and k_T are the rate constants for radiative decay, conversion into S_0, conversion into T_1^*, and the rate constant for the decay of T_1^* into S_0.

An important problem that cannot be solved starting from the kinetic data is the ratio of the cross sections of photoionization out of the S_1 and T_1^* states. A knowledge of this quantity permits calculating immediately the rates of all relaxation processes from the experimental dependence. In the work discussed (Dietz *et al.*, 1982a), the radiation of the short-wavelength ArF laser with photon energy 6.4 eV was especially chosen for photoionization from the excited states. For photoionization from the S_1 and T_1^* states, the excesses above the vertical ionization potentials were quite large: 2.3 eV and 1.1 eV, respectively (see Fig. 7.8). Taking into account the weak dependence of the efficiency of photoionization of polyatomic molecules far from the ionization limit gives a basis for assuming that under experimental conditions the cross sections of photoionization of S_1 and T_1^* are approximately identical. The experimental dependence (see Fig. 7.8) is then described by the expression

$$[S_1 + T_1^*] = \frac{[S_1(t = 0)]}{k_S - k_T} [(k_S - k_{isc} - k_T)e^{-k_S t} + k_{isc}e^{-k_T t}], \qquad (7.3)$$

with constants $k_S = 1.38 \times 10^7\ s^{-1}$, $k_{isc} = 8.5 \times 10^6\ s^{-1}$, and $k_T = 3.45 \times 10^9\ s^{-1}$. The rate of decay of the triplet state k_T turned out to be unexpectedly high, exceeding by more than two orders of magnitude the estimate made by Smyth *et al.* (1974).

The most striking results were obtained when studying highly excited molecular states in gas phase under collision-free conditions by means of subpicosecond laser pulses. The photoionization method was used to investigate the dynamics of excited states in *cis*-stilbene (Greene and Farrow, 1983). A 312.5-nm laser pulse 0.25 ps in duration and 10^{-6} J in energy excited the $S_1 \rightarrow S_N$ transition. The absorption of a second photon caused the ionization of the *cis*-stilbene molecule, its adiabatic ionization potential being 7.8 eV. The lifetime of the S_N excited state was measured by this method to be $\tau = 0.32$ ps.

A similar technique was employed to investigate the femosecond dynamics of Rydberg state relaxation in the benzene molecule by means of a laser pulse with a duration of 190 fs (Wiesenfeld and Greene, 1983; Greene, 1984). Figure 7.9a shows a diagram for the two-photon excitation of the benzene molecule at $\lambda_1 = 312.5$ nm, followed by its ionization upon absorption of a photon at $\lambda_2 = 625$ nm. Figure 7.9b presents the dependence of the ion yield on the delay of the λ_2 pulse relative to the λ_1 pulse. The lifetime of the $3R_g$ state of benzene was found from this dependence to be $\tau = 70$ fs. The measured lifetime was in good agreement with the linewidth of the two-photon absorption on the $3R_g 1_0^1$ transition, equal to 59 cm^{-1} (Whetten *et al.*, 1983). The photoionization method was also used to study the lifetimes (around 100 fs) of other Rydberg states in benzene and toluene. Combining the capabilities of the highly sensitive photoionization detection and femtosecond laser technology opens a wide field of investigations into the dynamics of highly excited molecular states.

7.3. Studies of States near Ionization Threshold

Until recently, energy states close to ionization threshold were the least studied. As the techniques for the multistep excitation of atomic and molecular states by means of tunable laser radiation and for the ionization detection of these states were developed, the situation changed radically. Highly

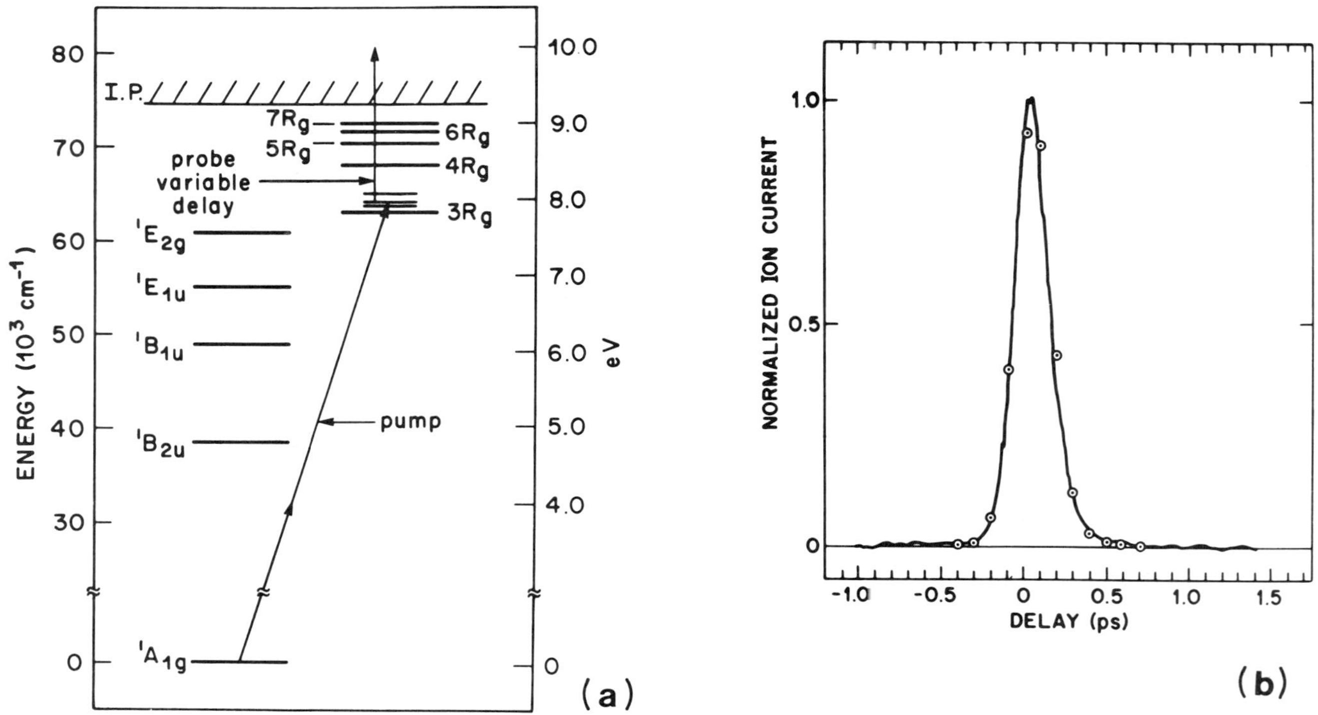

FIG. 7.9. Femtosecond photoionization spectroscopy of the nR_g Rydberg levels in benzene: (a) Benzene energy level diagram, excitation and ionization transitions. [From Whetten *et al.* (1983).] (b) Dependence of the ionization signal on the delay of the ionizing pulse and the lifetime of the $3R_g$ level ($\tau = 70$ fs) found from this dependence. [From Wiesenfeld and Greene (1983).]

excited states close to (below or above) ionization threshold are especially suitable for studies by the photoionization method. First, these states are necessarily reached when implementing the most optimal laser resonance photoionization spectroscopy schemes. Second, it is precisely these states that are the easiest to ionize, and hence detect, by the ionization technique. The flow of works in this area of systematic spectrosopic investigations grow rapidly, and these investigations perhaps merit a special treatment in a separate book. Here we restrict ourselves to the consideration of several examples giving a clear idea of the type and accuracy of the spectral data obtained by this method.

7.3.1. Rydberg States

This is a very active recent area of research in atomic and molecular spectroscopy, and most measurements here are carried out by means of the ionization technique.

7.3.1.1. Atoms. A brief introduction to the physics of Rydberg states in atoms has been presented in Section 3.3. For a more detailed description, the reader is advised to refer to the special monograph on this subject (Stebbings and Dunning, 1983). The technique of multistep excitation of Rydberg states (Svanberg *et al.*, 1973) by means of two or three laser pulses of suitable frequencies is fairly universal (applicable to practically any atom or molecule). Absorption of an even number of photons can lead to excitation of Rydberg states of the same parity as that of the initial state. This makes multiphoton excitation essentially different from one-photon excitation, which allows only states of the opposite parity to be excited through electric-dipole transitions.

Rydberg states can be ionized by various methods: with an electric field (in atomic-beam experiments), by collision (in experiments with thermonic gas diodes and ion cells), and with laser radiation. These methods have already been discussed in Section 3.3.

Experiments have been carried out on the studies of the Rydberg-state series of many elements. In the case of alkali metals that are most easily amenable to theoretical interpretation, they managed to measure by this method the hyperfine structure parameters and the Stark parameters of Rydberg states (Svanberg, 1977). In the more involved case of alkaline earth metals having two optical electrons, the study of Rydberg states has made it possible to experimentally verify the Lu–Fano quantum defect (Lu and Fano, 1970). These investigations have been carried out by the IBM group (Armstrong *et al.*, 1977, 1979; Esherick *et al.*, 1977; Wynne and Armstrong, 1979) and the SRI group (Cooke *et al.*, 1978). Observed in alkaline earth elements upon their two-photon excitation through a quasi resonant intermediate state (see Section 2.2) are two Rydberg series of the even states $^1S_0$

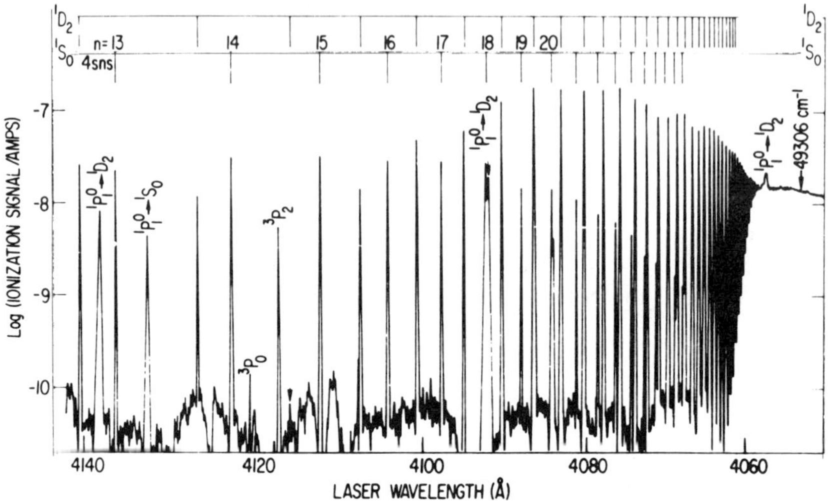

FIG. 7.10. Ionization spectrum in the case of two-photon excitation of even Rydberg states in the Ca atom that converge to an ionization limit of 49,306 cm^{-1}. [From Esherick *et al.* (1977).]

and $^1D_2$. They are clearly manifest in the ionization spectrum in the case of two-photon excitation of the Ca atom from the $^1S_0$ ground state, shown in Fig. 7.10. To observe the P^0 odd Rydberg states of alkaline earth atoms with the aid of optical-range lasers, use is made of three-step excitation (Wynne and Armstrong, 1979). Doubly excited autoionizing Rydberg states of Sr have been observed by Cooke *et al.* (1978).

In more complex atoms, several ionization limits for Rydberg states are observed. By properly selecting the intermediate states and the number of excitation steps, we can measure certain Rydberg series. In the absence of perturbations and with the appropriately selected ionization limit E_i of a series, the final term energies E_n of the series are described by the Ritz formula (3.11), which may be written in the form

$$\delta E_n = E_i - E_n = \mathrm{Ry}/(n^*)^2, \qquad n^* = n - \Delta. \tag{7.4}$$

In the Δ and δE_n coordinate system, this dependence must be a straight line. Figure 7.11 presents the results of such measurements by the photoionization method in an atomic beam of the $nf\,^2F^0$ series observed to occur in the Al atom upon its two-step excitation according to the $3p\,^2P^0 \rightarrow 3d\,^2D \rightarrow nf\,^2F^0$ scheme (Zherikhin *et al.*, 1984b). Figure 7.11b shows this dependence for the E_i value of Al, which was up to now believed to be correct. It can be seen that the curve greatly deviates from the straight line at large n. By adjusting

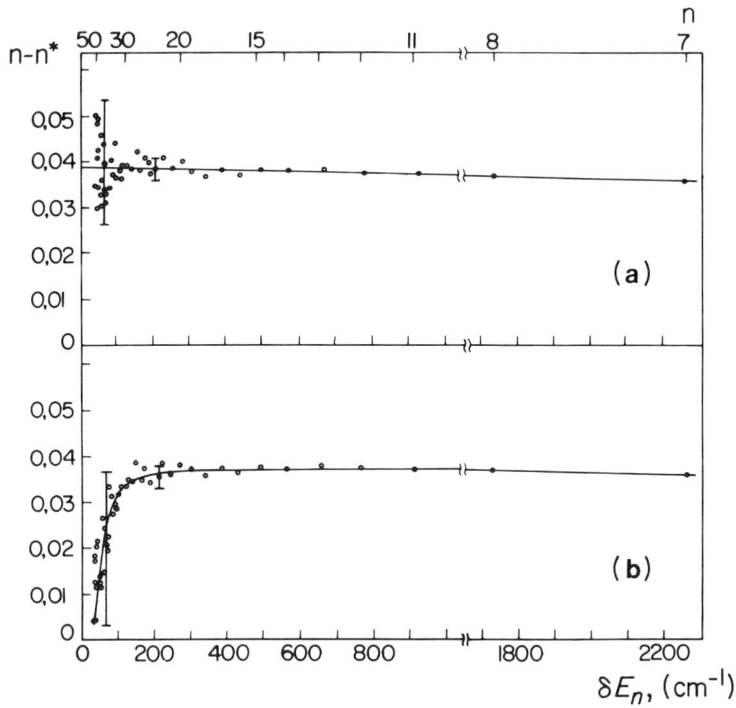

FIG. 7.11. Dependence of the quantum defect of the $nf\,^2F$ series of Al on the distance of the Rydberg level from the ionization limit E_i for two ionization limit values. (a) $E_i = 48{,}278.420\ cm^{-1}$ and (b) $E_i = 48{,}278.37\ cm^{-1}$. [From Zherikhin *et al.* (1984b).]

the value of E_i, it is possible to make the dependence a straight line again (Fig. 7.11a). Such a procedure provides a highly accurate method for measuring the ionization limits of atoms. The photoionization method was also employed to study the high-lying states of the following elements of the same subgroup of the periodic table: Tl, In (Mirza and Duley, 1978 a.b), and Ga (Young *et al.*, 1980).

A considerable number of works are devoted to studies of the Rydberg states in more complex atoms, those of lanthanides. Among the latter, it was Yb that had its highly excited states studied in detail. The experiments involved in these studies covered the two-step and two-photon excitation of the even series $6sns\,^1S_0$ (Camus *et al.*, 1980) and the three-step excitation of the odd series $6snp\,^{3,1}P^0$ (Bekov *et al.*, 1980b). The experimental data obtained were subjected to a detailed theoretical interpretation within the framework of the relativistic perturbation theory with a zero model potential

(Vidolova-Angelova *et al.*, 1981b, 1982). An analysis of highly excited Rydberg states in Yb has revealed that at large n the singlet–triplet splitting is greater than the distance between the adjacent levels of the given Rydberg series. This is substantially different from the situation with simpler atoms and apparently is characteristic of complex atoms. This effect is explained by the non-Coulomb nature of the frame potential. At large n, there also occurs a rapid disappearance of the fine level structure. Similar experiments and theoretical interpretation were also carried out for the Tm atom (Vidolova-Angelova *et al.*, 1984).

Extensive investigations into the Rydberg states of lanthanides and actinides by three-step ionization were carried out by the Livermore group in the course of their work on the laser separation of actinide isotopes (Solarz *et al.*, 1976; Paisner *et al.*, 1977). These investigations helped measure accurately some ionization potentials (see Section 7.3.2).

7.3.1.2. Molecules. Multistep excitation and ionization have opened up possibilities of studying systematically the Rydberg states in both simple and complex molecules. In contrast to atomic Rydberg states, a molecular Rydberg state (n, v^*, J^*) with the principal quantum number n, vibrational quantum number v^*, and rotational quantum number J^* may lie above the lowest (v^+, J^+) level of the molecular ion and therefore may ionize automatically. This is possible if vibrational–rotational energy is transmitted to the Rydberg electron. For this reason, research on autoionization of Rydberg states in molecules is of much interest in studying the nonadiabatic coupling of atomic and nuclear motions when the Born–Oppenheimer approximation holds no more. That is why Rydberg states in simple molecules have attracted interest and were studied in a number of works (Carlson *et al.*, 1980, 1981; Leutwyler *et al.*, 1982; Martin *et al.*, 1982; Bernheim *et al.*, 1982; Eisel and and Demtröder, 1982; Demtröder *et al.*, 1983). Earlier, in Fig. 5.4, we showed the spectrum of highly excited states of the K_2 dimer near the ionization limit, obtained by the method of two-step excitation and ionization. Studies of various Rydberg series allows determination of the adiabatic ionization potentials of molecules (see Section 7.3.2) and also of the vibrational and rotational constants of molecular ions. The study of Rydberg states of polyatomic molecules gives valuable information on dynamic effects associated with Jaln–Teller instability as demonstrated by Whetten *et al.* (1983) for benzene.

The use of ultrashort laser pulses makes it possible to effect multistep excitation of short-lived Rydberg states in polyatomic molecules. An example of such measurement for the benzene molecule has been illustrated in Fig. 7.9 of Section 7.2.2. This area of research has just started developing, and considerable progress can be expected in this direction in understanding the dynamics of highly excited electronic states near the ionization limit of polyatomic molecules.

7.3.2. Ionization Potentials

Measurements of Rydberg series limits provide a simple and accurate method for determination of the ionization potentials of atoms and molecules, as has been illustrated earlier with Al as an example (Fig. 7.11). In this area, the capabilities of the multistep ionization spectroscopy technique are out of any competition from other methods. Let us consider this approach for atoms as well as molecules.

7.3.2.1. Atoms. A good illustration of the capabilities of the photoionization method is Table 7.2 borrowed from works of the Livermore group (Solarz *et al.*, 1976) on the determination of the ionization potential of the U atom by various methods: surface ionization, electron impact, spectroscopic extrapolation, two- and three-step ionization, and finally, Rydberg series fitting. The last method proved the most accurate. Systematic studies by the photoionization method of Rydberg series and determinations on the basis of ionization potentials were carried out for almost all lanthanides (Paisner *et al.*, 1977). Of course, the Rydberg series of complex atoms do not necessarily lie exactly on a straight line as required by formula (7.4), which takes no

TABLE 7.2. Ionization Potential Values for U, Measured by Various Methods[a]

Ionization potential		Method	Reference
(cm^{-1})	(eV)		
49,000 ± 650	6.08 ± 0.08	Surface ionization	Bakulina and Ionov, 1959
49,000 ± 500	6.19 ± 0.06	Surface ionization	Smith and Hertel, 1969
49,300 ± 400	6.11 ± 0.05	Electron impact	Mann, 1964
49,200 ± 800	6.1 ± 0.1	Electron impact	Rauch and Ackerman, 1974
49,800 ± 600	6.05 ± 0.07	Spectroscopic extrapolation	Sugar, 1974
49,900 ± 16	6.187 ± 0.002	Two-step photoionization	Janes *et al.*, 1976
49,935 ± 20	6.1912 ± 0.0025	Three-step photoionization	Solarz *et al.*, 1976
49,958.1 ± 4	6.1941 ± 0.0005	Rydberg series fitting	Solarz *et al.*, 1976

[a] From Solarz *et al.* (1976).

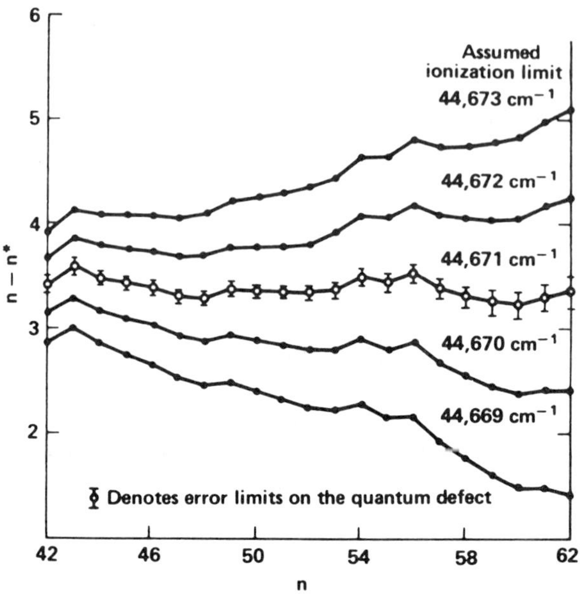

FIG. 7.12. Quantum defect $\Delta = n - n^+$ as a function of the principal quantum number n for the Ce atom for various ionization limit values. [From Paisner *et al.* (1977).]

account of level perturbation. Nevertheless, the observed dependences of the quantum defect $\Delta = n - n^*$ on the principal quantum number n allow them to be matched to horizontal straight lines, but only at a certain value of the ionization potential. This is illustrated well by Fig. 7.12 showing the dependences $n - n^* = f(n)$ for the Ce atom for several ionization limit values. Minimizing such a dependence, they obtain a fairly accurate value of the ionization potential. Table 7.3 lists ionization potentials of lanthanides, measured by the method of photoionization spectroscopy of Rydberg series. The major data obtained at the Livermore Laboratory (Paisner *et al.*, 1977) are supplemented with data borrowed from other works. Given for comparison purposes are data obtained by other methods.

7.3.2.2. Molecules. Multistep ionization also provides a possibility of accurately measuring the adiabatic ionization potentials of molecules. The method becomes especially effective when used in conjunction with the dynamic cooling of the molecules in supersonic jets, in which case the molecules are concentrated in the lower vibrational–rotational states, and the low temperature allows molecules with a low dissociation energy to be investigated. These possibilities were demonstrated in experiments with alkali metal dimers (Leutwyler *et al.*, 1981), van der Waals complexes (Gonohe *et al.*, 1984), and polyatomic molecules (benzene and benzene-d_6) (Grubb *et al.*, 1984).

TABLE 7.3. First Ionization Potentials of the Lanthanides[a]

Element	Electron impact[b]	Spectroscopic extrapolation[c]	Photoionization limit	Rydberg series fitting
Ce	5.44	5.47(5)[d]	5.537(4)	5.5387(4)
Pr	5.37	5.42(3)	5.464(6)	—
Nd	5.49	5.489(20)	5.523(3)	5.5250(6)
Pm	—	5.554(20)	—	5.582(10)
Sm	5.58	5.631(20)	5.639(3)	5.6437(10)
Eu	5.68	5.67045(2)[e]	5.666(3)	5.6704(3)
Gd	6.24	6.141(20)	—	6.1502(6)
Tb	5.84	5.852(20)	—	5.8639(6)
Dy	5.90	5.927(8)	5.936(3)	5.9390(6)
Ho	5.99	6.018(20)	6.017(3)	6.0216(6)
Er	5.93	6.101(20)	6.104(3)	6.1077(b)
Tm	6.11	6.18436(20)[f]	—	6.18436[g]
Yb	6.21	6.25394(20)[f]	—	6.25394[h]

[a] From Paisner *et al.* (1977).
[b] Ackermann *et al.* (1976).
[c] Reader and Sugar (1966); Martin *et al.* (1974).
[d] Errors are indicated in the parentheses (1 eV = 8065.479 cm^{-1}).
[e] Smith and Tomkins (1975, 1976).
[f] Camus and Tomkins (1969).
[g] Vidolova–Angelova *et al.* (1984).
[h] Vidolova–Angelova *et al.* (1981b).

Table 7.4 lists the results of experiments on measurements of the adiabatic ionization potentials of three alkali metal dimers, obtained by the method of two-step ionization in a supersonic jet. The supersonic jet method provides for measurements in collision-free conditions where the thermal and associative ionization as well as charge transfer effects are eliminated, considerable vibrational and rotational cooling ($T_{vib} \simeq 70$ K, $T_{rot} \simeq 30$ K), and enhancement of thc proportion of the dimer molecules. It can be seen from Table 7.4 that the accuracy of measurement of the ionization potentials of molecules comes to around 1 $cm^{-1} \simeq 10^{-4}$ eV.

Similar techniques are applicable to many molecules under the collision-free low-temperature conditions of a supersonic jet. A good example are experiments on the two-step photoionization of the van der Waals complexes of fluorobenzene with argon and CH_4 (Gonohe *et al.*, 1984). Figure 7.13 shows the results of such experiments with the fluorobenzene molecule and the fluorobenzene–$(Ar)_n$ complexes in which the frequency of the first laser was tuned to the (0, 0)-band of the transition to the S_1 state of each

TABLE 7.4. Ionization Potentials of Some Alkali Metal Dimers[a]

Na_2	NaK	K_2	Method[b]	Reference
4.9 ± 0.1	4.5 ± 0.1	4 ± 0.1	OPhI	Robbins *et al.*, 1967; Foster *et al.*, 1969
4.934 ± 0.011	4.52 ± 0.05	4.05 ± 0.05	OPhI	Hermann *et al.*, 1978
	4.57 ± 0.2		EII	Zmbov *et al.*, 1977
4.88898 ± 0.00016	4.41636 ± 0.00017	4.06075 ± 0.00015	TPhI	Leutwyler *et al.*, 1981

[a] From Leutwyler *et al.* (1981)

[b] OPhI, one-photon ionization; EII, electron-impact ionization; TPhI, two-photon ionization.

complex, while that of the second laser was varied. Observed for fluorobenzene-(Ar) and fluorobenzene–$(Ar)_2$ are red shifts of 200 cm^{-1} and 389 cm^{-1}, respectively. These shifts are evidence that the binding energy of the fluorobenzene cation in its ground state with Ar is higher than that of the neutral molecule in the ground state with Ar. This method is suitable for systematic studies of van der Waals complexes.

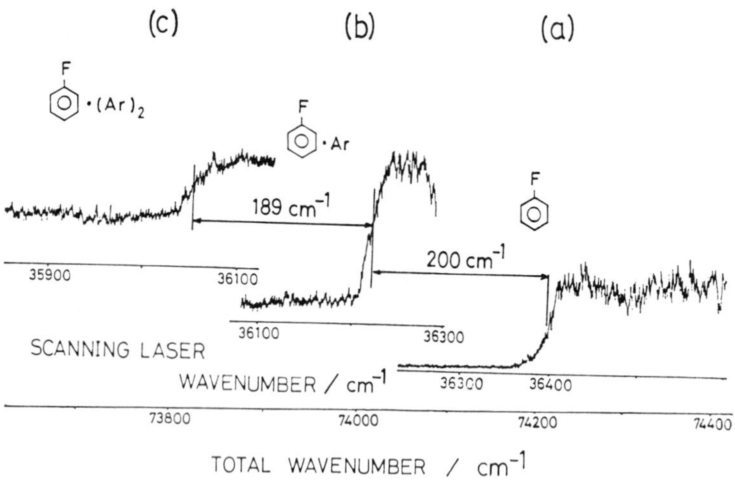

FIG. 7.13. Two-step, two-frequency photoionization spectrum of (a) the free fluorobenzene molecule, (b) the fluorobenzene-$(Ar)_1$ complex, and (c) fluorobenzene-$(Ar)_2$ complex in a supersonic jet. The frequency ν_1 is tuned to the (0, 0) band of the transition to the S_1 state of the molecule or complex. The frequency ν_2 of the second laser is variable. [From Gonohe *et al.* (1984).]

7.3.3. Autoionization States

Autoionization states are a natural object of investigation by the photoionization spectroscopy techniques involving multistep excitation. First, as is the case with Rydberg states, autoionization states are most convenient to excite through the absorption of several photons, since selection rules for one-photon transitions do not allow one-photon excitation of most autoionization states. Second, autoionization states usually rapidly decay with the formation of an "electron + ion" pair, and so it is only natural to employ ionization techniques to detect their excitation.

7.3.3.1. Atoms. A qualitative picture of the origin of autoionization states in a multielectron atom and the use of multistep excitation for their examination have been already considered in brief in Section 3.2.3. We have emphasized there the special value of doubly excited states which lead to the appearance of comparatively narrow autoionization states. This matter was studied in detail, both theoretically and experimentally, for the Yb atom having two $6s^2$ optical electrons (Bekov *et al.*, 1980a, 1981a). To illustrate, Fig. 7.14 shows an ionization spectrum recorded while varying the laser wavelength λ_3 at the third stage of excitation of the $4f^{14}6s6d\,^3D_1 \rightarrow 4f^{14}6s6p\,^3P_0^0$ transition in the Yb atom. The three-step excitation scheme

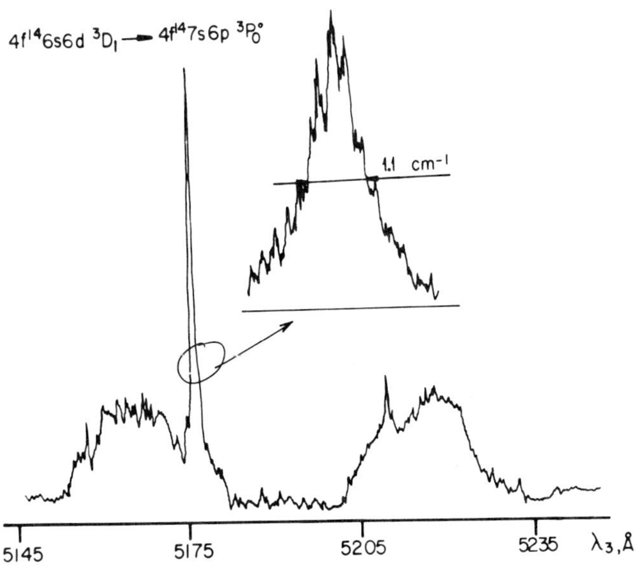

FIG. 7.14. Dependence of the ion signal on the laser wavelength λ_3 at the third stage of excitation of autoionization states from the $6s6p\,^3D_1$ intermediate state of Yb at a laser spectral bandwidth of $\Delta\nu_3 = 1\ \text{cm}^{-1}$ (Bekov *et al.*, 1980a, 1981a).

for Yb was shown in Fig. 3.6. The spectrum clearly shows a narrow autoionization resonance which we have managed to identify (see Table 3.1) and two broad resonances that are apparently due to the excitation of the 4f shell.

When searching for narrow autoionization resonances or other high-lying states in a complex atom, certain precautions should be taken in interpreting the narrow peaks that occur as λ_3 is being varied, with λ_2 and λ_1 kept constant. In spectra of complex atoms, a situation frequently occurs when λ_3, while being varied, becomes resonant with some transition from a low-lying atomic state, which then gets excited upon absorption of some of the photons λ_1, λ_2, or λ_3. This sort of "false" resonance can be discriminated by a standard procedure (Carlson *et al.*, 1976). First, the laser pulse at λ_3 can be delayed with respect to those at λ_1 and λ_2 so as to temporally separate the transitions $\lambda_1 \rightarrow \lambda_2$ and $\lambda_2 \rightarrow \lambda_3$. Second, radiation at λ_1 or λ_2 can be blocked to make sure that the absorption of all the three photons $\lambda_1 + \lambda_2 + \lambda_3$ is necessary. The persistence of the narrow resonance at λ_3 after these procedures is evidence of its true nature.

Narrow autoionization resonances were observed to occur also in other many-electron atoms such as Gd (Bekov *et al.*, 1978b), Lu (Bekov and Vidolova–Angelova, 1981; Miller and Nogar, 1982), and Eu (Zherikhin *et al.*, 1984a). Such resonances are apparently typical of many-electron atoms. Their systematic examination with the aid of photoionization spectroscopy in conjunction with multistep excitation has just started. Of course, the theoretical interpretation for atoms possessing three or more optical electrons will be more complex than for Yb.

The capabilities of multistep excitation manifest themselves most vividly in investigations of the doubly excited autoionization states in the so-called "planetary atoms" (Percival, 1977). The term refers to atoms in high-lying doubly excited states. Such states can hardly be excited through absorption of a single photon, because the cross section for two-electron transitions is extremely small. It is precisely in this case that the multistep excitation method proves indispensable, and it is this method that was used for the purpose in studying alkaline earth metal atoms (Gallagher, 1982; Freeman *et al.*, 1983). The essence of the method is explained in Fig. 7.15 with reference to the excitation of the 6p*n*d doubly excited autoionization states in BaI. The first two laser pulses excite one of the two optical electrons to the 6s*n*d Rydberg state. The third laser pulse excites the inner (relative to the "outer" Rydberg electron) electron to make the 6s → 6p transition, without affecting the Rydberg electron. The excitation at the third stage occurs on the strong resonance line of Ba^+, and so the cross section of such a transition is rather large (around 10^{-14} cm^2). The direct excitation of the Rydberg electron into the same energy range obviously has a very small cross section

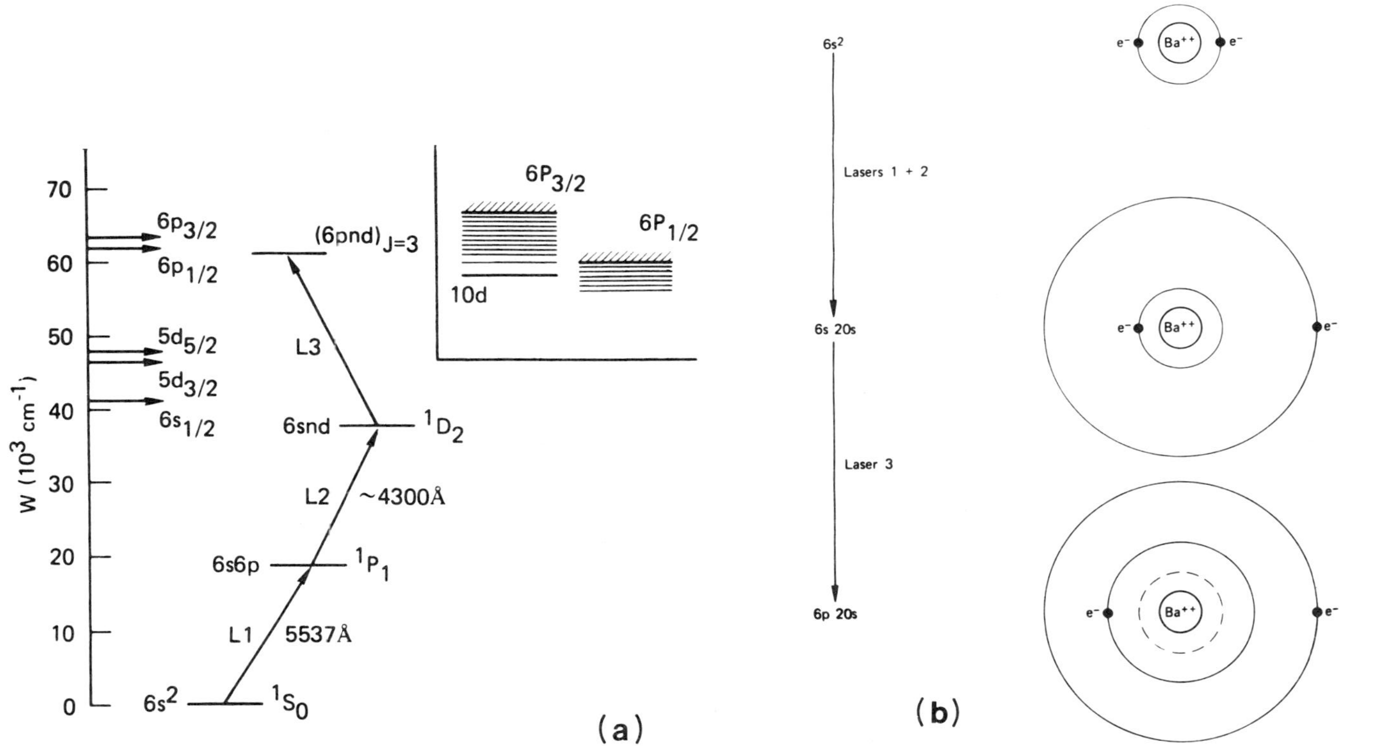

FIG. 7.15. Multistep excitation of doubly excited 6p*n*d states ("planetary atom" states) in BaI: (a) diagram of energy levels participating in three-step excitation and (b) states of the Ba atom in the course of excitation. [From Gallagher (1982).]

(about 10^{-20} cm^2) because of the great difference in spatial wavefunction oscillations between the initial and final states. As a result, both optical electrons of the Ba atom excited. The width of such autoionization states varies over a broad range (from a few fractions of a cm^{-1} to 100 cm^{-1}), depending on the principal quantum number and the character of the Rydberg state. As n increases, the electron–electron interaction weakens, this causing the decay rate and width of the autoionization state to decrease.

The width of autoionization states is commonly measured by varying the frequency of the narrow band laser employed at the laser excitation step (see Figs. 3.8 and 7.15b). In some cases, however, the width of an autoionization resonance can be measured by means of a wideband laser (Cooke *et al.*, 1982). This method is based on indirect measurement of the decay time of the autoionization state upon strong saturation of the transition from the Rydberg to the autoionization state. When this transition is highly saturated, there occur the depletion of the Rydberg state and ionization decay, even if the laser frequency is far beyond the autoionization transition band $\Gamma_{\text{q.i.}}$. As a result, the effective spectral bandwidth, Γ_{eff}, participating in the conversion of the Rydberg atoms to photoions is defined by the following expression (Cooke *et al.*, 1982):

$$\Gamma_{\text{eff}} = 2(\mu\mathscr{E}/\hbar)(\Gamma_{\text{a.i.}}\tau_p/\ln 2), \tag{7.5}$$

where μ is the dipole moment of the transition to the autoionization state, $\mathscr{E}$ the electric field strength of the light wave, and τ_p the laser pulse duration. When saturation is strong enough, Γ_{eff} can be made to exceed the laser bandwidth $\Delta\omega_L$, and hence can be measured. Then the true width $\Gamma_{\text{a.i.}}$ can be found using Eq. (7.5). It was exactly by this method that the widths of autoionization states with $n = 30$ to 40 in Ba were measured to be in the range 0.4–0.2 cm^{-1} (Cooke *et al.*, 1982).

7.3.3.2. Molecules. Autoionization states in molecules clearly manifest themselves when the latter are subjected to multistep or multiphoton excitation, and therefore they have already been discussed in Chapter 5. (Section 5.2.2, Fig. 5.4). In simple molecules, the vibrational limit turns them to autoionization states. Their study by the stepwise photoionization method has already been considered in Section 7.3.1 of this chapter. Here we only add that combining the techniques of multistep excitation by means of ultrashort laser pulses and ionization detection offers considerable scope for studying polyatomic molecules as well.

8. Laser Photoionization Atomic Spectral Analysis

The development of new analytical methods for measurement of ultralow contents of elements in different substances is very urgent for many trends of present-day science and engineering. This is because a wide range of problems currently faced in the processing of high-purity materials in geochemistry, toxicology, environment control, etc. could be solved by controlling the trace elements that may be present in amounts of 10^{-8}–10^{-11}%. In some cases, such sensitivity of analysis can be provided by conventional analytical methods or their modifications: atomic absorption and atomic fluorescence spectrometry, neutron activation analysis, spark mass spectrometry, and others. However, in most cases their sensitivity is limited to the level of 10^{-7}%.

The various laser spectroscopy techniques developed in recent years and described in monographs (Omenetto, 1979; Hieftje *et al.*, 1981; Kliger, 1983; Letokhov, 1986) are obviously of great interest as far as analytical applications are concerned. However, the most effective method for detecting atomic traces is the stepwise laser resonance photoionization (Bekov and Letokhov, 1983a,b). It is exactly the analytical applications of the resonance photoionization of atoms to which this chapter is devoted. Primary attention will be centered on the combination of atomic resonance photoionization with vacuum thermal vaporization and atomization of substances. Consideration will be given to a number of experiments on the trace element analysis of real substances—high-purity materials, natural samples, and biological objects.

8.1. Analytical Characteristics of the Photoionization Method

The attainment of the ultimate atomic detection sensitivity, i.e., the sensitivity required to detect single atoms in a sample, has been considered in detail in Chapter 4. Of course, so high a sensitivity is only necessary in solving unique scientific problems, for example, in nuclear physics and astrophysics. Some of them are considered in the next chapter. A much more moderate sensitivity is sufficient for analytical applications. For instance, Bekov *et al.* (1979) have successfully demonstrated the detection of Yb atoms in a tantalum sample at a level of 10^7 atoms (10^{10} atoms when measuring the hyperfine structure spectrum) even at a low laser-pulse repetition frequency when the fraction of the irradiated atoms is comparatively small ($\eta_{irr} = 10^{-5}$). The sensitivity of the method can be materially increased (10^3 times) by raising the laser-pulse repetition frequency to 10 kHz, with which $\eta_{irr} \simeq 10^{-2}$–$10^{-1}$. This has been demonstrated in the experiments on detecting Eu atoms in tantalum, described in Section 4.2.3.

The experimental data obtain make it possible to evaluate the capabilities of the stepwise laser resonance photoionization method in determining ultralow amounts of impurities in a sample. For example, in the case of vacuum evaporation and atomization of a sample with a mass of 10^{-3} g (total number of matrix atoms 10^{19}), 10^6 impurity atoms can easily be detected at a low laser-pulse repetition frequency. This corresponds to an impurity content of $10^{-11}\%$. In prospect, after optimizing the laser ionization process, e.g., by increasing the laser-pulse repetition frequency to 10^4 Hz, it will be quite possible to reach a *sensitivity* level of $10^{-14}\%$. So exceptionally low an impurity detection level is beyond the powers of the traditional analytical methods. This is an indication that the laser resonance photoionization method should be further developed to make it generally applicable to the detection of ultralow amounts of trace elements.

Under the conditions of real experiments on laser photoionization, the atoms to be studied are always surrounded by foreign atoms or molecules. Depending on the photoionization approach the concentration of foreign particles N_B may vary widely—from 10^{10} cm^{-3} and less in vacuum for the atomic beam, to 10^{19} cm^{-3} in the buffer gas medium. The interaction of these particles with the laser radiation inevitably gives birth to foreign ions, which can contribute to the photoionization signal. Let the photoionization *selectivity* S be defined as the ratio of the ionization yield of atoms A of interest to the ionization yield of foreign particles B, for example, the atoms or molecules of the matrix

$$S = Y_A / Y_B. \tag{8.1}$$

If no additional measures are taken to discriminate the foreign particle photoions relative to the photoions of the atoms to be detected, the selectivity value places an evident limitation on the minimal detectable concentration of the atoms to be analyzed: $N_A/N_B \gtrsim 1/S$.

The selectivity of the method of multistep resonant photoionization is many orders of magnitude higher than the selectivity of any spectroscopic method based on single-photon transitions as one resonant photon is absorbed. This is due to the fact that at multistep excitation there must be a resonance at each excitation step. Therefore, the selectivity of the process based on a sequence of several resonant processes will be apparently multiplied (Letokhov and Mishin, 1979). The ultimate selectivity of the multistep excitation of atoms, essential for the detection of extremely rare isotopes, is discussed later in the text (Chapter 9). Here we only note that there are three principal mechanisms responsible for the limitation of the selectivity of multistep excitation and ionization: (1) excitation of undesirable atoms B on the *wing of radiation spectrum* of the laser pulse tuned to resonance with the absorption line of the atoms A subject to detection; (2) excitation of the undesirable atoms B on the *wing of their absorption line* of the laser radiation tuned to resonance with the atoms A; and (3) *multiphoton nonresonance excitation* and ionization of the undesirable atoms B.

The contribution from each mechanism to the limitation of selectivity depends on the shift of the spectral lines ω_A and ω_B, the ionization thresholds and other quantum-level particulars of the atoms A and B, and finally, the intensity of the spectral wings and background noise of the monochromatic laser radiation. If the ionization potentials and spectra of the atoms A and B differ greatly, a very high selectivity of ionization detection of the atoms A can be realized. For example, Cs atoms in an argon atmosphere can be detected with a selectivity of up to 10^{19} (Hurst *et al.*, 1977c). On the contrary, when the spectral lines and ionization potentials of the atoms are close, the selectivity may be limited to a much lower level. Such a situation arises, for example, when detecting boron traces in germanium. In this case, we must resort to the special methods for improving selectivity, which are considered in Section 8.2.3.

The photoionization method of trace element analysis possesses such an important property as *versatility*, for any atom can be ionized in a resonance manner by means of a suitable scheme combining multistep and multiphoton excitation. General estimates as to this matter have been presented in Chapter 2, and various schemes of atomic excitation and ionization, including those for particular atoms, have been considered in Chapter 3. The most universal and easily realizable scheme is ionization via Rydberg states. This method is also most suitable for analytical applications. Indeed, by virtue of resonant interaction between the atom and the laser radiation at all the steps, the laser energetic parameters required will be minimal. As a result, the nonselective ion background inevitable in direct analytical experiments will be two or three orders of magnitude lower than in the case of stepwise ionization at the transition to the continuum.

The analysis of the spectrum of atomic energy levels for most of the elements (Chapter 3) shows that for 19 elements of the periodic table—H, He, C, N, O, F, Ne, P, S, Cl, Ar, As, Se, Br, Kr, I, Xe, At, and Rn—the wavelength of the first transition from the ground state lies in the region <2000 Å. Today it is rather difficult to generate tunable laser radiation in this region. It is possible to apply for these elements at the first step two- or three-photon excitation of the nearest levels by intense UV radiation (Young *et al.*, 1979). Unfortunately, in real analytical studies this gives birth to a considerable ion background as a result of nonselective photoionization of the matrix atoms or molecules with lower ionization potentials. It is impossible to detect the weak selective signal under such conditions without discriminating this background. For all other elements of the period table, resonant single-photon excitation can be realized at all steps using appropriate tunable lasers.

8.2. Technique of Photoionization Spectral Analysis

Of great importance for practical spectral analysis applications are the availability of laser equipment, the ease of atomization of substances, the minimization of background noise, and the matters of calibration. All these technical matters are briefly discussed.

8.2.1. Laser Equipment

The laser photoionization spectrometer makes use of tunable lasers which enable us to carry out resonant excitation of atomic transitions. At present, dye lasers excited by pulsed UV- and visible-range lasers—nitrogen laser ($\lambda = 337$ nm), excimer XeCl and KrF lasers ($\lambda = 308$ nm, 249 nm), and second- ($\lambda = 532$ nm) and third-harmonics ($\lambda = 355$ nm) of the Nd:YAG laser—are the most suitable for the photoionization of atoms in the region of 220 nm $< \lambda <$ 900 nm where absorption lines of most elements lie (Schäfer, 1977).

Dye lasers with transverse pumping by N_2 laser radiation have been rather widely used until recently (Hänsch, 1972). Despite the moderate power of its pulses, the N_2 laser can be applied for the simultaneous excitation of several dye lasers. Such a laser spectrometer was used in analytical experiments in which high sensitivity was achieved in detecting traces of some elements in real substances (see Section 8.3). As far as photoionization analytical applications are concerned, the models made by “Lambda Physik” and “Molectron" are the best of the variety of designed and produced dye lasers. They make it possible to apply any of the above-mentioned sources of pumping for simultaneous excitation of several dye lasers with a pulse repetition rate of up to 200 Hz. Dye lasers themselves are characterized by a high pulse power, sufficiently narrow line, and a wide range of wavelength tuning (217–970 nm). Thus, the laser systems considered are the optimal sources of multifrequency tunable radiation if there is no need for a high pulse repetition rate.

The most universal laser system for multistep resonant photoionization of atoms is a combination of several dye lasers pumped by an excimer XeCl laser ($\lambda = 308$ nm) or KrF laser ($\lambda = 249$ nm). The wavelengths of atomic transitions of more than 80% of the elements from the periodic table lie in the laser tuning range of such a system (217–970 nm). A relatively high pulse repetition rate (up to 200 Hz), the long-term stability of the frequency, and the energetic parameters of this system make it possible to realize a high absolute sensitivity of atom photoionization detection in prolonged analytical experiments. When the simplicity and mobility of laser systems are put in the forefront, it is advisable to apply the N_2 laser as a pumping source.

For a considerable number of elements (mainly heavy non-closed-shell elements) it is possible to choose a multistep photoionization scheme with laser wavelengths 530 nm $\lesssim \lambda_i \lesssim$ 750 nm. If in this case it is necessary to analyze a small absolute amount of a substance in a continuous atomic beam, when a high pulse repetition rate is required to intercept a significant part of the atoms to be detected, the Cu vapor laser is the most advisable to be used as a pumping source (see Section 4.2.2). If the mass of the substance to be analyzed is unlimited, it is better, of course, to apply a simpler laser system.

8.2.2. System of Substance Atomization

Atomization of substances can be accomplished in a gas or in a vacuum. The atomization in the medium of inert gas is widely used now in atomic–fluorescence spectroscopy. In most cases the process of atomization is performed in graphite crucibles, cells, and tubes heated electrically up to 3000°C. The main function of a buffer gas is to prevent thermal scattering of the substance and thus to create a considerable concentration of the atoms to be detected in the irradiation zone. This method is rather universal and is characterized by good speed of analysis. However, as for photoionization detection, the method has material shortcomings.

First, the atoms formed can be lost as a result of their interaction with atoms and molecules formed by the atomization of the base material with the microimpurities in the gas. The use of preliminary chemical separation of the matrix by means of different chemical reagents may lead to both an uncontrollable addition of the element with the reagents and a loss of the element in the process of chemical extraction of the matrix. Second, the detection limit of the chosen element in the substance will be defined by the content of this element in the inert gas. These shortcomings can be eliminated by excluding the interaction between the atoms of the element and the environment, for example, by atomization of the substance in a vacuum when direct analysis of the substance in its natural form is possible. Evaporation and atomization can be realized by thermal heating up to 3000°C, a laser radiation pulse, an electron or ion beam, an electric discharge, etc. The choice of one or another method of atomization is defined by a specific task. However, the simplest and most universal method, in our opinion, is thermal atomization in vacuum (Bekov and Letokhov, 1983a,b).

8.2.2.1. Thermal Atomization. This method is characterized by rather slow evaporation and atomization of a substance in order to form effusive atomic–molecular beams in a vacuum. In this case, the interference effects between the element and the matrix taking place at a high particle density during atomization in a buffer gas are essentially reduced due to a low vapor pressure of the substance in the crucible.

The main requirements in designing a vacuum atomizer are to achieve temperatures of up to 3000°C, to realize prompt replacement of samples and crucibles, to ensure its compatibility with the photoionization and ion detection systems as well as simplicity and reliability in work. For example, the atomizer used in our works (Bekov and Letokhov, 1983a,b) (Fig. 8.1) was based on a graphite crucible-tube heated by electric current. A special ion-suppressing system was placed above the outlet of the crucible to suppress a strong background of thermal ions and electrons. In the process of analysis, the sample of substance in its natural form with a mass of up to 200 mg was put into the crucible. In the analysis of solutions, first the moisture was

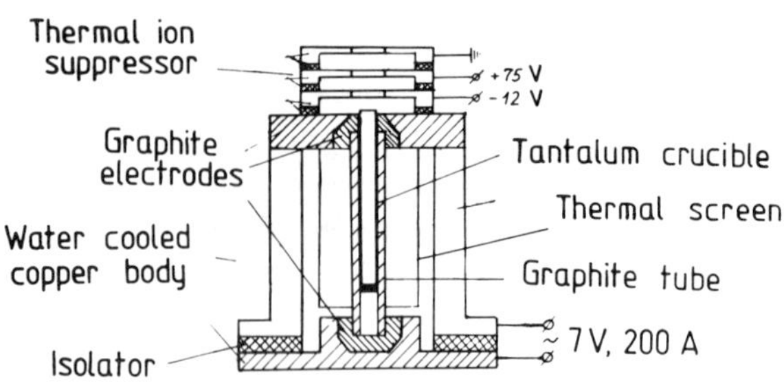

FIG. 8.1. Design of an electrothermal atomizer for creating collimated atomic beams in vacuum. [From Bekov and Letokhov (1983b).]

evaporated. Then the crucible was placed in the atomizer, and after the vacuum chamber was evacuated to a residual pressure of 10^{-6} Torr, the crucible was heated by steps to the working temperature. The evaporating atoms and molecules of the substance were formed by the cylindrical channel of the crucible as an atomic–molecular beam in which photoionization detection of the selected atoms was carried out.

8.2.2.2. Laser Evaporation. This method of evaporation and atomization of matter started to be used in analytical studies right with the appearance of powerful lasers. Evaporation by CW laser radiation is in fact a modification of thermal evaporation with some improvement of the locality of heating. When it is combined with photoionization detection of atoms, questions also arise about how to optimize the laser-pulse repetition rate, how to eliminate the thermal ion background, etc. In this respect it is more convenient to apply pulsed laser evaporation. It permits synchronization with the pulses of the detecting laser radiation that, in its turn, solves the problem of optimal consumption of the substance to be analyzed. The evaporation by laser pulse can be accomplished in a buffer gas or in a vacuum. The laser radiation intensity needed for effective evaporation and atomization of matter usually ranges from 10^7 to 10^9 W/cm^2. Such a wide range is due mainly to a different duration of the radiation pulse (10^{-8}–10^{-6} s), the pulse energy being approximately the same. Successful experiments have been conducted on the photoionization detection of atoms produced upon the pulsed heating and vaporization (ablation) of a substance with laser radiation (Beekman and Calcott, 1984).

8.2.2.3. Electron-Beam Evaporation. This type of evaporation has been applied for a long time in analytical studies as a means of producing the highest possible temperatures (above 3000°C), for example, to evaporate refractory metals and their oxides. The sample is heated in a vacuum by a continuous-power electron beam generated by an electron-beam gun. Many designs of electron guns now have been developed with electron energies of 5 to 30 keV, currents up to 1 A, and specific beam powers of 10^3 to 10^5 W/cm^2. With electron beam focused, it is possible to accomplish local heating of the sample with satisfactory spatial resolution.

Electron-beam evaporation is widely used in mass-spectrometric analysis of solids (Kuzmin *et al.*, 1970). The studies (Tzigel'man, 1972) proved that the intensity of a neutral component at such evaporation is 3 or 4 orders higher than that of an ionic component. With additional mass filtration of ions, this makes, in principle, the use of electron-beam evaporation in combination with laser stepwise photoionization of atoms promising for the analysis of matter.

8.2.2.4. Ion Sputtering. The sputtering of the surface of solids under the action of an accelerated ion beam is widely used in secondary-ion mass spectrometry to study the element composition of solids and particularly to detect traces of impurities in substances. One of the problems in this method of analysis is that the secondary-ion yield depends greatly on the matrix material, and the ion yield is small compared with that of the neutral component. The neutral-atom yield, as a rule, is 2–4 orders of magnitude higher than the ion yield and weakly depends on different matrix effects. All this points out that ion sputtering is promising as a method of atomization of solids. The ion background can be discriminated by using the methods of ion separation accepted in secondary-ion mass spectrometry. The application of laser radiation enables us to increase the ionization degree of the atoms of the selected element by several orders of magnitude in comparison with ion bombardment.

The ion-beam parameters may vary over a very wide range and are defined by a specific task. For example, an ion beam was applied with an energy of 5 to 30 keV and a direct current of 1 mA (Parks *et al.*, 1983). The beam was modulated in order to synchronize it with the laser pulses. This modulation provided a pulse repetition rate of up to 10 kHz with a pulse duration of 0.25 to 10 μs. The pulsed sputtering regime is more advantageous since the consumption of substance is minimal, and it is simpler to realize the vacuum conditions necessary for normal operation of the ion separation and detection systems (see Section 4.4.1).

8.2.2.5. Choice of an Atomization System. Each of the above-considered methods of atomization of matter has its advantages and disadvantages. Thermal atomization in a buffer gas is the simplest method of practical realization. However, this method cannot be used to realize an ultimate sensitivity of photoionization detection.

Pulsed laser evaporation and ion sputtering in a vacuum are best suited to local analysis of solids and thin films with a sensitivity of up to $10^{-10}\%$ and a high spatial resolution. This is due to the possibilities of the focusing of laser and ion beams and the application of ion-energy and mass-separation and filtration systems.

Thermal atomization in a vacuum combines the simplicity of experimental realization with the advantages of vacuum evaporation of substance and selective detection of ions. This method of atomization seems to be most universal, since it can be used for direct analysis of almost any class of substances, from high-purity materials to biological objects, with near-ultimate sensitivity. This method calls for a moderate vacuum (10^{-6} Torr) and is characterized by a rapid operation.

The most promising methods of atomization are ones that can be combined with laser stepwise atomic photoionization and make it possible to realize the ultimate sensitivity of photoionization detection. For this reason we have not considered flame atomization here, one of the oldest and most widely used methods, since the ion background of a flame, especially in the presence of a matrix, restricts the analysis sensitivity to a level from 10^{-5} to $10^{-7}\%$.

8.2.3. Registration System

There are two detectors operating in essentially different conditions that are sensitive enough to detect single ions and electrons (see Chapter 4). These are the proportional counter operating at near-atmospheric gas pressure and the secondary emission multiplier operating in a vacuum.

In the first case, the electrons formed by atomic photoionization are detected. The electron avalanche arising can be proportionally amplified by special choice of a gas and accelerating voltage, and with a gain factor of around 10^4 it is possible to detect single photoelectrons.

In detecting the ions by a secondary emission multiplier (SEM) in a vacuum, the scheme of ion detection depends essentially on the method of atomization. The common requirement in this case is the production of a vacuum no worse than 10^{-5} Torr for normal work of the SEM and free motion of the ions from the region of laser beam to the SEM. It is possible to use here both a unified vacuum system and differential pumping for producing a higher vacuum in the chambers of the ion source, the mass separator, etc.

To illustrate all this, we consider a scheme for ion detection at thermal atomization of matter in a vacuum, since it is this method that has been applied in most of the analytical photoionization experiments described below.

The atomic–molecular beam (Fig. 8.2) formed by atomization of matter is irradiated at a right angle simultaneously by all the laser beams in the region between the flat metal electrodes. An ionizing pulse of electric field ($\mathscr{E} = 10$–15 kV/cm, $\tau = 10$–20 ns) is fed to the electrodes 20–50 ns after the atoms are excited by laser to a Rydberg state. The ions formed by field ionization of Rydberg atoms gain in the field of this pulse a normal component to their thermal velocity in the atomic beam. The value of this component is almost two orders of magnitude higher than the average velocity of atoms in the beam, and so the ions move almost normally to the atomic beam axis. After passing through the slit in the electrode, the ions enter the SEM where they are detected. The estimation of the ion extraction efficiency from the interelectrode region shows that it is close to unity. Two types of SEM, 1° with discrete dynodes and 2° with a continuous channel dynode, were used in the experiments. The latter has better characteristics as far as the detection of single ions is concerned (the 1-type efficiency was 0.35–0.5 and less than 0.2 for the 2-type).

The signal from the SEM was processed in the standard manner: it is amplified, fed to the boxcar, and then registered by a recorder or by an amplitude analyzer.

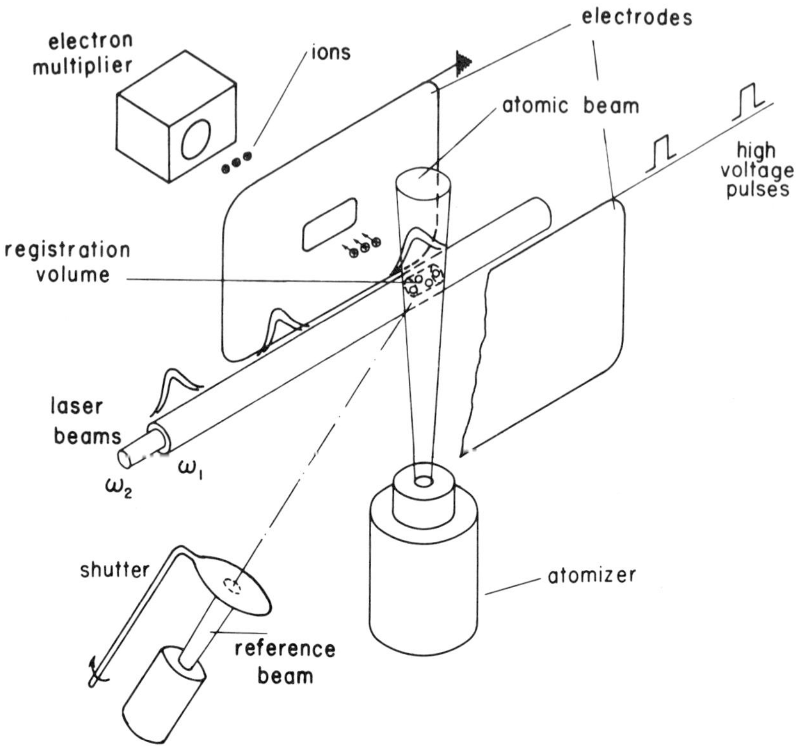

FIG. 8.2. Scheme of laser photoionization detection of atoms in an atomic–molecular beam at the thermal atomization of a substance in a vacuum. A reference atomic beam is used for calibration.

8.2.4. Discrimination of Nonselective Ion Background

In the direct analysis of matter, the atoms of the element to be detected are photoionized among a large number of atoms or molecules of the matrix. This gives rise to a nonselective ion background related in most cases to the processes of nonlinear interaction of the laser radiation with the substance.

The main sources of nonselective ions are the processes that limit the selectivity of photoionization detection (see Section 8.1). This is multiphoton nonresonant laser photoionization of unwanted atoms due to laser radiation absorption on the wings of absorption lines and laser spectral lines. Despite the small value of the cross sections of such processes, the ion background may be significant since the density of the matrix particles in the beam or in the vapor exceeds the density of the atoms to be detected by many orders of magnitude. Besides the ion background conditioned by laser radiation, there is a residual background from the thermal ions of the atomizer that have escaped the ion suppressor. As a result, the total nonselective ion background will restrict at a certain step the sensitivity of analysis. Consideration is given to the methods of separating the ions of a definite element from all the ions formed after the medium to be analysed was exposed to laser radiation.

8.2.4.1. Ion Mass Separation. The ions formed by atomic photoionization have almost the same thermal velocities as the primary atoms. This allows us to form from them by moderate electric fields a collimated pulsed ion beam which can serve as a good ion source for any ion mass separator. The mass resolution and transmission coefficient are the most important characteristics of the mass separator for detecting a small number of ions for analytical applications. Quadrupole and time-of-flight mass separators are most suitable for this purpose.

The basic advantages of quadrupole mass filters are near 100% transmission, which is of great importance for detecting weak ion flows, small dimension, which allows easy combining of the filter to the laser photoionization system, and reliability in operation. Even though quadrupole mass analyzers have a moderate resolution (10^3–10^4), in combination with selective laser photoionization they are very promising for application in analytical spectroscopy of atomic traces.

The time-of-flight mass separator naturally matches the scheme of atom ionization by an electric field via Rydberg states. As it has been said, under the action of a field pulse, the ions move at a right angle to the motion of the atoms in the beam. So in the detection scheme given in Fig. 8.2, the drift in the field-free region can be realized in the same vacuum chamber with the ion detector (SEM) sufficiently removed from the electrodes. It can be easily demonstrated that such a filter is linear with respect to ion mass, and with $\mathscr{E} = 10$ kV/cm, $\tau = 10$ ns, and $L = 50$ cm, the temporal interval between masses $\Delta m = 1$ amu is $\Delta T = 0.5\ \mu$s with the ion peak width being equal to 5 ns. Such an interval is convenient for time gating of signals from different masses.

8.2.4.2. Selection of Rydberg Atoms. Selective ionization of Rydberg atoms under the action of two electric field pulses with definite parameters is a simpler method for suppressing the nonselective ion background and separating the ions of the element under study (Bekov *et al.*, 1984b). The method is based on a sharp threshold dependence of the degree of ionization of Rydberg atoms on the amplitude of the electric field pulse and its weak dependence on pulse duration. For example, for the 17p state of aluminum, the critical strength is equal to 6.2 kV/cm with $\tau_1 = 10$ ns and 6.1 kV/cm with $\tau_2 = 30$ ns. It can be seen that the critical strengths differ slightly, while the pulse duration in the first case is three times shorter than in the second one.

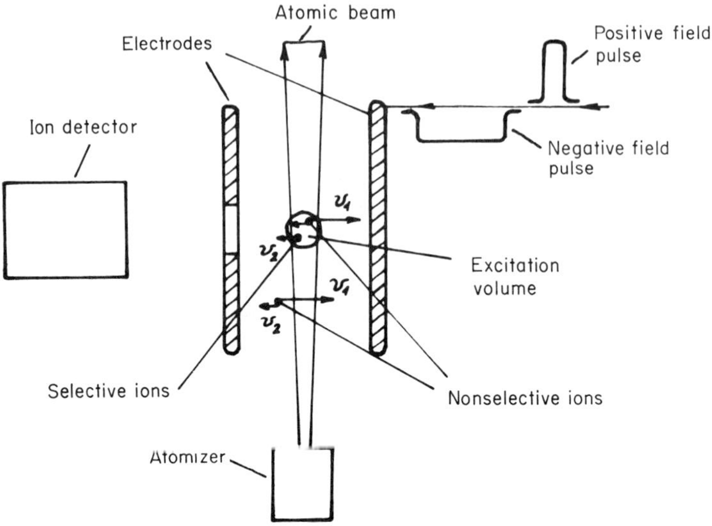

FIG. 8.3. Selection of Rydberg atoms at their two-pulse ionization by the electric field. First, a long negative pulse removes noise ions from the Rydberg atoms area; second, a short positive pulse ionizes Rydberg atoms only. [From Bekov *et al.* (1984b).]

Let the atoms of the selected element in the beam or in the vapor of different atoms (Fig. 8.3) be excited by laser radiation pulses to a Rydberg state in several steps. After the medium is irradiated by lasers, the region between the electrodes will have not only resonantly excited Rydberg atoms of the element being analyzed but also foreign ions: thermal ions and ions formed by laser multiphoton or multistep photoionization of the particles of the residual gas and the matrix. Then a negative electric pulse, with its strength being smaller than the critical one for an excited Rydberg state, is fed to the electrodes. Under its action all the ions between the electrodes acquire velocity components toward the electrode without a slit. The action of this pulse on Rydberg atoms is negligible. Then a positive electric pulse (the delays between the pulses and their durations are much shorter than the Rydberg state lifetime and equal to 10 to 50 ns), with its strength being larger than the critical one and sufficient for full ionization of Rydberg atoms, is fed to the electrode. The amplitudes $\mathscr{E}_{\mathrm{j}}$ and the pulse durations τ_{j} in this case should satisfy the condition

$$\mathscr{E}_{-}\tau_{-} > \mathscr{E}_{+}\tau_{+}. \tag{8.2}$$

The nonselective ions in this case, under the action of the second pulse, just decelerate their motion to the electrode without a slit. The ions of the element under analysis formed from Rydberg atoms will be accelerated in the field of the second pulse, pass through the slit in the electrode, and be detected by SEM. Thus, if only the ions of the selected element are extracted from the excitation zone, the detection selectivity will increase as against the selectivity in one-pulse ionization. According to simple estimations, the coefficient of suppression of nonselective background in two-pulse detection equals 3–4 orders and is limited by the background of the scattered ions outside the interelectrode space.

8.2.5. Calibration of an Analytical Spectrometer

The *ethalon calibration curve* is one of the most widely used methods for the relative calibration of analytical signals. It consists of constructing a dependence of the analytical signal on the concentration of an element in samples of the same substance with a known (standard) amount of this element. The standards are chosen so that the concentration to be found falls within the standard range in use. When the calibrating curve is linear over a wide range, it is possible, as a rule, to approximate the curve to the region of lower concentrations. This method is best suited to analysis of aqueous solutions of salts when the interference effects are minimal.

It is very difficult and sometimes impossible to choose or prepare standard samples with the same matrix content in analyzing complex objects. In these cases the *addition method* is advisable. Its essence lies in constructing a dependence of the analytical signal on the concentration of the element introduced into the substance in the form of a definite compound in a known amount and in approximating this dependence to the zero concentration level. It should be noted that this method becomes inadequate in the presence of strong effects from the interaction between the matrix and the introduced element, as well as when the chemical properties of the natural and introduced compounds of the element differ greatly.

These methods of calibration are applied in laser photoionization analytical spectroscopy as well. Atomization in a vacuum in this case allows the range of application of the addition method to be widened essentially, because in the process of such atomization all the interference effects between the element to be determined and the matrix are reduced considerably.

The use of a *reference atomic beam* is another method of calibration that can be applied in vacuum photoionization spectroscopy. Introduce an additional atomic beam of the element into an analytical vacuum chamber (Fig. 8.2). It can be formed, for instance, by thermal evaporation of the pure element from its metal or amorphous state in an additional oven. If this reference beam passes through the excitation zone, i.e., the region where the laser beams and the main beam intersect, the atoms of the elements are under the same conditions of laser excitation and field ionization no matter which

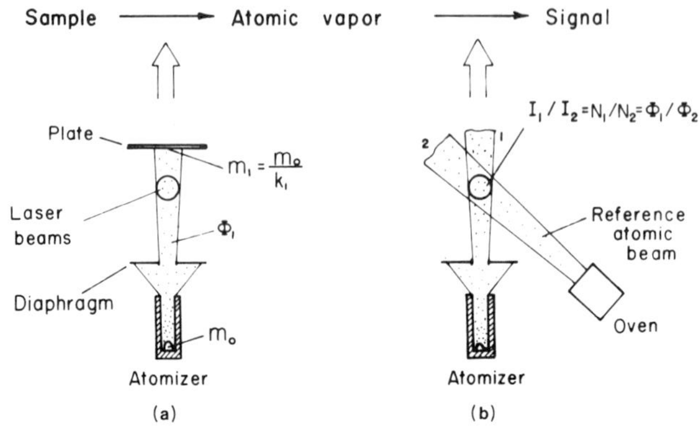

FIG. 8.4. Absolute calibration of a photoionization analytical signal by means of (a) the absolute measurement of atomic flow through the laser excitation volume and (b) comparison of ionization signals from the main and reference atomic beams.

beam they belong to (Fig. 8.4b). Then, if the system of ion registration operates in a linear regime, we can write

$$I_1(t)/\Phi_1(t) = I_2(t)/\Phi_2(t) = \bar{I}_2/\bar{\Phi}_2, \tag{8.3}$$

where $I_{1,2}$ are ion signals, $\Phi_{1,2}$ are the fluences of the atoms to be detected in the main and the reference beams, respectively. When the thermodynamic properties of the element are known, it is possible to calculate the fluence Φ_2 in a standard way (Ramsey, 1966), with the geometry and the temperature of the reference oven known. The material of the reference crucible in the experiment must, of course, be chosen so that the chemical interaction of the atoms of the element with the crucible walls is eliminated. Thus, if we know the ratio of the signals from the main and the reference beams, it is possible to determine the fluence of atoms of the element in the main beam. So, by using the reference beam, we can carry out an absolute calibration of the registration system.

The total analytical signal from the sample is proportional to the area $\int I_1(t)\,dt$ of the selective ion signal. Consequently, the atomic yield of the element from a sample will be expressed as

$$N = k_1 \int \Phi_1(t)\,dt = \frac{k_1\bar{\Phi}_2}{\bar{I}_2} \int I_1(t)\,dt, \tag{8.4}$$

where k_1 is the geometric factor that is equal to the ratio of the total fluence of atoms from the crucible to the fluence of atoms through the excitation zone. This coefficient is determined in model experiments by direct measurement of these fluences under the same conditions of atomization (Fig. 8.4a). Knowing the mass of the sample, i.e., the total number of its particles N_0, and the degree of atomization of the compounds of the element under the experimental conditions, we can calculate the relative content of the element from the formula

$$C = \frac{N}{N_0 \eta_{at}} = \frac{k_1 \bar{\Phi}_2}{\bar{I}_2 N_0 \eta_{at}} \int I_1(t)\, dt. \tag{8.5}$$

The calculation of Φ_2 in this formula is somehow connected with the value of the saturated vapor pressure of the element. However, existing pressure data are rather contradictory for a number of elements. The application of laser stepwise photoionization in the single-atom detection regime enables absolute calibration of the reference beam, too, (Bekov *et al.*, 1978d) (see Section 10.1.1). Such calibration makes it possible to choose the right values of vapor pressure from a variety of data and to use them for correct calculation of Φ_2.

8.3. Laser Photoionization Method in Analytical Experiments

In this section, consideration is given to the basic experiments on diagnostics and determination of impurities in substances that are in different states: from semiconductors to biological objects. Basic emphasis is placed on ultrahigh-sensitivity analysis of complex media.

8.3.1. Analysis of Solutions. Thermal Evaporation in Vacuum

The typical state of a subtance during its analysis by the methods of atomic absorption and atomic fluorescence is a specially prepared solution of the substance being analyzed. As will be seen, the photoionization method does not need such a solution since its high sensitivity enables detecting an extremely weak fluence of atoms evaporated as the sample is heated in a vacuum. Nevertheless, several experiments have been performed on the photoionization analysis of solutions.

The advantages of vacuum atomization were demonstrated in an experiment (Bekov *et al.*, 1983b) in which the detection limit of Yb was 5×10^{-2} μg/l or 1 pg in the sample. This value is smaller than the detection limits of the most sensitive analytical methods. An electrothermal atomizer was used to atomize ytterbium chlorides. The atoms of Yb were ionized through the autoionization state (AS) using the scheme in Fig. 3.17d:

$$\text{Yb: } 6^1\text{S}_0 \xrightarrow[\lambda_1]{555.6\text{ nm}} 6^3\text{P}_1 \xrightarrow[\lambda_2]{452.8\text{ nm}} 6^1\text{D}_2 \xrightarrow[(\sigma_3 = 10^{-16}\text{ cm}^2)]{452.8\text{ nm } (\lambda_3 = \lambda_2)} \text{AS}.$$

The results obtained were not ultimate. The sensitivity could be improved by two orders of magnitude by perfection of the atomizer design and increasing the laser radiation energy.

In analytical investigations, it is often necessary to measure with a high accuracy the isotope ratios in samples. Such tasks become more complicated if the interference of isobars of close elements takes place, because of their incomplete chemical separation. Such a situation takes place for Lu and Yb. These two elements are similar in their properties, and so the process of their chemical separation is very complicated. At the same time, laser photoionization in vacuum allows easy separation of these elements. The selectivity of Lu detection in the presence of Yb obtained by this technique was rather high (Miller *et al.*, 1982). A mixture of Yb and Lu chlorides was applied on a metal filament placed in the region of the ion source of the magnetic mass spectrometer. As the filament was heated to 700°C, the compounds of Lu and Yb were atomized, and the resulting atoms of Lu were ionized in two steps by a laser pulse running parallel to the filament. The laser wavelength ($\lambda = 452$ nm) was tuned to the transition of $^2D_{3/2}(5d6s^2) \rightarrow {}^2D^0_{3/2}(5d6s6p)$. Thus, the described atomization system and the ionizing laser radiation were the ion source for the mass spectrometer.

Under the condition of usual thermal ionization, the signal of Yb from an equimolar mixture of these elements is an order magnitude larger than the signal of Lu. At laser photoionization, due to the absence of accidental one- and two-photon resonances in Yb at $\lambda = 452$ nm, the selectivity of photoionization detection $S \geqslant 5 \times 10^4 = (N_{175}/N_{174})^{1/2}$, where N_{175} is the number of photoions, and N_{174} is the background signal for mass 174. The sensitivity of the method estimated under these conditions gives a value of the Lu content in the sample of less than 1 ng.

8.3.2. Analysis of High-Purity Semiconductor Materials

8.3.2.1. CdS Crystals. Separation of Surface and Bulk Impurities. The CdS crystals grown by the method of resublimation of CdS from the gas phase in an argon flow were analyzed for Na impurities (Akilov and Bekov, 1982). The Na impurities were detected in the following way. A sample of CdS crystal with a mass of several tens of milligrams was placed in a tantalum crucible with an outlet channel 1 mm in diameter and was heated in a vacuum ($p_{res} = 10^{-6}$ Torr) to a temperature of 800 to 1050°C. At this temperature, intense sublimation of the crystal occurs ($t_{melt} = 1475$°C), yielding various atoms and molecules—Cd, CdS, S_2, S_4, Na, etc. A fluence of atoms and molecules formed as a beam was irradiated by two dye lasers pumped by a N_2 laser. The wavelengths of the lasers were tuned for the two-step excitation of Na atoms to the Rydberg state 14d by the scheme of Fig. 3.16c:

$$3^2S_{1/2} \xrightarrow[\lambda_1]{589\ \text{nm}} 3^2P_{1/2} \xrightarrow[\lambda_2]{418\ \text{nm}} 14^2D_{3/2,5/2}.$$

Then the Rydberg atoms were electric-field ionized ($\mathscr{E} = 13$ kV/cm), and the ions were detected by an SEM (see Section 3.3.4). The reference beam method described in Section 8.2.5 was used to calibrate the registered ion signals.

Figure 8.5 presents the dependence of a Na-ion signal I_1 on evaporation time at $t = 940°C$ for a sample of CdS with a mass of 30 mg. The sharp maximum at the beginning of evaporation is conditioned by the evaporation of Na compounds from the polluted surface of the crystal. Similar dependences were obtained for several samples of CdS crystals in the temperature range from 800 to 1050°C. In some cases, the fluctuations of ion signal I_1 were stronger than in Fig. 8.5, which was probably connected with an inhomogeneous distribution of the impurity over the volume of the sample.

The content of the Na impurity in CdS was determined using formula (8.5) for the reference beam method. The value of $\int I_1(t)\,dt$ for each sample of CdS was determined from the area under the ion signal curve. At a definite temperature of the reference beam oven, the ion signal $\bar{I}_2$ was measured, and the fluence $\bar{\Phi}_2$ for this temperature was calculated from the beam geometry (Ramsey, 1966). For the atomizer used, the coefficient k_1 was measured, its value being $k_1 \simeq 50$. The degree of atomization of Na compounds in this experiment was taken to be equal to unity from the following considerations. Crystals with a regular structure and a low level of impurities are cha-

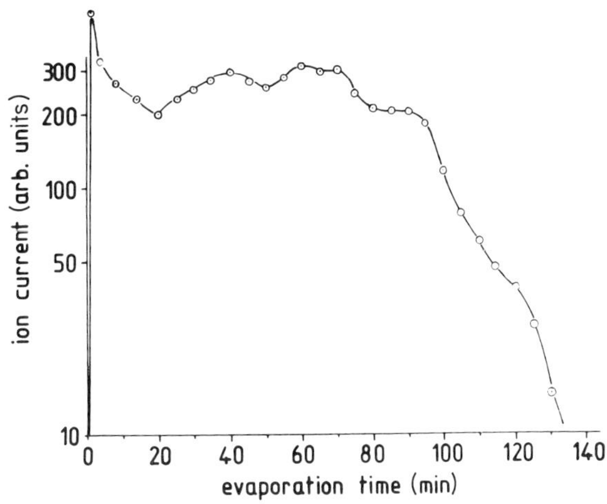

FIG. 8.5. Dependence of the Na ion signal on the evaporation time at a temperature $t = 940°C$ for a Cds sample with a mass of 30 mg. [From Akilov and Bekov (1982).]

racterized by the impurity molecules in them being in a dissociated state. When such crystals evaporate, the impurity molecule dissociates completely, giving rise to free atoms of the impurity element. Thus, the degree of atomization of the Na impurities in CdS crystals $\eta_{at} \simeq 1$.

The concentration values of Na impurities obtained in the experiments for different CdS samples ranged from 2×10^{-6} to $6 \times 10^{-5}\%$. In most cases, the signal-to-noise ratio was about 10^4, which allows us in principle, to reach a registration level below $10^{-9}\%$. The ion background, caused by nonselective multiphoton ionization of the atoms and molecules of the basic substance by the second-step laser radiation, is a factor limiting sensitivity. Nevertheless, the result obtained for Na—$2 \times 10^{-6}\%$ with a large signal-to-noise ratio (10^4)—is almost ultimate for most conventional analytical methods in the direct analysis of substance.

The experiment has shown that at vacuum atomization it is possible to separate surface and bulk impurities through careful evaporation of the substance below the melting point. In this case, first the contaminated surface layers evaporate, and under vacuum conditions their complete yield from the crucible takes place. Then the basic mass evaporates, and the bulk impurities are determined.

8.3.2.2. High-Purity Germanium. Thermal Evaporation in a Vacuum. Some tasks of the technology of high-purity materials, for example, the determination of aluminum, boron, and other group III acceptor elements in semiconductor materials, require control of impurities at a level of 10^{-8}–$10^{-10}\%$. The conventional analytical methods are not able to provide such sensitivity. The possibility of detection of sodium, aluminum, and boron in high-purity crystals of germanium using the method of laser stepwise photoionization in a vacuum was investigated (Akilov *et al.*, 1982, 1984; Bekov *et al.*, 1984b). Sodium was chosen as a model element because the scheme of its effective ionization via Rydberg states had been realized (Bekov *et al.*, 1978a,d). Under the conditions of saturation of all transitions and field ionization, every second atom of Na entering the laser beam became ionized. The experimental technique was similar to the one described in Section 8.3.2.1. The first photoionization experiments in a germanium atomic beam showed that, with the second-step energy fluence being sufficient to saturate the transition to state 3p → 14d, a significant ion background took place. It occured because of three-photon ionization of germanium atoms by laser radiation with $\lambda_2 = 418$ nm. To remove this background, the laser beam was defocused from a diameter of 0.5 to 1.5 mm. As a result, the ionization efficiency of the sodium atoms fell to 0.1, but the ion signal was decreased a little due to a multiple increase in the excitation zone volume. The signal of three-photon ionization, however, dropped by about two orders of magnitude because of a nonlinear dependence of the rate of such ionization on laser radiation intensity.

Figure 8.6 presents the time dependence of the Na ion signal at different temperatures. The increase of the photoionization signal observed with every increase of the crucible temperature (Fig. 8.6) seems to be caused by the atomization of more and more thermostable Na compounds or by the heating of new parts of the atomizer. It must be said that this signal is conditioned mainly by the Na compounds absorbed on the crucible surface, since it was observed every time the chamber lost sealing and the crucible was withdrawn into the air. At a temperature of 1750°C, there was a gradually decreasing background signal caused by the evaporation of Na from the surfaces of the oven components under the action of thermal radiation.

When a crystal of Ge with a mass of 5 mg, the content of Na being $2 \times 10^{-8}\%$ (Fig. 8.6c), was placed in the crucible, the amplitude of all the signals was increased due to an additional contribution of Na impurities

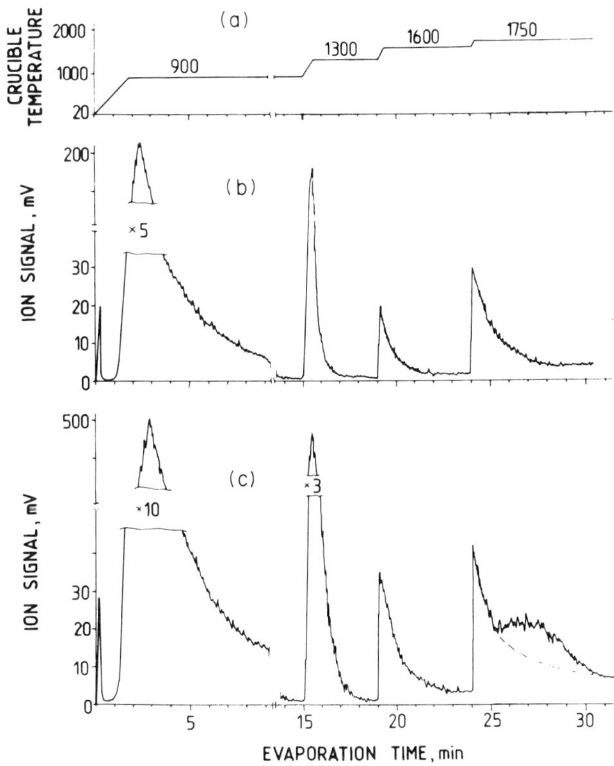

FIG. 8.6. Dependence of the Na ion signal on evaporation conditions. (a) Temperature–time dependence, (b) emtpy crucible, and (c) germanium sample $m = 5$ mg and $C_{Na} = 2 \times 10^{-8}\%$. [From Akilov *et al.* (1982).]

from the crystal surface. The yield of Na bulk impurities took place only at 1750°C when the diffusion rate of Na atoms from the Ge melt increased essentially. The absolute calibration of the registration system by the reference beam made it possible to evaluate from the additional ion signal the amount of Na in the sample. The result agreed quite well with the neutron activation analysis data. The detection limit of Na in Ge obtained in the experiment was about $5 \times 10^{-9}\%$.

The studies carried out show that the method of atomization concerned allows separate detection of surface and bulk impurities even at temperatures higher than the melting point of the substance, provided that the heating regime is chosen properly. This is probably due to the fact that when evaporating in a vacuum the flow of the basic substance from the surface of the melted material prevents the surface impurities from diffusing or being dissolved in the melt.

The next experiment was to determine the impurities of Al in Ge. The two-step scheme of Rydberg state excitation was also used for Al (Fig. 3.16c). The first-step laser radiation pulse with the wavelength $\lambda_1 = 3961.5$ Å excited the Al atoms to the $4^2S_{1/2}$ state from the $3^2P_{3/2}$ state, which is 112 cm^{-1} higher than the ground state. At a temperature of about 1000°C, the populations of the $3^2P_{3/2}$ and $3^2P_{1/2}$ (ground) states are made practically equal, and the laser excitation is more effective from the $3^2P_{3/2}$ state. The second-step laser radiation pulse with $\lambda_2 = 4474$ Å continued to excite the atoms to the Rydberg states $15^2P_{1/2,3/2}$, which were then ionized by an electric field pulse. The ionization efficiency was defined by the excitation efficiency at the last step and came to about 10%. The Al impurities were determined by a technique similar to the one for Na. First the annealing of the sample was carried out at 900°C ($t_{melt} = 937$°C). In contrast to Na, the stage of increasing the temperature to 1500°C was characterized by the absence of a surface aluminum signal. Only above 1500°C was it possible to detect an appreciable signal connected with the atomization of surface Al impurities.

Figure 8.7 shows how the Al ion signal depends on evaporation time at 1800°C for an empty crucible and for Ge samples differing in mass and surface area. As can be seen from Figs. 8.7b and c, there is no essential difference between the signals from samples of approximately the same mass but differing greatly in surface area. The signal from a sample (Fig. 8.7d)—the mass of which is ten times larger than, and the surface area is close to, that of the sample (c)—is much stronger than the signals from the samples (b) and (c). These results point to the fact that the recorded signal is caused mainly by the Al atoms from the volume of Ge rather than by the surface impurities. In the case shown in Fig. 8.7d, the selective signal-to-nonselective noise ratio was equal to about 10^2, which allowed achieving a detection limit of Al in Ge at about $10^{-9}\%$.

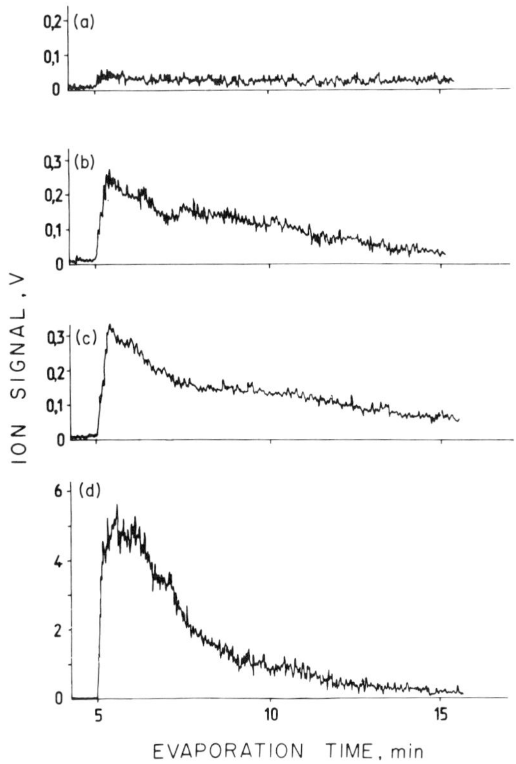

FIG. 8.7. Dependence of the Al ion signal on evaporation time at 1800°C for (a) an empty crucible; (b) germanium sample, $m = 2.75$ mg, surface area $S = 4\ mm^2$; (c) germanium sample, $m = 2.5$ mg, surface area $S = 18\ mm^2$; and (d) germanium sample, $m = 28.2$ mg, surface area $S = 20\ mm^2$. [From Akilov *et al.* (1982).]

8.3.2.3. Traces of Boron in High-Purity Germanium. Rydberg Selection. The first experiments on direct laser photoionization analysis of Ge for boron impurities revealed the presence of a strong nonselective ion background caused by Ge atoms. The analysis of this signal showed that there was two-step photoionization of Ge atoms by the laser radiation used to ionize B atoms. This process limits the selectivity of detection of B impurities in Ge to the level $S = 3 \times 10^7$. Actually, Fig. 8.8 presents the simplified energy level diagrams of boron and germanium. It also shows the transitions used for the photoionization of boron atoms through Rydberg states. It can be seen that the germanium atom has an absorption line $\lambda_1(\mathrm{B}) = 249.8$ nm, very close to the boron line $\lambda_1(\mathrm{Ge}) = 249.68$ nm, that is conditioned by the intercombination transition ${}^3\mathrm{P}_0 \rightarrow {}^1\mathrm{P}_1^0$. The absorption of the first-step laser UV radiation at the wing of the Ge line causes the Ge atoms to be

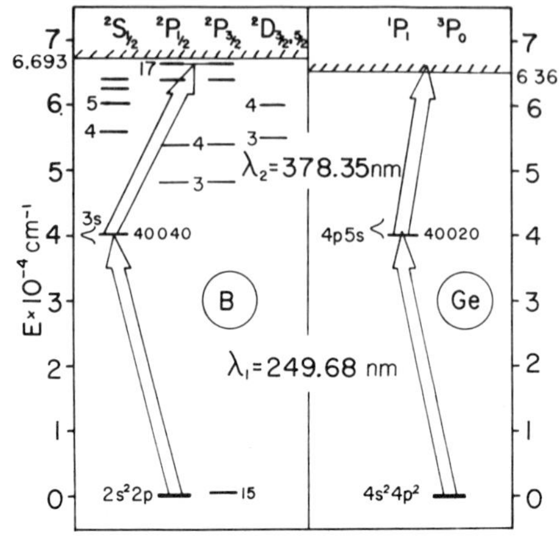

FIG. 8.8. Diagram of the two-step resonant ionization of B atoms in the presence of a great number of Ge atoms which can be ionized also via the quasi-resonant 4p5s state. [From Bekov *et al.* (1984b).]

excited to the $4s^24p5s\,^1P_1^0$ state. The quantum energy of the second-step laser $\lambda_2(B) = 378.3$ nm used to excite the Rydberg 17p boron state is sufficient to photoionize the germanium atoms from the $^1P_1^0$ level. It is also possible to photoionize them by a pulse at $\lambda_1(B)$, but due to the small energy of this laser pulse, the contribution of such a process will be small compared with $\lambda_2(B)$.

To increase the detection selectivity, the method of ion background discrimination considered in Section 8.2.4b was used. The boron atoms were ionized according to the scheme

$$2^2P_{1/2} \xrightarrow[\lambda_1]{249.68\ \text{nm}} 3^2S_{1/2} \xrightarrow[\lambda_2]{378.35\ \text{nm}} 17p \xrightarrow[\text{pulse}]{\mathscr{E}\text{-field}} B^+.$$

The energy parameters of laser radiation in the experiment (Bekov *et al.*, 1984b) provided on ionization efficiency of the B atoms of about 5%. The electric field pulses had the parameters $\mathscr{E}_- = 5$ kV/cm, $\tau_- = 30$ ns, $\mathscr{E}_+ = 7$ kV/cm, and $\tau_+ = 10$ ns. The technique of ion detection and the basic elements of the experimental setup were similar to those described before.

In the experiment, a Ge sample with a mass of about 10 mg was put into a graphite crucible annealed before at $t \simeq 2500$°C. The germanium was then evaporated in a vacuum ($p = 2 \times 10^{-6}$ Torr) at about 1700°C. Since the thermodynamic properties of Ge and B differ greatly, it was mainly Ge evaporation at such a temperature. This led to a gradual increase of B concentration in the remaining Ge melt. After 20–30 min the crucible temperature was increased to 2200°C for intense evaporation of the rest of the germanium and boron. Figure 8.9 shows the dependences of the B ion signal on the evaporation time at $t = 2200$°C for an empty crucible and two

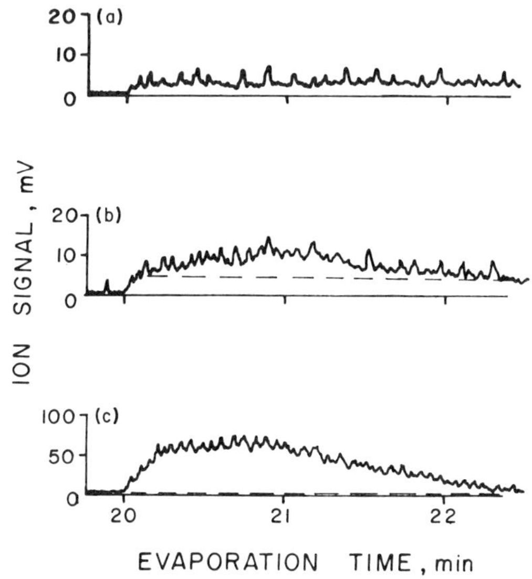

FIG. 8.9. Dependences of the B ion signal on evaporation time at 2200°C. (a) Empty crucible background; (b) germanium sample, $m = 9.5$ mg, $C_B = 2 \times 10^{-8}\%$; (c) germanium sample, $m = 10$ mg, $C_B = 2 \times 10^{-7}\%$. [From Bekov *et al.* (1984b).]

Ge samples with a mass of about 10 mg containing B impurities of 2×10^{-8} and $2 \times 10^{-7}\%$ (these concentration values were obtained from electrophysical measurements of the corresponding samples). The comparison of the signals in Fig. 8.9b and c shows that the ratio of B ion yields in these cases is close to 10, i.e., it coincides with the results of preliminary certification. Thus the recorded signal is conditioned by the bulk impurities of B rather than by its surface impurities. The absolute calibration of the registration system was done as before by the reference beam of B atoms. This allowed estimating the integral signal for case (c). It was found to be consistent with the complete yield of B from the Ge sample. All this confirms the correctness of the analysis carried out.

The B impurities left in the crucible after annealing were a factor limiting the ultimate sensitivity of analysis in the experiment. Since B is widely spread in nature, it is necessary to choose graphite with a low content of B for crucibles. An enhancement of the ionization efficiency of boron atoms by increasing the energy of laser pulses will enable under the some conditions of atomization realizing the detection limit of B in Ge at $10^{-9}\%$.

The experiments carried out show that the method of laser stepwise atomic photoionization in combination with atomization of substances in a vacuum holds much promise for determining the content of impurities in high-purity substances. The method allows direct analysis of natural samples without

TABLE 8.1. Results of Laser Photoionization Analysis of Germanium at Thermal Atomization in a Vacuum

Element	Concentration (wt.%)	Detection limit (wt.%)
Na	2×10^{-8}	5×10^{-9}
Al	10^{-7}	10^{-9}
B	2×10^{-8}	5×10^{-9}

special preparation. The experiments illustrate the possibility of separate registration of surface and bulk impurities since the diffusion of surface impurities inside the melted sample decreases considerably at vacuum evaporation. The results of the photoionization determination of Na, Al, and B impurities in Ge are given in Table 8.1

8.3.2.4. Silicon. Laser Evaporation in Buffer Gas. The pulsed laser evaporation (ablation) of Si was used together with synchronized laser photoionization of Na atoms in the resulting cloud of substance (LARIS) (Mayo *et al.*, 1982). The processes of evaporation of substance, atomic photoionization, and electron detection were carried out in a buffer gas that contained 90% argon and 10% methane. This gas was a working medium of the proportional counter that was used to detect the resultant photoelectrons. Evaporation was performed by a pulse of flash lamp-pumped dye laser with the parameters $\lambda = 590$ nm, $\tau_p = 0.6$ μs, and $E = 1$–10 mJ. Only the central part of the beam with a Gaussian intensity profile was used, and this allowed focusing the radiation onto a spot with a diameter of up to 30 μm. The intensity from 10^7 to 10^8W/cm^2 in the spot was optimal for evaporation.

After the evaporated plume of substance was thermalized as a result of collisions with the buffer gas particles, the Na atoms were photoionized according to the scheme (Fig. 3.17a):

$$3^2S_{1/2} \xrightarrow[\lambda_1]{589.6\text{ nm}} 3^2P \xrightarrow[\lambda_2]{568.8\text{ nm}} 4^2D \xrightarrow{\lambda_1,\lambda_2} Na^+.$$

The estimated ionization efficiency was close to 1. The sensitivity of the photoionization system allowed the detection of a concentration of Na atoms in the buffer gas of no less than 10 atoms/cm^3. Several samples of Si differing in purity were investigated. The experiments have shown that at intensities of 10^7–10^8 W/cm^2, 10^{-8} cm^3 of silicon crystal evaporate, which corresponds to 5×10^{14} Si atoms. The density of silicon atoms in the gas plume was estimated in this case to be 10^{14} cm^{-3}. In the process of detection, one point was probed by several evaporating pulses. The first few pulses served to clean any accidental surface contamination, and the others were determined. For purest silicon, the minimal signal corresponded to 500 Na atoms in the excitation volume or 5×10^{11} cm^{-3} in the volume of Si. When the laser result was obtained, it was assumed that (1) the influence of surface contamination was eliminated, (2) the plume of substance was fully thermalized by the moment of photoionization, and (3) the concentration of reactive species in the buffer gas was not high enough to cause any significant decay of the concentration of the resulting Na atoms.

It can be seen that very essential assumptions on the character of all processes are needed in order to interpret quantitatively the obtained results. The simplest way to their elimination is to use standard samples of Si with a definite concentration of Na. However, even in this case of evaporation in a buffer gas, the interference effects arising between the atoms of the element to be analyzed, the matrix, and the impurities in the buffer gas complicate considerably the calibration of the signals and their interpretation. Under such conditions, the approximation of the calibration curve to the low-concentration region may turn out to be incorrect.

8.3.2.5. Silicon. Ion Sputtering in a Vacuum. Another version of the photoionization method for the analysis of solids (called SIRIS by the authors) that is free of the shortcomings listed above uses ion sputtering in vacuum (Parks *et al.*, 1983). A pulsed beam of argon ions with the parameters $E = 5.30$ keV, $\tau_p = 2\ \mu$s, and $f = 30$ Hz was used in the analytical experiment for sputtering silicon. It was formed from a continuous ion source with a special system of modulation and filtration. The ion beam intersected the target plane at 60° with respect to the normal, thus forming a 2.6×3.3-mm^2 spot on the surface. The target system only provided the escape of neutral particles into the region of detection. The resultant atomic–molecular cloud was irradiated by laser radiation tuned for stepwise photoionization of the atoms of the element under study. The resulting ions were then subjected to mass filtration using a mass spectrometer with double focusing that comprised an electrostatic sector for analyzing the ion energy and a magnetic sector for mass analysis. After magnetic separation, the ions were detected by a channel electron multiplier equipped with a special conversion electrode. To suppress the extraneous background, time gating of the electronics was used.

The potential of the method was studied by detecting the impurity of gallium in silicon with a concentration of 10^{-4}%. The atoms of Ga were photoionized according to the following scheme (Fig. 3.18b):

$$4p^2P^0_{1/2,3/2} \xrightarrow{2\lambda_1} 4d^2D_{3/2,5/2} \xrightarrow{\lambda_2} Ga^+.$$

The wavelength λ_1 corresponded to three first-step transitions: 5750.16 Å, 5890.07 Å, and 5888.99 Å. The spectrum of the ^{69}Ga and ^{71}Ga isotopes was recorded in the experiment without extraneous mass peaks. At the same time, analysis of silicon by the method of secondary-ion mass spectrometry was performed. In this case, there was observed an additional mass peak of GaH^+ formed in the process of ion sputtering.

Although this experiment was preliminary and qualitative in nature, nevertheless, it has demonstrated the wide potential of the new method which enables us now to realize a detection sensitivity of impurities on a level of about $10^{-7}\%$. The realization of all the technical reserves of the SIRIS apparatus will allow the sensitivity of 1 part in 10^{-12}.

8.3.3. Analysis of Objects of Marine Environment

The next important step in the investigations into the analytical capabilities of the photoionization method were the experiments on the detection of trace elements in a complex matrix such as natural water. For this reason, we discuss them in greater detail.

8.3.3.1. Aluminum in Natural Waters. The development of new highly sensitive methods of detection of microelements, and aluminum particularly, in natural waters is dictated by the following reasons. Aluminum, one of the most important elements in the lithosphere, is used in the lithology and chemistry of the ocean as an indicator of terrigenous runoff. Besides, there are some results (Mackenzie *et al.*, 1978) relating aluminum dissolved in seawater to certain biological processes. Thus the correct quantitative estimation of the participation of aluminum in the processes taking place in the ocean needs a universal method that allows determining aluminum in waters with different mineralization and a different content of organic substance that can form with aluminum-soluble metallo-organic complexes.

At present there is not a direct method and a universally adopted technique for the determination of dissolved aluminum in natural waters. There are a lot of methods that differ from each other by the application of different chemical reagents, which may contaminate the samples and act as an undesired background. The absence of reliable data on dissolved aluminum in sea and ocean waters and the presence of just single data for interstitial waters point to the fact that these methods are complicated and unsatisfactory. The merits of the method of laser stepwise atomic photoionization revealed by analysis of impurities in high-purity substances (see Section 8.3.3) stimulated the application of this method to the analysis of natural waters, which are more complex objects (Bekov *et al.*, 1983a).

The procedure of aluminum determination in natural waters consists of the following. A sample of water with a volume of 40 μl was introduced into a tantalum or graphite crucible and evaporated in the air at 90°C to a dry residuum. Then the crucible with the contents was placed in an atomizer, the vacuum chamber was evacuated to a residual pressure of 10^{-6} Torr, and the crucible was heated by steps to 1750°C. The further technique of aluminum signal detection was analogous to the one described in Section 8.3.2.

Figure 8.10 shows a typical dependence of the ion signal on evaporation time at 1750°C for 40 μl of seawater. During 0–5 min, the crucible was heated stepwise to 1500°C. The step heating was due to the fact that at fast ashing of the dry residuum of seawater, considerable gasing and vacuum deterioration take place. In this temperature range the aluminum ion signal was almost absent. To discriminate the selective signal of Al from the total ion signal, the first-step laser wavelength was periodically tuned off by 3–5 cm^{-1} from the Al transition $3^2P_{3/2} \xrightarrow{\lambda_1} 4^2S_{1/2}$. The dips in Fig. 8.10 correspond to the detuning of λ_1, which causes the ionization efficiency of Al atoms to drop by more than two orders of magnitude (in the experiment the ionization efficiency was equal to about 0.1). Such a procedure allowed discriminating the nonselective ion background of the seawater matrix from the ion signal conditioned only by aluminum. The nonselective background is probably

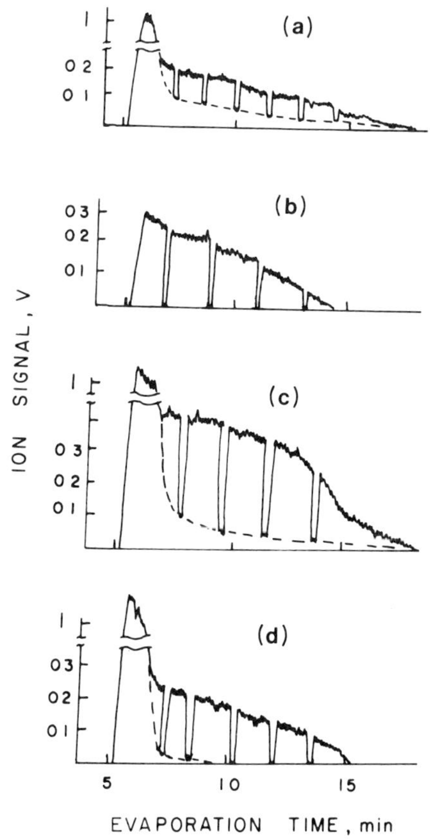

FIG. 8.10. Ion signal dependence on evaporation time at 1750°C for 40 μl of sample. (a) Seawater; (b) $AlCl_3$ calibrating solution with Al concentration of 100 μg/l; (c) mixture of seawater (9 parts) and $AlCl_3$ solution with Al content of 1000 μg/l (1 part); and (d) aqueous solution with a concentration of NaCl of 2% (2×10^7 μg/l) and Al of 100 μg/l. [From Bekov *et al.* (1983a,c).]

conditioned by the thermal ionization of easily ionized seawater components and the laser photoionization of the compounds constituting the seawater matrix. The yield of Al in this case is defined by the area lying between the curve of the total ion signal and the dashed line.

In quantitative measurements, the universally accepted technique is to construct an analytical calibration curve. The calibration solutions in this case must be prepared on the basis of the matrix being analyzed. First the aluminum yield from solutions with Al content known as investigated. The solutions were prepared from aluminum chloride and deionized water. The typical time dependence of the ion signal for the calibrating solution of $AlCl_3$, with the Al content being 100 μg/l, is presented in Fig. 8.10b. As can be seen, for a "pure" solution the ion signal is conditioned only by aluminum. Its total yield is defined by the area under the ion signal curve.

With the use of the additions method, it was shown that the seawater matrix composition does not distort the results of the determination of aluminum content. For this purpose, a solution was prepared that consisted of nine parts of seawater and one part of a calibrating solution of $AlCl_3$ with an Al content of 1000 μg/l. This corresponded to an additional concentration of Al in seawater of about 100 μg/l. The dependence of the ion signal on evaporation time for 40 μl of such a mixture in the same conditions is shown in Fig. 8.10c. The total signal of aluminum defined by the "selective" area under the signal curve is equal to the sum of the selective signal of aluminum from seawater and the Al signal from the solution with Al concentration of 100 μg/l. The nature of the nonselective ion signal in seawater was studied using an aqueous solution that contained 2% of NaCl (this concentration of NaCl is typical of seawater) and aluminum with a concentration of 100 μg/l (Fig. 8.10d) which shows that the yield of Al here is similar to that in the solution of $AlCl_3$, i.e., NaCl does not affect the yield of Al. The nonselective ion signal conditioned by the ionization of NaCl differs from the nonselective ion signal for seawater in time behavior. This signal in the interval 5–7 min is caused by the ionization of NaCl, and the prolonged nonselective signal is defined by a more complex composition of the seawater matrix.

On the basis of an $AlCl_3$ solution in deionized water, an analytical calibration curve was constructed. It turned out to be linear in the concentration range from 5 up to 10^3 μg/l (Fig 8.11). Above 10^3 μg/l, the calibration curve becomes nonlinear because of saturation of the registration system. The lower limit of the calibration curve was defined by the Al content in deionized water (5 μg/l). The detection limit of Al in these experiments, determined by extrapolating the analytical curve to the background level, was equal to 0.01 μg/l. An analytical curve was constructed on the basis of seawater with aluminum chloride additions. The curves turned out to be parallel, which indicates that it is possible to use the calibration curve on the basis of $AlCl_3$

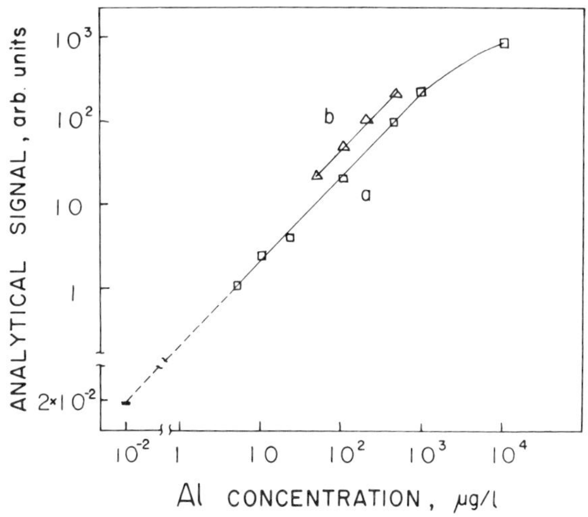

FIG. 8.11. Analytical calibration curves for the photoionization analysis of Al traces in seawater, constructed on the basis of deionized water (curve A) and seawater (curve B) by means of the additions method. [From Bekov *et al.* (1983a,b).]

solutions in quantitative measurements of aluminum in the seawater. The nonselective ion background affects only the value of the minimum detectable concentration of Al in seawater, since it is difficult to discriminate a small selective signal against the background of a strong nonselective one. In the experiments, this value was equal to 1 μg/l, which corresponded to the absolute amount of Al in the sample of 40 pg.

The value of the relative standard deviation was determined from 20 parallel measurements carried out for seawater. It was equal to 5.6%. The largest contribution to this value was made by the volume fluctuations of the sample introduced in the crucible.

Also, the concentrations of dissolved aluminum in the pore waters of the Indian Ocean were determined. The pore water was squeezed out of sediment monoliths (10–15 cm below the bottom) that were picked with a bottom scoop and filtered through 0.45-μm membrane filters. The content of dissolved aluminum increases in the regions of subsea ridges, which is conditioned by the formation of metalliferous sediments there. Such an increase is particularly characteristic in the "triple point" region, i.e., the point of intersection of three submarine ridges, where the process of subsea orogenesis is probably not finished yet. The results obtained agree well with the known data (Caschetto and Wollast, 1979a).

The content of dissolved Al in the surface waters of the Mediterranean has also been measured. The value obtained (6.5 ± 0.7) μg/l is close to the average concentration of Al in the Mediterranean Sea (Caschetto and Wollast, 1979b).

8.3.3.2. Ruthenium in the Ocean. The distribution of the rarest elements in the ocean is of interest not only for oceanology but also for geology, geochemistry, and cosmochemistry. Its study enables us to understand the processes of lithosphere formation. Among these elements is ruthenium, which has not been studied yet in ocean objects because of its ultralow content and the absence of rather sensitive analytical methods for Ru detection.

For the lack of a sufficient amount of analytical data on ruthenium, its abundance in the earth's crust has not finally been ascertained and, according to the estimations, ranges from 10^{-8} to $10^{-9}\%$. The best results, with detection limits of 10^{-6}–$10^{-7}\%$ in determining the ruthenium content in natural objects, were obtained by conventional analytical methods only after preliminary chemical or fire assay concentration of ruthenium from large samples 100–500 g in mass (Jäger, 1978). A complex multistage procedure of chemical treatment of samples and the volatility of ruthenium oxides make the method of chemical concentration labor consuming and not quite reliable. More convenient are the methods of fire assay concentration of the platinum metals and ruthenium, particularly, into copper, lead, or tin alloys.

Laser photoionization spectroscopy with vacuum thermal atomization is very promising for the analysis of such alloys. Concentration in this method allows solving the problem of sample representativeness. The point is that if the sample has a typical mass of 10–100 mg used in vacuum atomization, and the content of element in the substance is small, the amount of element in samples of such a mass fluctuates over very wide ranges. A comparatively large sample is needed to exclude these fluctuations. For example, a sample 10–50 g in mass is representative for geological objects and concentrations of 10^{-9}–$10^{-11}\%$.

Laser photoionization spectroscopy with vacuum thermal atomization in combination with fire assay concentration was used to detect ruthenium traces in a wide class of ocean species (Bekov *et al.*, 1984a, 1985).

In connection with a wide-ranging variety of objects and an ultralow content of ruthenium in them, fire assay concentration to a lead alloy was used. This procedure comprised a fusion of 25 g of the substance in powder form with lead oxide and additional fluxes at 1000°C (Haffty *et al.*, 1977). In this case, ruthenium is collected into a metal lead alloy (lead button) 30–40 g in mass. After mechanical separation of slags, the lead button was reduced to 50–100 mg by oxidation in magnesite cupells in an air flow at a temperature of 900–1000°C. Such a reduction of the lead button does not lead to losses of ruthenium (Kolosova, 1982). The concentration factor obtained in this way ranged from 10^2 to 10^4. In the analysis of waters, the moisture was first evaporated, and then the fire assay enrichment of the dry salt residuum was carried out.

The lead bead produced after cupellation was put into the atomizer crucible and after the vacuum chamber was evacuated, the crucible was heated to 900°C. At this temperature the lead was evaporated, while the ruthenium was concentrated in the rest of the lead alloy because of its extremely low volatility. In this way the matrix was evaporated for 15 to 40 min, depending on the lead mass. The sample remainder was atomized by the stepwise increase of the crucible temperature to 2400°C.

The photoionization detection of the ruthenium atoms in the beam formed was carried out at the field ionization of the Rydberg state with the effective principal quantum number $n^* = 16$. The three-step excitation of Ru was performed according to the scheme (Fig. 8.12):

$$4d^7 5s(^5F_5) \xrightarrow[\lambda_1]{393.6\ \text{nm}} 4d^5 5s5p(^7D_4^\circ) \xrightarrow[\lambda_2]{633.05\ \text{nm}} 4d^7 6s(^5F_5) \xrightarrow[\lambda_3]{567.3\ \text{nm}} 4d^7 np \qquad (n^* = 16).$$

The technique of ion detection is similar to that described above for aluminum.

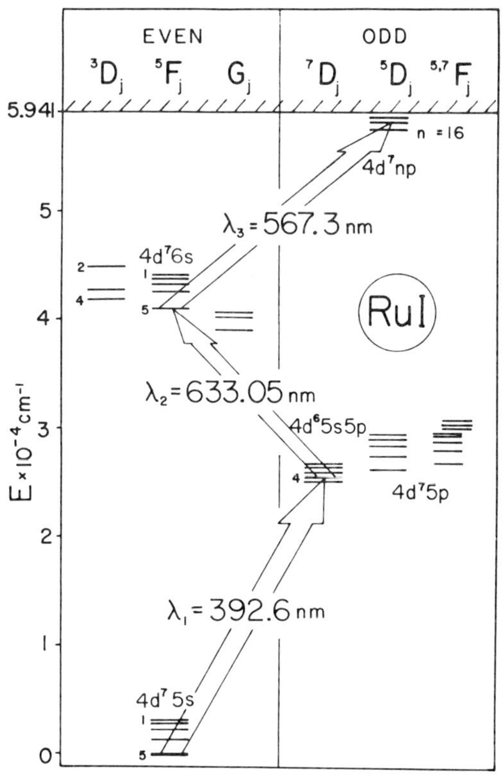

FIG. 8.12. Simplified energy level diagram of the Ru atom and laser transitions for the excitation of Rydberg states. [From Bekov *et al.* (1984a).]

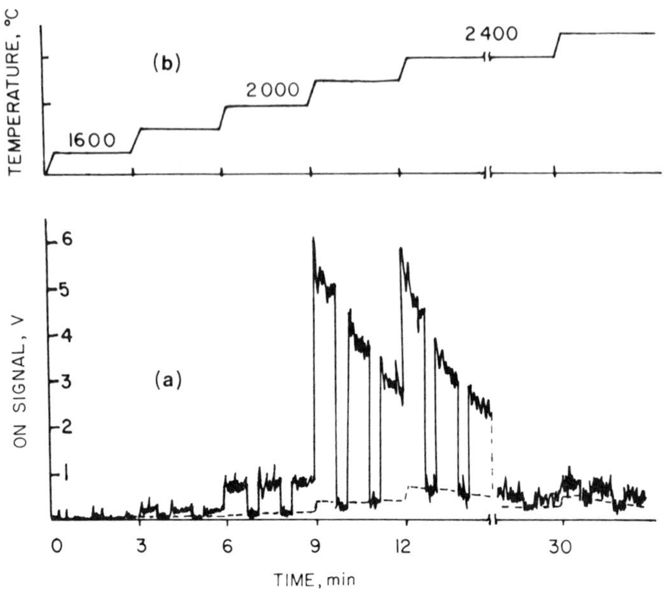

FIG. 8.13. Measurement of an ultralow concentration of Ru. (a) Typical dependence of the Ru ion signal on the evaporation time and atomizer temperature for a lead bead of 50–100-mg mass and (b) time dependence of the atomizer temperature. [From Bekov *et al.* (1984a).]

Figure 8.13 presents a typical dependence of the Ru ion signal on evaporation time and atomizer temperature. The content of ruthenium in the sample is defined by the integral selective signal. It was discriminated from the total ion signal by tuning off the frequency of one of the lasers from the resonance with the corresponding transition. The ion signal was calibrated with a calibration curve constructed on the basis of copper standards that contain a definite amount of ruthenium. The possibility of such calibration was confirmed by the coincidence of the analytical photoionization signals from copper and lead alloys in which ruthenium was collected from the same amount of standard ruthenium-bearing ore.

Table 8.2 describes a number of objects investigated and the results obtained, which are the first data on the content of ruthenium in these objects. Ferromanganese nodules are an interesting group of ocean formations enriched in iron, manganese, and heavy metals (Mero, 1965). We can see that they are characterized by rather a high content of ruthenium.

Of particular interest is the high content of ruthenium in the sample of aerosol taken in the central part of the North Atlantic at the latitude of the Sahara. This concentration is almost three orders higher than in ocean water and four or five times higher than in pelagic bottom sediments. If such

TABLE 8.2. Ruthenium Content in Objects of Ocean Environment[a]

Sample	Location	Ru concentration[b] ($10^{-10}\%$)
Clayey-siliceous ooze	Southern Indian Ocean	150
Metalliferous sediment	East Pacific Rise	1700
	Red Sea, Discovery Deep	7600
Ferromanganese module	Northern Pacific	3500
	Southern Pacific	1600
Phosphorite (Pleistocene)	Mid-Pacific Mountains	170
Fish bones	Chile shelf	56
Ocean water	Bay of Biscay	1.3
Aerosol	Northern Atlantic, off the Sahara coast	910

[a] From Bekov *et al.* (1984a).
[b] The values are given with an accuracy of $\pm 15\%$. The detection limits are equal to $10^{-10}\%$ for solid samples (25 g) and $3 \times 10^{-12}\%$ for seawater (1 1).

enrichment in ruthenium is characteristic of ocean aerosols as a whole, this could be an argument in favor of the hypothesis on cosmogenic origin of a considerable part of the ruthenium in the ocean, since many cosmogenic materials (dust, meteorites, etc.) are enriched in ruthenium (Crocket, 1972; Wedepohl, 1968).

The measurements performed show that the combined photoionization method has undoubted promise, and since there are no fundamental limitations on the fire assay concentration of other platinum metals, the method can be applied to detect traces of any platinum-group elements in natural materials.

8.3.4. Analysis of Biological Objects

One of the urgent analytical tasks for biology and medicine is to determine trace amounts of metals in biological objects. Such analysis results in determining normal concentrations of metals in the organism as well as correlation between their deviations from normal level and different functional disorders and diseases.

Biological objects are especially interesting for the application of the photoionization method since they reveal its essential merit: insensibility to other elements except the element being analyzed. This means that no preliminary treatment of the sample is needed.

This was confirmed by the experiments on photoionization detection of Al traces in human blood (Bekov *et al.*, 1984c). The choice of Al is connected not only with the fact that it can be easily detected, but also that it is one of the metals of interest for toxicology. The role of this element in the metabolism of living organisms remains to be determined. The point is that it is difficult to produce "aluminumless" diets, because of the element's wide abundance in nature. Some lung and nervous system diseases were confirmed to be caused by an increased content of aluminum in an organism (Berman, 1979).

The analytical procedure of direct determination of the Al content in blood consisted in the following: a sample of blood of 40 μl in its natural form was introduced into a tantalum crucible and was dried in the air at a temperature of 90–100°C for 3 to 5 min. The processes of ashing and atomization of the dry residuum took place in a vacuum chamber. It is necessary to find heating regimes such that ashing does not cause the vacuum to be deteriorated below 10^{-4} Torr. At the same time, this process must go quickly enough in order not to lead to significant thermal evaporation of the residuum without its atomization. The crucible temperature during ashing in the experiment was increased to 1500°C in five steps for 10 min. The technique of the detection of Al atoms is described in detail in the foregoing sections.

Figure 8.14a shows a typical time dependence of the ion signal during stepwise heating of the crucible. The procedure of wavelength detuning described in the previous section was used to discriminate the selective signal

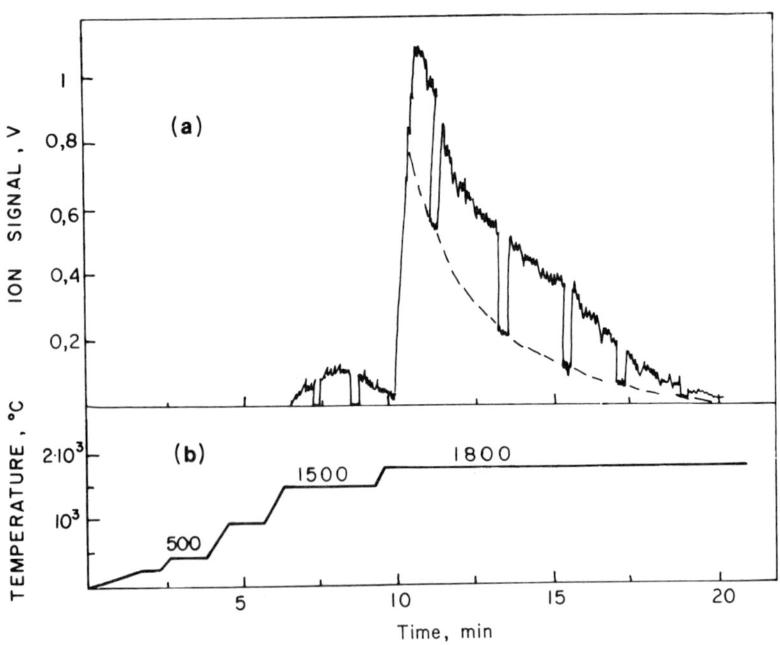

FIG. 8.14. Measurement of Al traces in blood. (a) Dependence of the Al ion signal on evaporation time for 40 μl of human blood and (b) stepwise increase of crucible temperature. [From Bekov *et al.* (1984c).]

of aluminum. The total Al signal for a sample was defined by the "selective" area under the signal curve (the difference between the total signal and the background) (Fig. 8.14a). The value of Al concentration corresponding to such a signal was determined by the calibration curve constructed on the basis of $AlCl_3$ aqueous solutions. The possibility of such calibration was examined by the additions method. In this case, 40 μl of blood and 40 μl of $AlCl_3$ solution with an Al concentration of 100 μg/l were introduced into the crucible. The aluminum signal from such a mixture within the measurement error (about 10%) turned out to be equal to the sum of signals from the mixture components at their independent analysis. This fact proved the absence of effects of the blood matrix on the aluminum yield at thermal atomization in a vacuum. The results of measurements of the aluminum content in five blood samples lie in the range of 230 ± 50 μg/l.

8.4. Comparison of Analytical Methods for Atomic Traces

The experiments on laser photoionization detection of element traces in different media described in the above section point conclusively to the merits of vacuum atomization of matter. The results obtained under nonoptimized experimental conditions are ultimate for the most sensitive conventional analytical methods. The comparative characteristics of different analytical methods are given in Table 8.3. The method of laser photoionization spectroscopy has reserves to reduce the atomizer design, increase the laser photoionization efficiency and selectivity, remove the nonselective ion background, and so on.

This is borne out by the recent successful experiments on the resonance photoionization detection of plutonium traces (Krönert *et al.*, 1984). Knowledge of the ecological behavior of radiotoxic materials is essential to the minimization of the hazard associated with the use of atomic energy. Of

TABLE 8.3. Comparison of Various Methods of Trace Element Determination

Method	Detection limit in aqueous solutions (%)	Detection limit in real matrix (%)	Element selectivity
Atomic absorption spectrometry	10^{-4}–10^{-9}	10^{-4}–10^{-7}	Moderate
Neutron activation analysis	10^{-5}–10^{-9}	10^{-5}–10^{-9}	Moderate
Spark mass spectroscopy	10^{-5}–10^{-8}	10^{-5}–10^{-7}	High
Laser fluorescent spectroscopy	10^{-6}–10^{-11}	10^{-5}–10^{-8}	High
Laser stepwise photoionization spectroscopy	10^{-11}–10^{-14}	10^{-8}–10^{-12}	Ultrahigh

special interest is the behavior of actinides, plutonium in particular, in the biosphere, for example, in the vicinity of atomic power plants. The alpha-spectroscopy and mass-spectrometry techniques currently employed for Pu detection are not sensitive enough. The method of three-step photoionization of Pu atoms, in conjunction with the mass-spectrometric detection of the photoions produced, is ideally suited for this purpose. The use of high-repetition frequency dye lasers pumped by a Cu laser provides for the necessary Pu detection sensitivity.

The experiments performed confirm the universal character of the method of laser photoionization spectroscopy in combination with vacuum thermal atomization of a substance and open up wide prospects for using it as a new analytical method. Besides, laser stepwise photoionization of atoms in a vacuum can be combined with other methods of atomization and allows direct combination with a mass spectrometer and different methods of discriminating selective ions.

Considered in the chapter has been primarily the most difficult case of detecting trace atoms in a condensed matrix, where it is necessary to overcome the problem of maintaining the accuracy of analysis when atomizing the sample. Of course, it is much simpler to analyze gas media. Let us mention some work in this direction: spatial diagnostics of a Ga atomic beam by the two-step photoionization method (Tursunov and Eshkobilov, 1982), and detection of hydrogen atoms in a flame by the method of four-photon ionization at 365 nm (Tjossen and Cool, 1983). In the case of gas media, the method makes it possible to detect molecular traces as well. The photoionization detection of molecular traces has already been discussed in various sections of Chapter 6.

In conclusion, we should emphasize again that the merits of the laser photoionization method—(1) the possibility of single-atom detection in the volume of interaction with laser radiation, (2) the possibility of direct analysis of a substance in its natural form, (3) the removal of uncontrollable contamination by atomizing the substance in a vacuum, (4) the possibility of discriminating a selective signal against the background in one measurement, and (5) separate detection of surface and bulk impurities in solid samples—enable it to be used in determining traces of most elements in almost any matrix.

9. Laser Photoionization Detection of Rare Isotope Atoms

The present chapter is concerned with the possibilities of detecting with a high selectivity rare atoms (isotopes, isomers) against the background of huge numbers of atoms of the same element. This problem is of considerable interest in nuclear physics, astrophysics, and particle physics. To begin with, let us consider the essence of the problem with reference to detection of rare cosmogenic isotopes.

9.1. Problem and Solution Techniques

Consider the requirement imposed on the laser photoionization technique and other possible laser techniques for the detection of very rare isotope atoms.

9.1.1. Rare Cosmogenic Isotopes and Methods for Their Detection

There are a fairly large number of rare isotopes of cosmic origin, particularly those formed in the upper atmosphere as a result of nuclear reactions under the effect of cosmic rays. They include such isotopes as ^{10}Be, resulting from the interaction of galactic cosmic rays with N and O nuclei; ^{14}C, formed in the reaction between secondary neutrons and N; and ^{26}Al, produced as a result of fission of the Ar nucleus. These isotopes form in the upper atmosphere, precipitate, and accumulate on the earth's surface and ocean bottom. The rate of their precipitation in the ocean can be considered to remain constant during a long period of time exceeding their half-life $T_{1/2}$. Figure 9.1 lists some rate isotopes of cosmic origin along with their concentrations.

The best known radioisotope among these is radiocarbon, ^{14}C, which is used for estimating the age of objects of organic origin (Libby, 1980). Radiocarbon, which is formed in the upper atmosphere at a concentration of $^{14}C/^{12}C = 10^{-12}$, is involved in the earth's biochemical life cycle. After

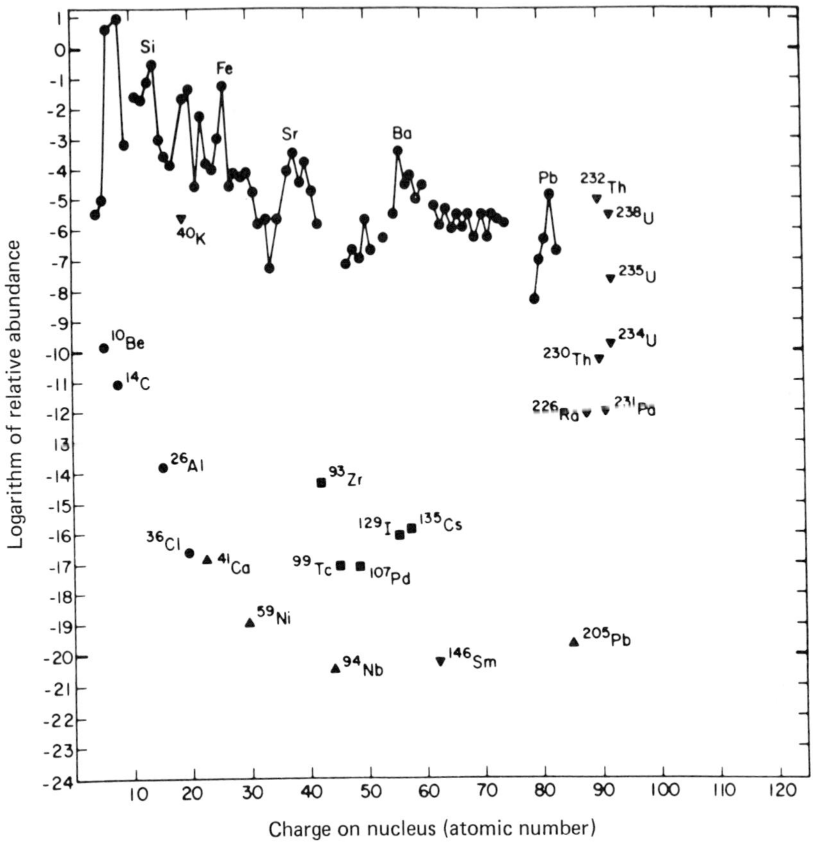

FIG. 9.1. Relative abundances of the stable elements in the earth's crust shown together with the relative abundances of some long-lived radioisotopes: (●), radioisotopes, generated primarily by cosmic rays; (▼), survivors from element formation prior to the formation of the solar system; (▲), elements created by neutron capture; and (■), elements from the spontaneous fission of uranium-238. [From Litherland *et al.* (1981).]

an organism has ceased to be living and participating in the carbon cycle, its content of ^{14}C decreases exponentially in accordance with the 5730-year half-life of this radioisotope. To date organic archaeological objects of events of 50,000 years ago, for example, it is necessary to detect ^{14}C in relative concentrations as low as 10^{-15}. By utilizing other, longer-lived isotopes, radioisotope dating can be extended to cover millions of years in the past.

At present, there exist two universal methods for detecting cosmogenic radioactive isotopes in low concentrations. The first and most popular one consists in measuring the specific radioactivity of the sample under analysis and comparing it with that of a specimen of zero age. This standard quantity for ^{14}C, for example, is well known and amounts to 15.3 β-decay events per minute per gram of the natural mixture of carbon isotopes. To realize this method requires fairly large sample (around 5 g) and a long observation time (approximately 1 day). Therefore a major part of a very valuable sample must frequently be sacrificed in order that its age can be determined. Serious measures should also be taken to ensure proper protection against background activity. The nuclear method of detecting rare isotopes is disadvantageous in that it depends on radioactive transformations of the isotopes for its operation, which occur extremely seldom. Therefore, to have a reasonable observation time (a few days), the sample must, in principle, contain a large number of the rare isotope atoms of interest.

The second method for detecting rare isotopes consists of utilizing an accelerator as a high-resolution mass spectrometer (Litherland *et al.*, 1981). The principal difficulty in implementing this method is the need to suppress background noise due to abundant isotopes and isobaric atoms, such as ^{14}N in the case of ^{14}C. The method handles much smaller samples (down to 5 mg), but the sample is completely lost as a result of measurements. The cost of the equipment required by the method is rather high.

It is clear that the shortcomings of both generally accepted methods open up a wide field of application for laser techniques, for these are, in principle, capable of tackling the very difficult task of detecting a few rare isotope atoms against the background of 10^{10}–10^{20} atoms of the most abundant isotopes of the same atomic species. Taking as an example the detection of ^{10}Be in polar ice, Table 9.1 gives an idea of the parameters of the known methods and the requirements of the potential laser methods for detecting rare isotopes.

TABLE 9.1. Sensitivity of Various Methods in Detecting ^{10}Be ($T_{1/2} = 1.5 \times 10^6$ years, $^{10}Be/^{9}Be = 10^{-10}$) in Polar Ice

Method	Sample weight (kg)	Number of ^{10}Be atoms	Experiment
Ultralow radioactivity measurement technique	10^6	10^{13}	McCorkell *et al.*, 1967
Ultrasensitive accelerator mass spectrometry	1	10^7	Beer *et al.*, 1983
Potential laser technique for detecting single rare isotope atoms	10^{-3}	10^4	—

9.1.2. Possible Laser Techniques. Limitations and Ways to Overcome Them

All the existing laser spectroscopy methods whose ultimate sensitivity lies at a level of single atoms (for reviews, see Letokhov, 1980, 1981a; Balykin *et al.*, 1980b) can, in principle, be employed to effect a highly selective detection of rare isotope atoms. Figure 1.2b–d presents simplified schemes of the three main laser techniques for detecting single atoms. These techniques take advantage of the effects due to resonant interaction of the atom with photons: (1) spontaneous reradiation of many photons absorbed from the laser beam, (2) photoionization of the atoms as a result of absorption of a few photons, and (3) changes of the atomic coordinate and velocity as a consequence of reradiation of a large number of photons.

The main question relating to the applicability of these methods to detecting rare isotopes is their ultimate selectivity S, i.e., the ability to detect a small number (N_A) of the rare isotope atoms A in the presence of a much greater number (N_B) of the main isotope atoms B:

$$S = N_{\mathrm{B}}/N_{\mathrm{A}}. \tag{9.1}$$

The selectivity of these techniques stems from the presence of a small isotope shift $\Delta = \omega_{\mathrm{A}} - \omega_{\mathrm{B}}$ of the spectral line of one or several consecutive resonant transitions of the atom from its ground state to an excited state. The fact that the width of any spectral line is finite naturally limits the selectivity because of the overlapping of the wings of the close spectral lines of atoms A and B, but the character of the limitation largely depends on the particular technique used.

With the fluorescence technique, the fluorescence excitation line usually has a Lorentzian shape, i.e., the probability that a photon will be scattered by the atoms B when the laser frequency coincides with the center of the spectral line of the rare atoms A is defined (in excitation conditions far from saturation) by the expression

$$W_{\mathrm{scat}}^{\mathrm{B}} = \sigma_0 I \mathscr{L}(\Delta/\Gamma) \simeq \tfrac{1}{2}(\mu_{12}\mathscr{E}/\hbar\Delta)^2, \tag{9.2}$$

where I is the laser radiation intensity (in photons/cm^{2}s), σ_0 the cross section of the radiative transition $1 \rightarrow 2$ at a maximum, Γ the natural half-width or the rate of spontaneous decay of the atom into its initial ground state, μ_{12} the dipole moment of the transition $1 \rightarrow 2$, $\mathscr{E}$ the electric field strength of the light wave, and $\mathscr{L}(x) = 1/(1 + x^2)$ the Lorentzian function. By virtue of Eq. (9.2), the selectivity of the fluorescence detection of rare atoms is limited to the level of

$$S_{\mathrm{fl}} = W_{\mathrm{scat}}^{\mathrm{A}}/W_{\mathrm{scat}}^{\mathrm{B}} = (\Delta/\Gamma)^2, \qquad (\Delta \gg \Gamma). \tag{9.3}$$

For typical isotope shift and radiative linewidth values, $S_{\mathrm{fl}} = 10^4$–10^6, i.e., it is much lower than the required values indicated in Fig. 9.1.

The photoionization technique detects excited atoms by their subsequent transition into an ionized state. Its selectivity is therefore governed by the probability of the atomic excitation and not by that of photon reradiation. The selectivity of the method depends materially on the type of subsequent ionization of the excited atoms A. In the case of nonresonant simultaneous ionization, where the difference between the energy of the first absorbed photon and the atomic excitation energy $\hbar\Delta$ can be compensated for the second absorbed photon, the selectivity is reduced to the former level defined by Eq. (9.3). This matter is discussed in detail in Section 9.2 for various schemes of multistep resonance ionization of atoms.

It follows from the above simple estimates that none of the laser techniques in its simplest version can provide for the very high detection selectivity required. However, each of the techniques can be modified so as to ensure a substantial increase in selectivity. The methods that are possible here may be divided into two groups:

(1) Methods based on a repeated resonance interaction of the atom with the laser light, wherein the atom reradiates a large number of photons. In this case, there occurs what may be called a "selectivity accumulation" on account of an effect that is possible only when a single atom reradiates a large number of photons following its repeated resonance excitation. This effect is made use of in the method of "fluorescent bursts" (Greenlees *et al.*, 1977; Balykin *et al.*, 1977, 1980a) and that of laser deceleration and cooling of atoms (Balykin *et al.*, 1984). It is possible to effect a multiple isotope-selective ionization of atoms with laser radiation and thus improve the selectivity of ionization detection of rare isotopes. This possibility is discussed in Section 9.3.

(2) Methods based on a multistep resonance excitation of the atom in a multiple-frequency laser field, wherein use is made of the isotope shift on several consecutive resonant transitions. As a result of such a multistep resonance excitation, the selectivities S_i attained at each excitation and ionization stage are multiplied (Letokhov and Mishin, 1979). The idea of such a method is illustrated in Fig. 9.2. For example, in the case of a three-step excitation in which use is made of the isotope shift at each step, the net selectivity of optical excitation is

$$S_{\text{opt}} = S_1 S_2 S_3. \tag{9.4}$$

At moderate selectivity values, $S_i \simeq 10^4$–10^6, the net ionization selectivity can reach the exceptionally high values of $S_{\text{opt}} = 10^{12}$–10^{18}. If the given energy level diagram does not allow realization of several successive excitation steps up to the ionization limit, we can, in principle, stimulate with laser radiation transitions from excited states downward (N-type multistep excitation scheme), as shown in Fig. 9.2b. Such an excitation method widens

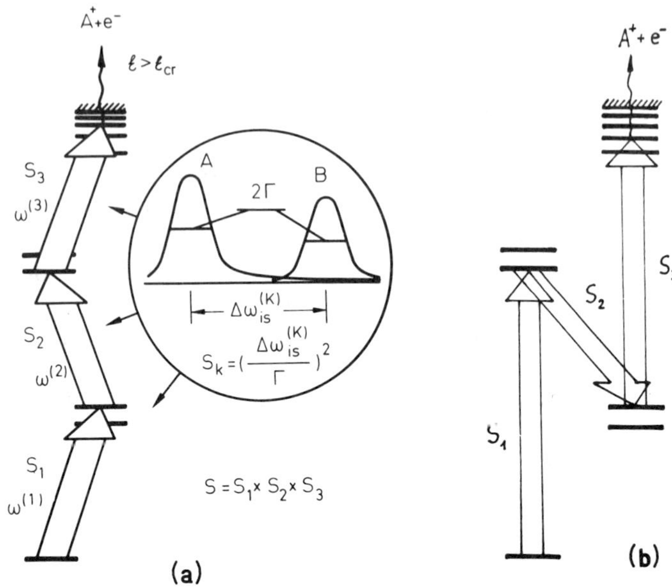

FIG. 9.2. Selectivity multiplication in the course of multistep excitation taking advantage of selectivity at each excitation step. (a) Successive upward excitation and (b) successive excitation and deexcitation under the action of laser radiation.

the possible choice of intermediate transitions involving an isotope shift and the possible choice of laser radiation wavelengths. Ways to detect rare isotopes, based on such a multistep technique, are discussed in Section 9.4.

9.2. Selectivity of Multistep Ionization

The problem of the selectivity of multistep excitation and ionization of free atoms reduces essentially to the calculation of atomic excitation with light pulses of frequencies detuned from exact resonances. Such a calculation was, in fact, made specially for the purpose (Makarov, 1983). Presented below is a summary of the results of this analysis for several schemes differing in the number of resonance excitation steps.

For the sake of definiteness, all the schemes under consideration are assumed to be subject to the following conditions:

(1) The laser pulse duration satisfies the condition

$$\tau_p \gtrsim 1/\Gamma, \tag{9.5}$$

i.e., the time of spontaneous decay of the intermediate level is longer (but not necessarily much longer) than the pulse duration.

(2) The pulses are of smooth shape, and the isotope shift is much greater than the spectral pulse width:

$$\Delta \gg 1/\tau_p. \tag{9.6}$$

(3) At all excitation stages, the rates of excitation of the rare isotope are of the order of the reciprocal of the pulse duration at these stages, i.e.,

$$(\mu\mathscr{E}/2\hbar)\tau_p \simeq 1. \tag{9.7}$$

Subject to these conditions, the probability of ionization of the rare isotope A, $W_A(\Omega_i = \omega_{kn}^{(i)})$, whose successive quantum transition frequencies $\omega_{kn}^{(i)}$ are in exact resonance with the exciting laser frequencies Ω_i, is in the range 0.1–1. The exact figure depends on the shape of laser pulses, their overlapping, and the degree of saturation of the quantum transitions. These parameters are usually adjusted in the course of the experiment so as to achieve the maximum ionization efficiency. It is necessary to calculate the probability of excitation of the abundant background isotope B, $W_B(\Omega_i = \omega_{kn}^{(i)} + \Delta_i)$, subject to condition (9.7), Δ_i being the isotope shift at the ith resonance excitation stage, Ω_i the laser field frequency at the ith stage, and $\omega_{kn}^{(i)}$ the transition frequency of the atom A at the ith stage. The ratio between the ionization probabilities W_A and W_B defines the selectivity of detecting isotopes atoms, given by Eq. (9.1)

9.2.1. Two-Step Ionization: One Resonance Step, One Isotope Shift

Let us begin with the simplest case of two-step photoionization with only one resonance stage in which an isotope shift occurs (Figs. 3.17a and b). The following two processes lead to the ionization of the background isotope B: (1) direct two-photon ionization and (2) population of an intermediate level on account of a process whereby two photons are absorbed from the first laser field and a photon of necessary energy obeying the law of conservation of energy is spontaneously emitted (Fig. 2.7). In the general case, the following estimate holds for the two processes:

$$W_B \simeq (\overline{F^2 V}/\Delta_1^2)\tau_p^{(1,2)} + (\overline{F^4}/\Delta_1^4)\Gamma\tau_p^{(1)}\overline{V}\tau_p^{(2)}, \tag{9.8}$$

where $F = \mu_{12}\mathscr{E}/2\hbar$, $\tau_p^{(i)}$ is the laser pulse duration at the ith stage, and the overscore denotes time averaging:

$$\overline{F^2 V}\tau_p^{(1,2)} \equiv \int F^2(t)V(t)\,dt; \qquad \overline{F^4}\tau_p^{(1)} \equiv \int F^4(t)\,dt, \tag{9.9}$$

the quantity V describing the rate of ionization of excited atoms from the intermediate state 2 by means of either laser radiation on a transition to the continuous spectrum ($V = \sigma_i I_2$) or an electric field if the state 2 lies close to the ionization limit.

The first term in Eq. (9.8) is written in the general form so as to allow for the possible overlapping of two pulses, for it is only in the case of such overlapping that the direct two-photon ionization of the rare isotope is possible. Consider two cases differing in the degree of overlapping of the pulses

(1) No Delay Between the Pulses. In this case, the main contribution to the loss of selectivity is from the first term of Eq. (9.8), which may be reduced to the form

$$W_B = (\overline{F^2 V}/\Delta_1^2)\tau_p = 1/\Delta_1^2\tau_p^2 \simeq (\Gamma/\Delta_1)^2[1/(\Gamma\tau_p)^2], \tag{9.10}$$

where we take $\Gamma\tau_p \simeq 1$ and $V\tau_p \simeq 1$. The selectivity of two-step ionization with a single resonance stage is governed by the amplitude of the Lorentzian wing of the absorption line of the background isotope B. Physically, it is explained by the fact that the energy difference $\hbar\Delta_1$ can be compensated for at the second, nonresonance excitation step.

(b) The pulses are separated in time by an amount $\Delta\tau$ satisfying the condition

$$1/\Gamma \gtrsim \Delta\tau \gg \tau_p^{(1,2)}, \tag{9.11}$$

so that the first term in Eq. (9.8) can be disregarded. In such ionization conditions, the following estimate is valid:

$$W_B \simeq (\overline{F^4}/\Delta_1^4)\Gamma\tau_p^{(1)}\overline{V}\tau_p^{(2)} \simeq (\Gamma/\Delta_1)^4[1/(\Gamma\tau_p^{(1)})^3], \tag{9.12}$$

where $\overline{F}\tau_p^{(1)} \simeq 1$ and $\overline{V}\tau_p^{(2)} \simeq 1$. In this case, the probability of ionization of the background isotope is much lower because the law of conservation of energy prohibits the direct absorption of the resonant photon. The excitation of the intermediate state is possible only on account of two-photon absorption, followed by spontaneous reemission of a photon of shifted frequency (Fig. 2.7). It is for this reason that the fourth power of the ratio between the linewidth Γ and the isotope shift Δ_1 appears in the first multiplier of Eq. (9.12). The second multiplier in Eq. (9.12) appears as a result of the exponential decay of the intermediate level by the moment the second pulse arrives.

Thus, the maximum selectivity of two-step ionization is achieved when irradiation is effected by temporally separated laser pulses, and the direct two-photon ionization of the background isotope is thus suppressed.

9.2.2. Three-Step Ionization: Two Resonance Steps, One Isotope Shift

Let us now consider the case of two-step resonance excitation with an isotope shift occurring only at a single, e.g., the first, step. In this case, even if two pulses act simultaneously, the absorption on the line wing of the first transition must be impeded, because the photon of the second resonance stage cannot compensate for the difference of the energies of the first transition and the first photon. It might be expected that in this case the estimate (9.12) for W_B will hold true even for overlapping pulses.

The ionization of the background isotope atom comes mainly as a result of the following three processes (Fig. 9.3): (1) direct three-photon ionization, (2) population of an intermediate level on account of the process that has already been considered in Section 9.2.1 (i.e., absorption of two photons from the first pulse of frequency Ω_1 and reemission of a spontaneous photon at a shifted frequency), and (3) population of an intermediate state on account of a process whereby two photons of frequencies Ω_1 and Ω_2 are absorbed and a photon with such energy as is necessary to satisfy the law of conservation of energy is spontaneously reemitted. In the general case, the following estimate for the contribution from all the three processes is valid:

$$W_B = (\overline{F_1^2 F_2^2 V/\Delta_1^4})\tau_p^{(1,2,3)} + (\overline{F_1^4/\Delta_1^4})(\Gamma_1 \tau_p^{(1)})(\overline{F_2 \tau_p^{(2)}})\,\overline{V\tau_p^{(3)}} + (\overline{F_1^2 F_2^2/\Delta_1^4})(\Gamma_2 \tau_p^{(1,2)})(\overline{F_2 \tau_p^{(2)}})(\overline{V\tau_p^{(3)}}). \tag{9.13}$$

In most cases, $\Gamma_2 < \Gamma_1$ and therefore the last term in Eq. (9.13) is of minor importance.

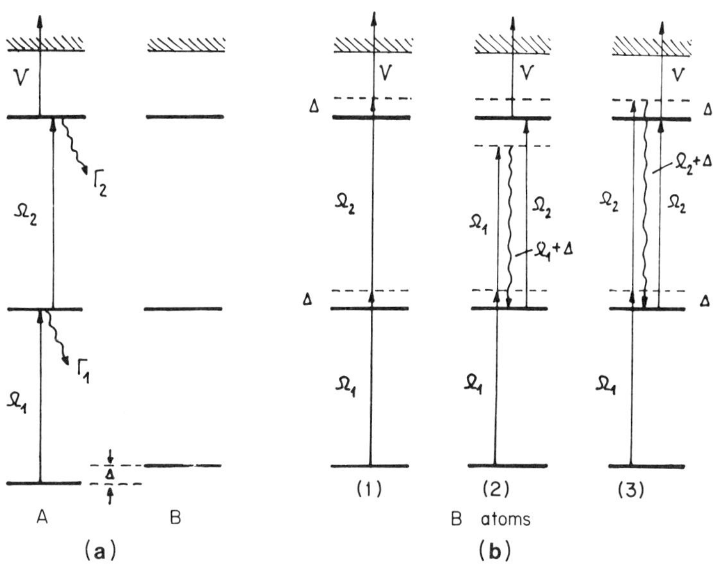

FIG. 9.3. Estimation of the selectivity of ionization of atoms A in the presence of atoms B, the process involving two resonance excitation stages with an isotope shift occurring only at one of them (the first). (a) Stepwise ionization diagram and (b) three different ionization mechanisms for the B atoms that limit the selectivity of the step-wise ionization of the rare atoms A.

As can be seen from the general expression (9.13), the greatest contribution is provided by the effect of overlapping all three pulses. If the overlapping is only partial, the second term may become equal to the first, and if there is no overlapping at all, the second term dominates. In the case in which there is no delay between the pulses (complete overlapping), formula (9.13) takes the form

$$W_{\mathrm{B}} \simeq (\overline{F_1^2 F_1^2 V / \Delta_1^4}) \tau_{\mathrm{p}}^{(1)} \simeq (\Gamma_1/\Delta_1)^4 [1/(\Gamma_1 \tau_{\mathrm{p}}^{(1)})^4]. \tag{9.14}$$

In all other cases in which there are delays between the pulses, the following estimate holds:

$$\begin{aligned} W_{\mathrm{B}} &\simeq (\overline{F_1^4}/\Delta_1^4)(\Gamma_1 \tau_{\mathrm{p}}^{(1)})(\overline{V \tau_{\mathrm{p}}^{(3)}}) \\ &\simeq (\Gamma_1/\Delta_1)^4 [1/(\Gamma_1 \tau_{\mathrm{p}}^{(1)})^3]. \end{aligned} \tag{9.15}$$

So, two-step excitation with two resonance steps provides for a substantial enhancement of selectivity, even where an isotope shift is taken advantage of only at a single step, as compared with single-step resonance excitation for which the probability of ionization of the background isotope is described by expression (9.10).

9.2.3. Three-Step Ionization: The Resonance Steps, Two Isotope Shifts

The diagram of transitions and isotope shifts for this case is similar to the one shown in Fig. 9.3, the only difference being that use is made here of two isotope shifts. For convenience, the isotope shift for the two-photon transition is denoted as Δ_2. The isotope shift for the second resonance transition will then be $\Delta_2 - \Delta_1$. In addition to the condition (9.5), i.e., $\Delta_2 \tau_{\mathrm{p}} \gg 1$, the following condition is taken to be satisfied:

$$|\Delta_1 - \Delta_2| \tau_{\mathrm{p}} \gg 1. \tag{9.16}$$

Four processes contribute to the ionization of the background isotope: (1) three-photon ionization; (2) excitation of the first level on account of the processes that have already been considered in Section 9.2.2, followed by two-photon ionization; (3) excitation of the first level on account of the processes considered in Section 9.2.2, followed by excitation of the second level as a result of similar processes, with subsequent one-photon ionization; (4) excitation of the second level on account of processes whereby 1 (2) photon(s) are absorbed from the first laser field, 2 (1) photons are absorbed from the second laser field, and a photon of necessary energy is spontaneously emitted. In the general case, we may write

$$\begin{aligned} W_{\mathrm{B}} \simeq{}& (\overline{F_1^2 F_2^2 V / \Delta_1^2 \Delta_2^2}) \tau_{\mathrm{p}}^{(1,2,3)} + (\overline{F_1^4}/\Delta_1^4)(\Gamma_1 \tau_{\mathrm{p}}^{(1)}) [\overline{F_2^4}/(\Delta_1 - \Delta_2)^4] \\ &\times (\Gamma_2 \tau_2^{(2)})(\overline{V \tau_{\mathrm{p}}^{(3)}}) + (\overline{F_1^4}/\Delta_1^4)(\Gamma_1 \tau_{\mathrm{p}}^{(1)}) [\overline{F_2^2 V}/(\Delta_1 - \Delta_2)^2] \tau_{\mathrm{p}}^{(2,3)} \\ &+ [1/\Delta_1^2 \Delta_2^2 (\Delta_1 - \Delta_2)^2] [(\overline{F_1^4 F_2^2} \Gamma_1 \tau_{\mathrm{p}}^{(1,2)} + \overline{F_1^2 F_2^4} \Gamma_2 \tau_{\mathrm{p}}^{(1,2)})]. \end{aligned} \tag{9.17}$$

Consider several different limiting cases:

(a) *No delays between the pulses.* The main contribution to W_B is from the first term of (9.17), as in formula (9.13):

$$W_B \simeq (\overline{F_1^2 F_2^2 V}/\Delta_1^2\Delta_2^2)\tau_p \simeq (\Gamma_1^2/\Delta_1\Delta_2)^2[1/(\Gamma_1\tau_p)^4]. \tag{9.18}$$

In this case, the excitation selectivities attained at both resonance stages are simply multiplied as ordered by the simple formulas (9.3) and (9.4).

(b) *All the laser pulses are temporally separated*, except in compliance with the condition (9.11). The main contribution to (9.17) is then made by the second term:

$$\begin{aligned} W_B &\simeq (\overline{F_1^4}/\Delta_1^4)(\Gamma_1\tau_p^{(1)})[\overline{F_2^4}/(\Delta_1 - \Delta_2)^4](\Gamma_2\tau_p^{(2)})(\overline{V}\tau_p^{(3)})] \\ &\simeq (\Gamma_1/\Delta_1)^4[1/(\Gamma_1\tau_p^{(1)})^3][\Gamma_2/(\Delta_1 - \Delta_2)]^4[1/(\Gamma_2\tau_p^{(2)})^3]. \end{aligned} \tag{9.19}$$

In this case, the excitation selectivities attained at the first two resonance stages are multiplied, each of the selectivities being defined by an expression similar to (9.12).

(c) *The second and third pulses are delayed with respect to the first.* In this case, the main contribution to W_B is from the third term of (9.17):

$$\begin{aligned} W_B &\simeq (\overline{F_1^4}/\Delta_1^4)(\Gamma_1\tau_p^{(1)})[\overline{F_2^2 V}/(\Delta_1 - \Delta_2)^2]\tau_p^{(2,3)} \\ &\simeq (\Gamma_1/\Delta_1)^4[1/(\Gamma_1\tau_p^{(1)})^3][\Gamma_2/(\Delta_1 - \Delta_2)^2(1/\Gamma_2\tau_p^{(2,3)}). \end{aligned} \tag{9.20}$$

In this case, the excitation selectivity is somewhat reduced.

(d) *The third pulse is delayed with respect to the first and second ones.* In this case, the main part is played by the fourth and last term in (9.17):

$$W_B \simeq \Gamma_1^6/\Delta_1^2\Delta_2^2(\Delta_1 - \Delta_2)^2[1/(\Gamma_1\tau_p^{(1,2)})^5] \tag{9.21}$$

Thus, increasing the number of steps of resonant, isotope-selective excitation and temporally separating all the laser pulses, both exciting and ionizing, allow the multiplication of the selectivities of excitation of a rare isotope in the presence of a predominant number of abundant isotope atoms to be realized.

9.3. Methods for Highly Selective Ionization Detection of Rare Isotopes

Let us now consider the various actual approaches to the realization of the highly selective ionization detection of rare isotopes that have already been discussed briefly in Section 9.1.2.

9.3.1. Multiple Repetition of Resonance Ionization

One of the obvious ways to improve the selectivity of ionization detection of isotope atoms is to repeat the laser resonance ionization process again and again. This trend has developed as a result of the combination of several ideas put forward as early as the first years of the development of the resonance ionization detection technique.

Because the ionization technique is destructive, it was suggested that this drawback should be eliminated by repeating the process (Letokhov *et al.*, 1977, p. 183):

> If there is no need to extract from the mixture the atom being detected, we can imagine a certain modification of the photoionization method wherein each atom will participate in ionization many times. Indeed, the ionized atom can be neutralized and the selective ionization process repeated once more, and so on. In this way, each atom can, in principle, be made to yield a large number of photoelectrons, which will lower the atomic detection limit.

This idea was discussed once more (Balykin *et al.*, 1980b) from the standpoint of the realization of a cyclic laser resonance conversion of atoms to ions and then back to atoms.

That it is possible to neutralize ions by implanting them into a foil, liberate them subsequently by heating the foil, and then detect the neutral atoms thus produced by the three-step resonance photoionization method was demonstrated in experiments with Yb isotopes (Bekov *et al.*, 1979).

Figure 9.4 shows the scheme of ionization detection of rare isotopes that was suggested on the basis of the above ideas and experiments (Letokhov,

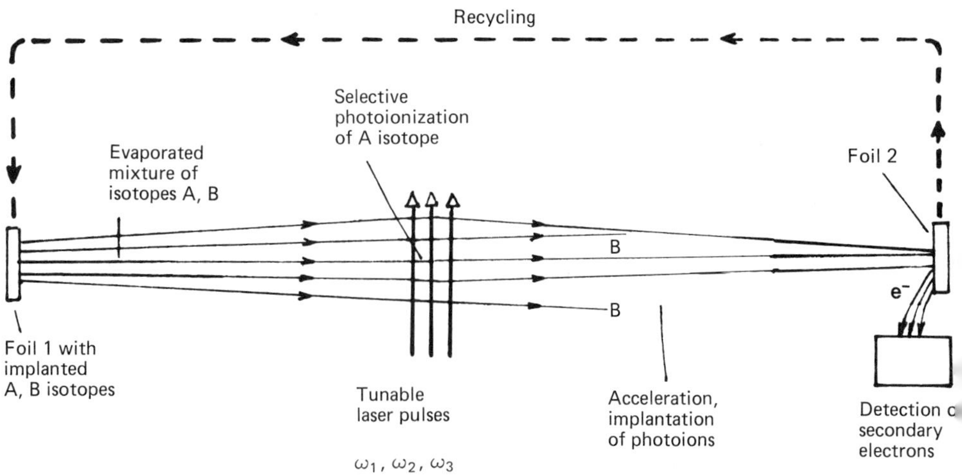

FIG. 9.4. Idea of enhancement of the selectivity of ionization detection of rare isotopes B in mixture with A isotopes abundant by repeating the laser ionization process.

1981b). Let sample 1 contain a mixture of isotopes A and B. By vaporizing them and then subjecting them to resonance ionization, we can increase the ratio of their concentrations in the photoion stream by S_{opt} times and then implant the ions produced into foil 2. The selectivity of such a single-shot process is equal to the ratio of the net ionization efficiencies Y_i for the atoms B and A:

$$S_{opt} = ([A]/[B])_2/([A]/[B])_1 = Y_i(A)/Y_i(B). \tag{9.22}$$

By repeating this process m times, it is possible to ensure the selectivity

$$S = (S_{opt})^m. \tag{9.23}$$

Such a scheme is suitable for detecting rare isotopes only if the atoms are highly effectively collected after each cycle. Indeed, let the net ionization efficiency $Y_i(A)$ differ greatly from 100% and the nonionized, neutral atoms [their proportion is equal to $1 - Y_i(A)$] be lost irretrievably. In this case, the sensitivity of ionization detection is unavoidably reduced to a level of $[Y_i(A)]^m$. For example, if $Y_i(A) = 0.1$ and the process is repeated three times, the loss of 99.9% of the rare isotope atoms A is inevitable. There are several possible ways to overcome this difficulty.

First, we can try to achieve a 100% ionization efficiency by using maximum efficient quantum transition schemes, intense laser radiation, etc. Of course, a loss of laser ionization selectivity is in this case inevitable. However, the selectivity can be improved by placing in the way of the photoions a mass separator with a separation selectivity of S_{mass}. The total selectivity of the single-stage process will then be $S = S_{opt}S_{mass}$, and if the process is repeated m times, it will be increased accordingly:

$$S_m = (S_{opt}S_{mas})^m. \tag{9.24}$$

This method is applicable in cases where the atomic ionization is not selective ($S_{opt} = 1$). The whole selectivity is then accumulated on account of the repetition of the mass separation process, the laser radiation carrying out only the function of high-efficiency ionization. The method is useful when the isotope shift and hence the isotopic selectivity are very small, particularly in detecting rare isotopes of noble gases (Chen *et al.*, 1980, 1984).

Second, it is possible to try to eliminate the loss of nonionized atoms by employing suitable experimental geometries. In particular, this objective is attained when detecting rare gas atoms in a closed space, because such atoms do not enter into chemical reactions with the walls of the containing vessel, and their adsorption on certain spots on the walls can be controlled through cooling. In this case, the same atom can enter the irradiation region again and again, and so a high detection efficiency can be achieved even at a low ionization yield per pulse.

A laser photoionization detector for rare isotopes of noble gases, which uses both of the above possibilities of photoionization detector modification, is now being developed at the Oak Ridge laboratory (Chen *et al.*, 1980, 1984). Figure 9.5 shows a schematic diagram of a photoionization detector for Kr atoms. The atoms are accumulated on a cold (15 K) substrate. A pulsed laser heats the substrate and liberates the atoms into a vacuum. The Kr atoms entering the irradiation region are subjected to three-step photoionization ($\lambda_1 = 116.5$ nm, $\lambda_2 = 558.1$ nm, $\lambda_3 = 1064$ nm) according to the scheme shown in Fig. 3.17b. Such a stepwise ionization is effected without taking advantage of isotopic selectivity, i.e., it provides for selection by the atomic charge Z only. To provide for isotopic selectivity, use is made of mass separation (A-selection) by means of a quadrupole mass filter. The separated photoions are detected with the aid of an electron multiplier. The net per-pulse ionization efficiency for Kr atoms comes to 3×10^{-5}. However, non-ionized atoms can return onto the cold substrate and take part in the subsequent detection cycle. At a laser pulse repetition frequency of 10 Hz, it takes approximately an hour to warrantably detect the desired ^{81}Kr atoms against the background of all Kr atoms. The possibility of selectively detecting

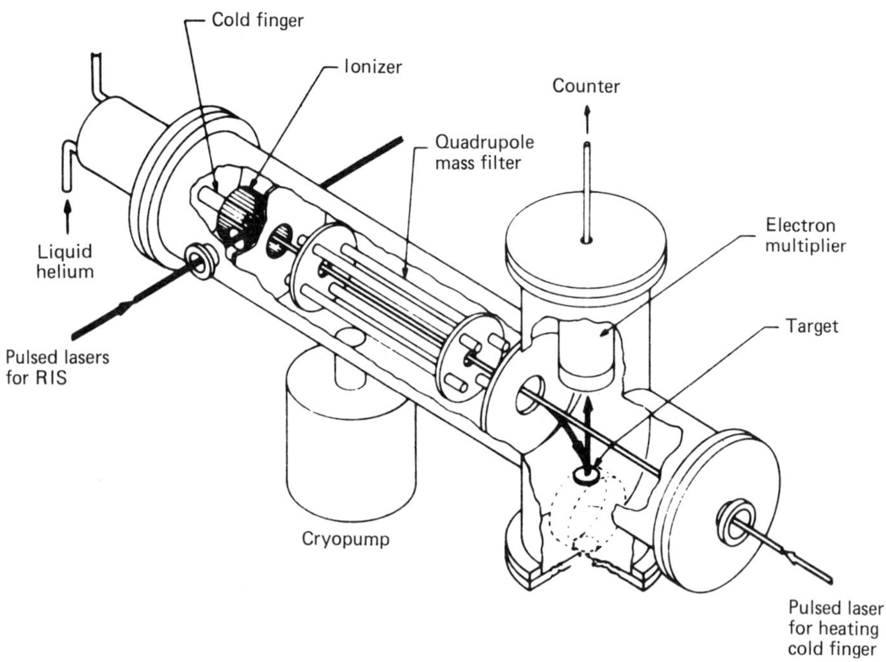

FIG. 9.5. Schematic of the experiment on sorting out and counting noble gas isotopes. [From Chen *et al.* (1984).]

10^3 atoms of ^{81}Kr amongst 10^7 atoms of ^{80}Kr or ^{82}Kr has been demonstrated. As a matter of fact, one stage of photoionization detection of isotopes has thus been realized. The ^{81}Kr atoms implanted into the substrate of the electron multiplier can obviously be liberated, and the detection process can be repeated to improve selectivity in accordance with the idea illustrated in Fig. 9.4.

Subject to discussion are other schemes of ionization detection of rare isotopes, based on the above-mentioned ideas of improvement of selectivity on account of mass separation and repetition of the ionization process. For instance, it has been suggested that isotope-selective ionization and mass separation should be combined with the fluorescence detection of the mass-separated photoions (Keller *et al.*, 1981). The combination of isotope-selective ionization with mass separation has been proposed with a view to developing an ultrasensitive mass spectrometer (Lucatorto *et al.*, 1984). In this proposal, to improve the isotopic selectivity of the excitation of ^{90}Sr possessing a small isotope shift, provision has been made for the use of two-photon excitation with counter-propagating beams, which eliminates the Doppler broadening (see Section 2.2.4) (Vasilenko *et al.*, 1970). Finally, they have proposed to realize the idea of a cyclic interaction of an atom with ionizing laser radiation, illustrated in Fig. 9.4, in a ring-shaped ion collector (Snyder *et al.*, 1985). It has also been proposed that laser selective photoionization should be effected with transverse irradiation of the atomic beam (the Doppler shift being eliminated by two-photon ionization in counter-propagating beams), acceleration of the ions produced, and their subsequent conversion into neutral atoms in a gas cell. In this case, undesirable isotopes are removed in each cycle, and selectivity is thus accumulated. Estimates have shown it possible to obtain a detection selectivity of 10^{15} for ^{90}Sr against the background of ^{88}Sr at an atomic flux of 10^{13} atoms/s and a detection efficiency of better than 10%.

9.3.2. Multistep Selective Resonance Ionization

To implement the idea of excitation selectivity multiplication in accordance with Fig. 9.2, it is necessary to select such a sequence of resonance transitions as makes two or three intermediate levels suffer isotopic displacement (for brevity, this term will be used to denote any disturbance of the atomic spectrum by the nucleus that causes a level shift). Since only the S and $P_{1/2}$ electron orbits penetrate the nucleus, the intermediate terms should contain such electrons.

With some atoms, such as the alkaline metals, only the first step of excitation is isotope-selective, since only the ground state S is shifted. The resonant state P is shifted by one order less. Practically, there is no shift in the rest of the states. However, the isotopical shift of many other atoms is observed on the first as well as on the second and third steps. Examples of such atoms are the elements of the lanthanide and actinide groups, for example, Yb, Gd, and U. For the sequence of transitions in the ytterbium atoms, $6s^2(^1S_0) \rightarrow 6s6p(^3P_1) \rightarrow 6s7p(^3S_1) \rightarrow 6snp(^3P_1)$, the shift of the $^1S_0$ and $^3S_1$ terms is the most important. The values of the isotopic shift are large enough to attain a total photoexcitation selectivity greater than 10^{15}! (Letokhov and Mishin, 1979).

The idea of multiplication of isotopic selectivity at each excitation stage is difficult to realize for the most interesting cosmogenic isotopes (Fig. 9.1), because it is hard to find for them a series of consecutive upward transitions with noticeable isotope shifts, such shifts being characteristic of the ground state only.

A universal way to overcome this difficulty and make the method of multi-step resonance photoionization really applicable to the detection of rare isotopes was suggested (Kudriavtzev and Letokhov, 1982). The idea of the method is based on collinear stepwise photoionization of a beam of accelerated atoms. It is common knowledge that acceleration of atoms in the form of ions under a given potential difference U and subsequent neutralization of the ions into atoms lead to the bunching of the longitudinal ionic velocities, hence to the narrowing of the Doppler width of all the spectral lines of the given atomic species (if viewed in a collinear fashion) as compared with the initial Doppler width $\Delta\nu_D^{(0)}$ at an ion source temperature of T (Kaufman, 1976; Anton *et al.*, 1978; Kluge *et al.*, 1979):

$$\delta\nu_D(U)/\delta\nu_D^{(0)} = \tfrac{1}{2}(kT/eU)^{1/2}. \tag{9.25}$$

This relation can be simply derived from the following considerations. Since the ion energy width remains constant throughout the course of electrostatic acceleration, i.e.,

$$\delta E = \delta\, \tfrac{1}{2}\, Mv_z^2 = Mv_z\, \delta v_z, \tag{9.26}$$

an increase in the longitudinal velocity component v_z must cause a decrease in the velocity interval δv_z. In other words, faster atoms span a given acceleration interval more rapidly, and so the constant force exerted by the electric field imparts a lower velocity to them than to slower atoms. At $U = 10^4$ V, the narrowing factor reaches 10^3, which has been observed experimentally (Anton *et al.*, 1978). What is important is that the atoms in this case gather in a smaller volume of the phase space, and their Doppler-free laser excitation is therefore effected without any loss in sensitivity.

Along with the narrowing of the Doppler width, there also occurs the Doppler shift of all the spectral lines of the accelerated atoms, which depends on the mass of the ion. As a result, there occurs an artificial "mass" shift on any spectral transition of the atom:

$$\Delta\nu_{sh}/\nu_0 = (1/c)(\sqrt{2eU})(1/\sqrt{m_1} - 1/\sqrt{m_2}). \tag{9.27}$$

This shift at $U = 10$ kV is approximately an order of magnitude greater than the ordinary isotope shift $\Delta\nu_{is} \simeq m_e(\Delta m/m^2)\nu_0$. What is most important is that such a kinematic isotope shift occurs on any spectral transition of the atom, thus making it possible to realize in a natural way the idea of selectivity multiplication in multistep isotope-selective excitation.

To illustrate, let us consider the possibility of detecting ^{26}Al by the stepwise ionization technique (Kudriavtzev and Letokhov, 1982). Natural aluminum has only one stable isotope, ^{27}Al. The equilibrium content of radioactive ^{26}Al is 10^{-14}. Consequently, to detect ^{26}Al, the ionization selectivity must be much in excess of 10^{14}. Figure 9.6a shows schematically the energy levels of the Al Atom, along with the transitions that can be used for the purpose of multistep ionization. The transition wavelengths lie in the spectral region covered by dye lasers. Ionization is effected through the 3s3p^2 ^{3}P autoionization state. Figure 9.6b illustrates the difference between the absorption spectrum on the 3p → 4s transition of Al atoms accelerated to an energy of 10 keV and that of thermal Al atoms at $T = 1500$ K. For the thermal atoms, the isotope shift (590 MHz) is smaller than the hyperfine splitting (750 MHz) and much smaller than the Doppler width (4.1 GHz). For the accelerated atoms, an additional isotope shift of $\Delta_1 = 12$ GHz develops, and the Doppler width becomes less than the radiative width.

Following Section 9.2.3 let us estimate the selectivity of such a three-step ionization scheme with two resonance stages and isotope shifts occurring at each of them. According to Eq. (9.18), the selectivity in the case of irradiation with concurrent laser pulses is defined by the product of the selectivities attained at the first and second stages:

$$S = S_1 S_2, \qquad S_1 = S_2 \simeq (\Delta_1/\Gamma)^2(\Delta_2/\Gamma)^2. \tag{9.28}$$

A simple estimate gives $S_1 = S_2 = 0.95 \times 10^7$. The linewidth on the transition to the autoionization state comes to 230 GHz. This value is much greater than the isotope shift, and so there is no selectivity at all at the third stage ($S_3 = 1$). In accordance with Eq. (9.28), the total selectivity is $S \simeq 0.9 \times 10^{14}$. By employing separation by masses with the aid of a mass spectrometer having a selectivity of only $S_{mass} \simeq 10^2$, we can achieve the necessary net selectivity (around 10^{16}) to detect ^{26}Al against the background of ^{27}Al. A serious problem in realizing such a high selectivity will be in obtaining a sufficiently intense ion flux. At an ion current of $I = 10^{-4}$ A, about 600 ^{26}Al ions can, in principle, be obtained during an observation time of 10^3 s.

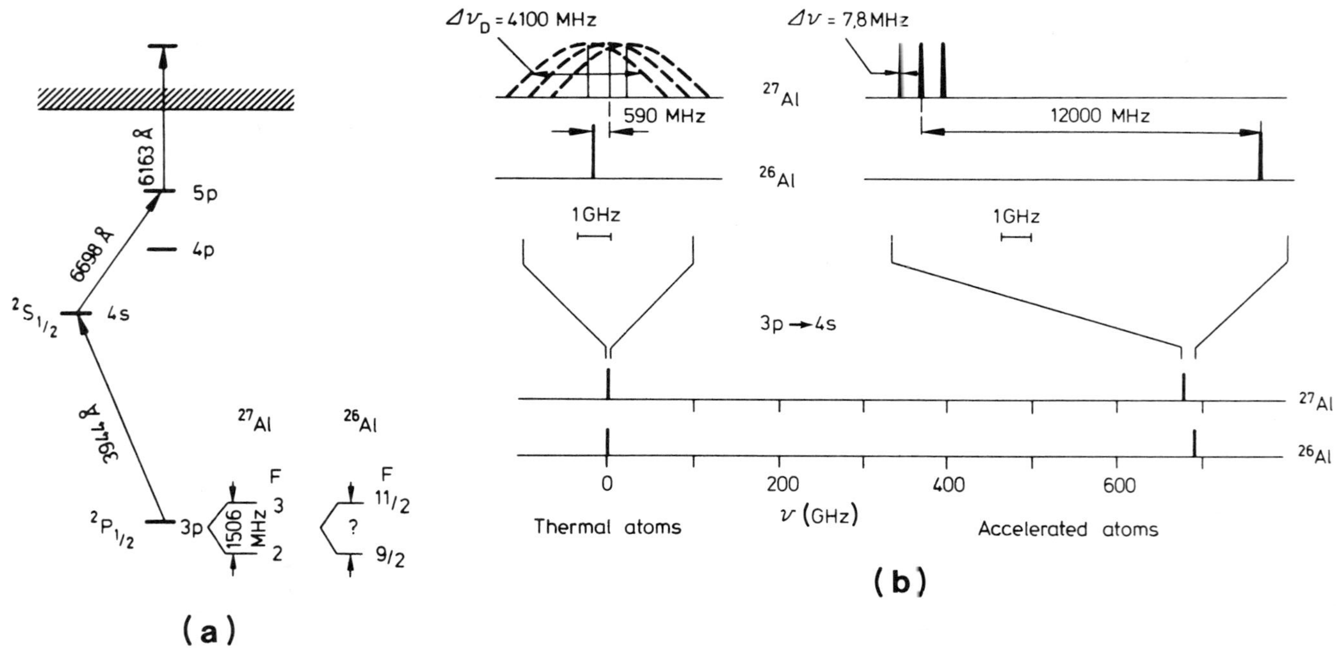

FIG. 9.6. (a) Diagram of the energy levels in Al and transitions that can be used for detecting ^{26}Al. (b) Structures of the absorption lines of the ^{26}Al and ^{27}Al isotopes on the 3p → 4s transition for thermal atoms and for atoms accelerated to 10 keV, observed collinearly. [From Kudriavtzev and Letokhov (1982).]

The first steps have been taken on the way to the realization of such a laser ionization detector for rare isotopes, as exemplified by the highly selective collinear ionization of ^{40}K in a beam of accelerated K atoms (Kudriavtzev *et al.*, 1985). The potassium atom has two stable isotopes, ^{39}K and ^{41}K, with a relative concentration of 93.2% and 6.8%, respectively, and a radioactive isotope, ^{40}K ($T_{1/2} = 1.3 \times 10^9$ years), with a relative concentration of 0.012%. When K atoms are accelerated at 4 keV, their isotope shift is increased, for example, on the $4P_{3/2} \rightarrow 21D$ transition by 20 times (from 200 MHz to 4.1 GHz), whereas the Doppler width is reduced by a factor of $2(eU/kT)^{1/2}$ and amounts merely to 5.9 MHz at a source temperature of 1073 K.

Figure 9.7 shows a schematic of the experimental setup. Metallic potassium was first atomized in an atomizer and then ionized on the surface of the tantalum foil of an ionizer that was heated to 400°C. The ions thus produced were accelerated in the electric field ($U_A = 4$ kV) of the accelerator and then collected into a parallel beam by a single lens. The ions then entered a 10-cm long recharge cell containing potassium vapor at 160°C. In the cell, 80% of the ions were converted to neutral atoms, the remaining 20% being removed by means of a capacitor filter. The effective atomic current in the beam amounted to 3 nA at a beam diameter of 3 mm. The fast atoms then passed through a field-free region where they were subjected to two-step isotope-selective excitation. In the first experiment, the atoms of all three isotopes were excited from the $4S_{1/2}$ ground state to the $4P_{3/2}$ first excited state by means of a wideband first-step dye laser ($\lambda_1 = 765$ nm, $\Delta\nu_1 = 0.8$ cm^{-1}). Then the atoms of the desired isotope were excited to the 21D Rydberg state with the aid of a narrowband second-step dye laser ($\lambda_2 = 460$ nm, $\Delta\nu_2 = 0.015$ cm^{-1}) and then ionized in the dc field of the ionizer. The ions of the desired isotope thus produced were diverted to a secondary electron multiplier for detection.

Exact tuning to resonance with the absorption lines of the different isotopes was effected by varying not the laser frequency, but the kinetic energy of the atoms. For this purpose, retardation system 4 with an applied positive retarding voltage of $U_R = 0$–320 V was installed upstream of the recharge cell. With the potassium atoms possessing a kinetic energy of 3.8 keV, a 1-eV change in the energy corresponds to a 40-MHz change in the resonance frequency at the second step. As the retarding voltage was varied, the different potassium isotopes became successively tuned to resonance with the laser radiation. The laser radiation at λ_1 and λ_2 was directed opposite to the atomic beam by means of mirrors. In the first experiments, the pulse repetition frequency of the dye lasers was not very high (20 Hz), and so not all of the desired isotope atoms could be ionized.

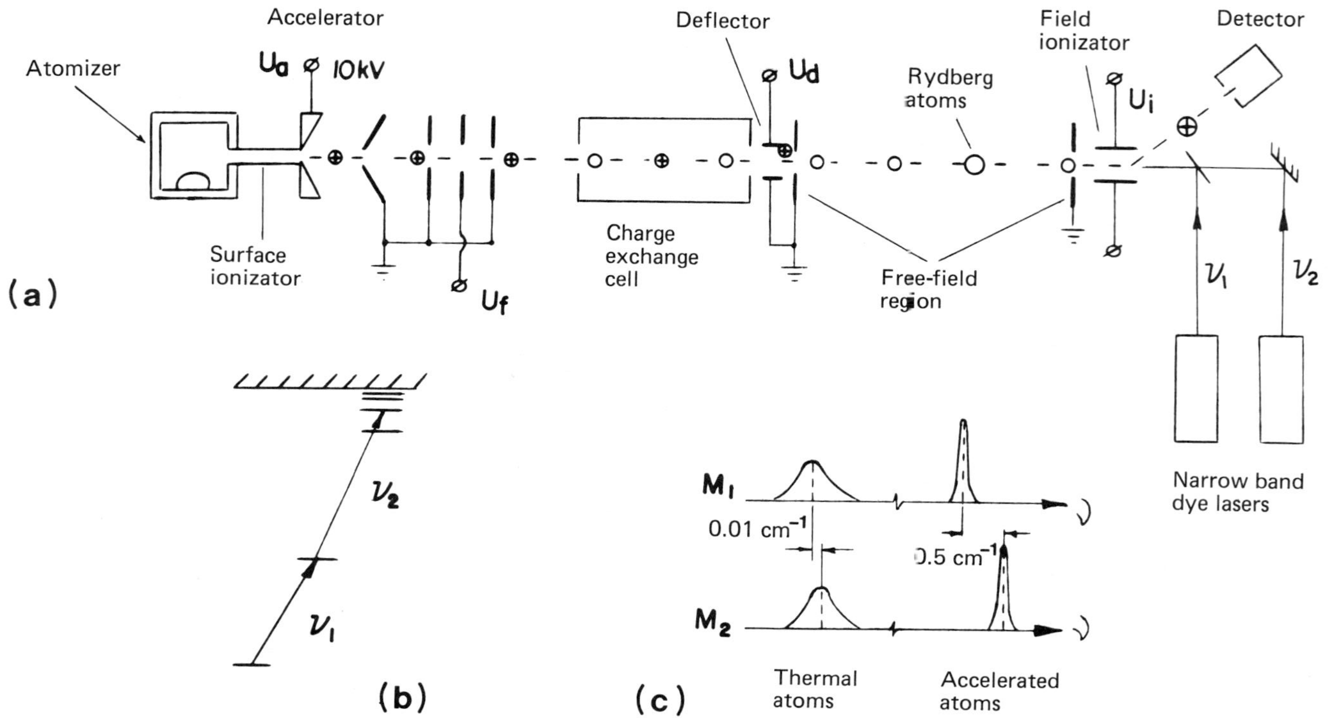

FIG. 9.7. Schematic of the experimental setup for the isotope-selective collinear two-step ionization of accelerated K atoms. [From Kudriavtzev *et. al.* (1985).]

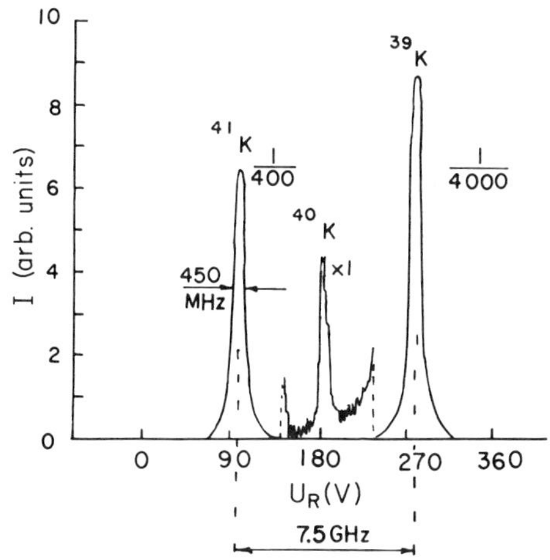

FIG. 9.8. Photoion yield for different potassium isotopes as a function of the retarding voltage in the ionization of accelerated atoms in a beam. [From Kudriavtzev *et al.* (1985).]

Figure 9.8 shows the ionization spectrum of potassium atoms accelerated to 3.8 keV. Corresponding to the higher retarding voltage is the lighter potassium isotope ^{39}K. The ratio between the intensities of the peaks for ^{39}K, ^{40}K, and ^{41}K is equal to the ratio of the natural concentrations of the isotopes. The width of the lines observed in the ionization spectrum is equal to 450 MHz and is accounted for by the effective laser radiation linewidth at the second stage. The distance between the lines for ^{39}K and ^{41}K, equal to 7.5 GHz (185 V), corresponds to the value calculated from Eq. (9.27). The ionization selectivity defined as the ratio of the signal from the ^{39}K isotope to the background noise signal is equal to 10^5. In the experiments, it was ensured by isotope-selective excitation at the second step only.

The next steps on the way to the detection of still more rare isotopes will be to improve selectivity on account of the narrowing of the laser radiation spectrum, introduce a delay between the first- and second-stage pulses, add isotope-selective excitation at the first step, and increase the laser pulse repetition frequency.

9.4. Detection of Rare Isotopes in a Radiochemical Neutrino Detector

A natural field of application of laser techniques for detecting single atoms are the numerous experimental nuclear and particle physics tasks in which extremely rare atoms need to be detected among other atoms. For instance, the possibility was discussed of applying laser stepwise photoionization to the detection of single atoms with an excited (isomeric) nucleus and superheavy (transuranic) atoms (Letokhov, 1976a), superdense nuclei (Letokhov, 1977a), nuclear shape isomers [^{m}Am-224 in particular (Balykin *et al.*, 1978)], nuclei formed upon neutrino capture (Hurst *et al.*, 1979a), and so on. Many applications of the photoionization technique in experimental weak-interaction physics (detection of solar neutrinos, double beta-ray decay) and particle physics (search for fractional charges, superheavy atoms, magnetic monopole, etc.) were discussed at the Second International Conference on Resonance Ionization Spectroscopy (Hurst and Payne, 1984).

The requirements for the photoionization technique and the conditions of its application in all these tasks are rather different. Besides, all experiments of this sort are still at the stage of discussion and realization. So considerable progress can be expected in the development of the technique exactly in connection with these interesting applications. For this reason, it would be advisable to restrict ourselves here to the discussion of the most obvious application that is now being intensively developed, namely, counting single atoms in radiochemical solar neutrino detectors.

According to the standard solar model, the neutrino flow onto the Earth appears as a result of several nuclear reactions of the proton–proton and carbon–nitrogen cycles inside the Sun (Bahcall *et al.*, 1982), which are summarized in Table 9.2. The table also lists various possible radiochemical neutrino detectors sensitive to different regions of the neutrino energy spectrum, i.e., to different sources of neutrinos.

The best known neutrino detector is the Cl–Ar type which is sensitive to high-energy neutrinos, accounting for a comparatively small proportion of the total solar neutrino flux. This type of radiochemical neutrino detector

TABLE 9.2. Solar Neutrino Sources and the Radiochemical Neutrino Detector

Source	Neutrino energy (MeV)	Neutrino flux (cm^2/s)	Detector
p–p	0–0.42	6.1×10^{10}	$^{71}Ga(\nu, e^-)^{71}Ge$
pep	1.44	1.5×10^{8}	$^{71}Ga(\nu, e^-)^{71}Ge$
^{7}Be	0.862	4.3×10^{9}	$^{81}Br(\nu, e^-)^{81}Kr$
^{8}B	0–14	5.6×10^{6}	$^{37}Cl(\nu, e^-)^{37}Ar$ $^{97,98}Mo(\nu, e^-)^{97,98}Tc$
^{13}N	0–1.2	5×10^{8}	—
^{16}O	0–1.73	4×10^{8}	—

has been realized in extremely delicate experiments with 615 tons of C_2Cl_2 on the bottom of a 1.6-km deep mine (Davis *et al.*, 1968; Davis, 1978). Single ^{37}Ar atoms produced in the target decayed on account of the beta-process, their half-life being $T_{1/2} = 35$ days, so that only 18 atoms of ^{37}Ar were contained in the 400 m^3 of the target under steady-state conditions. All these atoms could be extracted during 20 days and collected in a miniature proportional counter containing pure argon with a few ^{37}Ar atoms. The beta-decay of ^{37}Ar gives rise to Auger electrons with an energy of 2.8 keV, and it was these electrons that were registered by the proportional counter. The total efficiency of extraction and detection of the ^{37}Ar atoms amounted to 40–45%. This was the most sensitive radiochemical experiment ever to be realized. The measured rate of production of ^{37}Ar in the target was 0.39 ± 0.05 atom/day, which was by a factor of 3 to 4 lower than predicted by the standard solar model (Bahcall, 1982). This discrepancy came to be known as the "solar neutrino anomaly" and gave impetus to the realization of other radiochemical detectors (see Table 9.2).

Kuz'min (1965) proposed a Ga detector for low-energy neutrinos. The detector is based on the reaction $^{71}Ga(\nu_e, e^-)^{71}Ge$ which involves only 39.6% of ^{71}Ga contained in the natural mixture of ^{69}Ga and ^{71}Ga. In 50 tons of Ga, there occurs only one ν_e capture a day. The half-life of ^{71}Ga is $T_{1/2} = 11.8$ days, and so the steady-state concentration of ^{71}Ge in 50 tons of Ga comes only to 17 atoms. The radiochemical Ga–Ge detector can operate on the same principle as the Cl–Ar detector: (1) chemical extraction of ^{71}Ge from Ga (the process is considered to be capable of being conducted with an efficiency of more than 90%) and (2) registration of the events of radioactive decay of ^{71}Ge in the Davis proportional counter (Bahcall *et al.*, 1978; Zatsepin, 1978). The radioactive decay of ^{71}Ge proceeds via two channels:

$$^{71}\text{Ge} \begin{cases} \to (41\%) \to {}^{71}\text{Ga}^+ + e^- \quad (10.4\ \text{keV}) \quad (\text{I}) \\ \to (57\%) \to {}^{71}\text{Ga}^+ + e^- \quad (1.2\ \text{keV}) \quad (\text{II}) \end{cases} \tag{9.29}$$

thus giving rise to Auger electrons 10 keV in energy (channel I) and Auger electrons of low energy (channel II).

The decay events of the first type can be registered perfectly in the Davis proportional counter. However, those giving rise to low-energy electrons are much more difficult to register reliably because of the background noise. If these latter events were disregarded, the size of the Ga target would have to be increased 2.5 times to have the necessary signal-to-noise ratio, which would be extremely costly. Another way was suggested to surmount this difficulty, based on the discrimination of the events giving rise to low-energy electrons by means of concurrent laser detection of the resulting Ga atoms (Letokhov *et al.*, 1982).

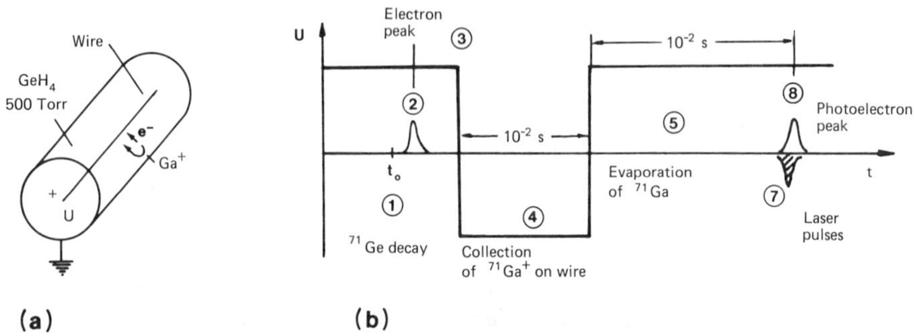

FIG. 9.9. Laser detection of the ^{71}Ge atoms produced in the radiochemical Ga detector for neutrinos. (a) Geometry of the miniature proportional counter and (b) sequence of events in the laser discrimination of low-energy electron peaks by checking on the formation of Ga atoms.

The idea of the laser radiochemical Ga detector is explained in Fig. 9.9. The Davis proportional counter is filled with GeH_4 extracted from the irradiated Ga target. The cylindrical counter has a small diameter (e.g., 5 mm) and in its center is a 5-μm filament. The beta-decay of ^{71}Ge gives rise to either K or L electrons. The K electrons, with an energy of 10 keV, are discriminated electrically. The low-energy L electrons, which are difficult to discriminate in this way, are discriminated by irradiating the counter with laser pulses capable of ionizing the Ga atoms resulting from the decay of ^{71}Ge. It is supposed that the extracted Ge can be completely freed from the Ga of the target. The decay of ^{71}Ge gives rise to $^{71}Ga^+$ ions and not to neutral atoms. It has therefore been suggested that to detect Ga, use should be made of the sequence of actions illustrated in Fig. 9.9b. Let the proportional counter give at the moment t_0 an electron signal due to a decay event of type II or due to noise. At that moment the polarity of the counter is reversed so as to collect the Ga^+ ions on the filament and thus neutralize them. The filament then can be heated in a pulsed manner to evaporate the neutral ^{71}Ga atoms. Once the atoms have diffused over a sufficient distance, two tunable pulsed lasers are switched on to effect their resonance photoionization. The photoionization signal can be used to judge the nature of the initial electron pulse which occurred at the moment t_0. Preliminary experiments on the photoionization detection of Ga atoms were already carried out (Letokhov *et al.*, 1982; Kirsten, 1984), although this is apparently the simpler part of the whole experiment. There are a number of difficulties on the way to the realization of such a laser–radiochemical detector. Some of these were discussed by Letokhov *et al.* (1982).

A ^{81}Br–^{81}Kr radiochemical detector is now being developed for detecting solar neutrinos with a maximum energy of 470 keV, which is most sensitive produce ^{81}Kr with a very long half-life ($T_{1/2} = 2 \times 10^5$ years), and proportional counters are unfit for detecting such isotopes. A natural way out of this situation is the use of laser resonance ionization to count single ^{81}Kr atoms (Hurst *et al.*, 1980, 1984b). With the target size being the same as in the case of the Cl–Ar detector, it is expected that around two ^{81}Kr atoms/day will be obtained for such bromine compounds as $CHBr_3$ and CH_2Br_2. In this way, it will be possible to obtain a few hundred ^{81}Kr atoms during an irradiation period of half a year. It was demonstrated that such a quantity of Kr atoms can be fully extracted by the same method as that used with the Cl–Ar detector (Hurst *et al.*, 1984a).

The laser detection of ^{81}Kr can be effected by a combination of the resonance photoionization of Kr atoms (Z-selectivity) with the mass separation of the photoions produced (A-selectivity). This technique has been described in Section 9.3.1, and a schematic diagram of the photoionization detector has been shown in Fig. 9.5. The detected photoions are then accelerated to 10 keV and implanted into a silicon target. In such experiments, the detection selectivity is limited by the selectivity of the mass filter used (around 5×10^3). The sample obtained after extracting ^{81}Kr from the irradiated target is expected to contain about 10^8 ^{82}Kr atoms in addition to 500 ^{81}Kr atoms. Therefore, the silicon target will have implanted into it, alongside the 500 ^{81}Kr atoms, also about 2×10^4 ^{82}Kr atoms. To improve selectivity, it is contemplated that the process will be repeated at least once more, as described in Section 9.3.1 (see Fig. 9.4). For this purpose, the silicon target will be subjected to pulsed laser annealing, as a result of which the Kr atoms will come out of the target into an evacuated volume. It is expected that after the second cycle, the silicon target will have implanted into it 500 ^{81}Kr atoms as before and only 40 ^{82}Kr atoms. If necessary, the process can be repeated again, the number of photoions (which must belong to the ^{81}Kr atoms only) being registered. Based on preliminarily experiments, including the photoionization detection of 10^3 Kr atoms, a conclusion was drawn that the ^{81}Br–^{81}Kr laser–radiochemical detector for solar neutrinos was quite feasible (Hurst *et al.*, 1984b).

Also, the possibility is now being discussed of detecting the flow of solar neutrinos issued by ^8B with the aid of the $^{97,98}Mo(\nu, e^-)^{97,98}Tc$ reaction (Cowan and Haxton, 1982). The resulting ^{97}Tc and ^{98}Tc isotopes have a long half-life (2.6×10^6 and 4.2×10^6 years, respectively), and so it is natural to employ the highly selective stepwise laser photoionization method for their detection (Cowan *et al.*, 1984).

10. Applications in Molecular, Chemical, and Statistical Physics

Photoionization spectroscopy is effective in all experiments in which very small partial concentrations of atoms and molecules of a given sort must be measured. Such measurements can be carried out (1) with selection of a type of atom or molecule, (2) with selection of a quantum state of an atom or molecule, (3) with time resolution, (4) with spatial resolution, and (5) with concentrations of up to single particles in the volume observed when statistical effects become essential. A combination of these properties makes photoionization spectroscopy a very useful addition to the available spectroscopy applied in molecular, chemical, and statistical physics. This chapter is briefly concerned with some of the applications of the photoionization method.

10.1. Measurement of Ultralow Concentrations of Atoms and Molecules

The photoionization method can be applied to measure low concentrations of atoms and molecules. In the second case, it is possible to ionize molecules directly or ionize the atoms formed by the photodissociation of the molecules. Such measurements can be taken both in vapor and in molecular beams. These possibilities are considered separately below.

10.1.1. Atoms in Vapor. Absolute Measurement of Partial Pressure

Resonant photoionization of atoms can be realized for atoms in a buffer gas and in low-pressure atomic vapors. In both cases, we can attain a sensitivity at single-atom level in a required (irradiated) volume, and absolute calibration of the number of ionized atoms can be accomplished.

10.1.1.1. Atoms in a Buffer Gas. The first experiments on the ionization of Cs atoms in the buffer gas of a proportional counter (Hurst *et al.*, 1977a) show that the method of resonant ionization is ideal for measuring ultralow partial concentrations of definite atoms. For precise measurement of a concentration of atoms N_a, we should know the following quantities: (1) the volume of the irradiated region where resonant photoionization is carried out, (2) the photoionization yield Y_i, and (3) the detection efficiency of the resultant photoelectrons (photoions). Experiments must be performed with laser beams of definite geometry. The ionization yield is determined beforehand, for example, by choosing a mode of saturation for the exciting and ionizing steps, when $Y_i \simeq 1$ (see Chapter 3) and the resonant atoms in the irradiated volume are fully converted to photoions and photoelectrons. If the ground atomic state, however, is split into several sublevels, each of them populated according to the Boltzman distribution at a given temperature T, it is necessary to calculate the relative thermal population of the ground state sublevel from which the resonant excitation and ionization of the atom occurs.

Finally, the photoelectron signal in the proportional counter can be subjected to analog and digital measurement with 100% efficiency (Hurst *et al.*, 1978a). Analog measurement is applied in experiments when large numbers of photoelectrons are formed and the signal is proportional to their number. Digital measurement is used when just one or several photoelectrons are formed and their number is measured precisely. For absolute calibration an x-ray source can be used. It produces a definite average number of electrons in the proportional counter during a pulse (Hurst *et al.*, 1978a).

10.1.1.2. Atoms in Low-pressure Vapor. Photoionization experiments in low-pressure vapor with a sensitivity of several atoms are possible with the technique of extraction and detection of photoions from a quasi-closed cavity (Andreyev *et al.*, 1985). Such an approach in the photoionization detection of single atoms is considered in Section 4.2. It was successfully used in experiments with ultralow-pressure Sr vapors with sufficiently high ionization efficiency (17%) (Andreyev *et al.*, 1986). In other words, every sixth atom of Sr in a quasi-closed heated cavity was converted to a photoion that could be recorded. Such a high sensitivity of measurement of partial pressure (about 10^{-15} Torr) was attained with three-step photoionization of Sr atoms by dye-laser radiation with a moderate peak power and a high pulse-repetition rate (10 kHz). Due to this high laser pulse-repetition rate, an atom randomly moving in the quasi-closed cavity was able to fall within the irradiation region many times and, finally, be ionized. It is this factor that leads to high total probability of ionization of atoms in a closed heated cavity.

This method of measurement is useful because it can be applied for direct measurement of the equilibrium pressure of atoms in the region of ultralow concentrations, up to several atoms in the volume under study.

10.1.2. Atoms in a Beam. Measurement of Beam Geometry

It is much easier to measure the concentration of atoms in a beam than in low-pressure vapor, since in the first case it is simple to detect the resulting photoions (see Section 4.2). Figure 4.3a shows some dependences of the concentration of Yb atoms in a beam on the temperature of the source measured by this method, which is rather universal and can be applied to many elements refractory elements, in particular). However, the results of measurement must be extrapolated to the case of equilibrium pressure of the atomic vapor very carefully, and only when there is no uncontrollable attenuation of the atomic beam as it leaves the source. In the low-temperature range, the measured pressure is at variance with the results of equilibrium pressure extrapolation (the dashed curve in Fig. 4.3a).

In the measurement of ultralow concentrations, when there is about one atom in the irradiated region, strong fluctuations are inevitable in the number of detectable atoms obeying the Poisson distribution (Fig. 4.3b,c):

$$P(N_i) = (\bar{N}_i^{N_i} e^{-\bar{N}_i})/(N_i!), \qquad (10.1)$$

where $\bar{N}_i = Y_i\bar{N}_a$ is the average number of ions formed by pulsed irradiation of an atomic beam, Y_i is the average ionization yield per pulse, and $\bar{N}_a$ is the average number of atoms in the region under irradiation. Figure 10.1 shows a measured dependence of the relative fluctuations of the number of ions $\delta = \overline{[(\Delta N_i)^2]}^{1/2}/\bar{N}_i$ on the average number N_a of atoms in an experiment with a Yb beam (Bekov *et al.*, 1978d). This figure also presents a dependence for the Poisson distribution that looks like $\delta(N_a) \approx 1/\bar{N}_a^{1/2}$.

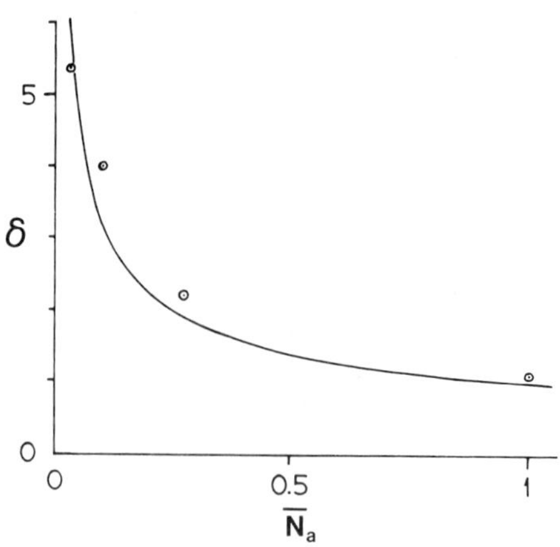

FIG. 10.1. Dependence of the relative fluctuations of the ionization signal on the average number of atoms $\bar{N}_a$ in the region of detection. [From Bekov *et al.* (1978d).]

These observations lead to a simple method of absolute calibration in measuring ultralow concentrations of atoms. When the irradiation volume V_{irr} is reduced to $(Y_i\overline{N_a})^{-3}$, one photoion, on the average, is formed in the irradiated region, and so strong fluctuations in their number can be observed. If the photoionization yield Y_i is known, its value is large enough, and its fluctuations are unessential (all the excitation steps are saturated), the fluctuations observed will be directly associated with a small number of atoms in the volume. The measurement of this limit of strong fluctuations gives at once an absolute value of partial atomic pressure:

$$\overline{N_a} \simeq (Y_i V_{irr})^{-1}. \tag{10.2}$$

This method of absolute calibration is very handy for experiments with beams and low-pressure vapors, particularly for analytical applications (Chapter 8).

Focusing laser pulses onto a certain point of the atomic beam makes it possible to measure the concentrations of atoms with a spatial resolution, i.e., to carry out three-dimensional measurement of the atomic beam geometry. Such an experiment was performed with an atomic beam (Tursunov and Eshkobilov, 1982). This method is of interest for the diagnostics of atomic beams in molecular epitaxy setups.

10.1.3. Molecules in Vapor

The molecular vapor pressure can be measured through direct photoionization as well as by photodissociation with a known yield and subsequent photoionization of the resultant atoms. The second method was demonstrated in experiments with RbI and KI molecules (Young *et al.*, 1979). The photodissociation of these molecules by UV laser pulses (N_2 laser, $\lambda =$ 337 nm) gives rise to atoms of alkali elements in the ground state which can be easily detected by two-step photoionization. If the photodissociation yield is unknown, the temperature dependence of pressure is usually extrapolated to a region of higher temperatures where the temperature dependence is known from independent measurements. This method makes it possible to obtain a vapor pressure curve in a very wide range, from 10^3 to 10^{-10} Torr. The case of such measurement for RbI molecules is shown in Fig. 10.2 (Capelle *et al.*, 1984). In this experiment, the equilibrium partial pressure of RbI vapor was measured in the buffer gas of a proportional counter (10% CH_4 + 90% Ar, 200 Torr). The photodissociation yield of RbI molecuules was about 10^{-5}, and the volume of the irradiated region 0.04 cm^3. So the limit of pressure measurement lay in the region of 10^{-10} Torr when just several photoelectrons were formed by the photoionization of Rb atoms in the exposed region.

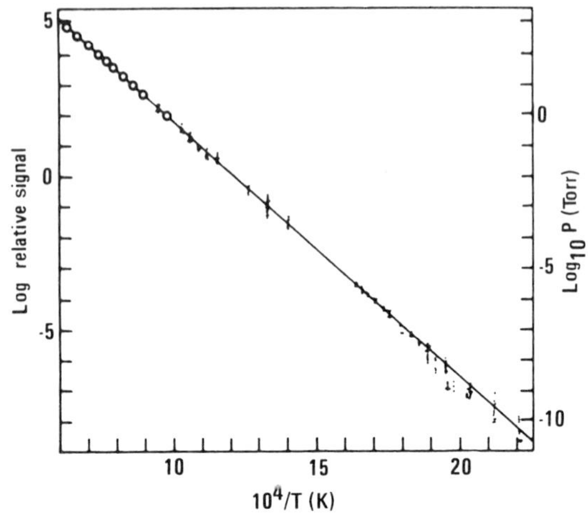

FIG. 10.2. Temperature dependence of the vapor pressure of RbI molecules measured by the method of RbI photodissociation and subsequent resonant ionization detection of Rb atoms in a buffer gas. [From Capelle *et al.* (1984).]

The method of resonant photoionization can be evidently applied for the systematic measurement of temperature dependences of vapor pressure in the region of ultralow concentrations that has not been studied yet.

10.2. Detection of Primary Products of Photodissociation

One of the most important problems of gas-phase photochemistry and the kinetics of gas-phase chemical reactions is direct detection of the primary products formed in the course of both monomolecular and bimolecular processes. In this case, method is needed that allows selective detection of atoms and radicals with high sensitivity and time resolution under conditions free of collisions between the resulting primary products and other particles and between each other. For this purpose, different methods are used: laser excitation of the resultant product fluorescence (Zare and Dagdigian, 1974; Reisler *et al.*, 1980), mass spectrometry in molecular beams (Schulz *et al.*, 1979), etc. Much progress has been made in studying the kinetics of gas-phase chemistry and photochemistry, but the sensitivity and time resolution of the existing methods are still far from the required values. Therefore, the application of the photoionization detection of atoms and radicals for this purpose seems to be highly promising. This has been demonstrated in a number of experiments, especially in studying the primary products of IR multiphoton dissociation of molecules. All these possibilities are considered in more detail below.

10.2.1. Detection of Atoms

The photoionization method can be used for the detection of atoms formed in any process of molecular dissociation: (a) UV photodissociation, (b) IR multiphoton dissociation, and (c) IR + UV photodissociation. This method enables atoms to be detected in certain quantum states. That is the advantage of the photoionization method over, say, the method of mass spectrometry.

10.2.1.1. UV Photodissociation. The first experiments in this direction were carried out with the CsI molecule (Grossman *et al.*, 1977a,b). This molecule has a wide photoabsorption band in the range from 3100 to 3400 Å which leads to photodissociation yielding Cs and I atoms in ground states. It is especially convenient to detect Cs atoms by means of resonant two-step photoionization in a one-frequency laser pulse with $\lambda = 4593$ Å. Such experiments were performed with CsI molecules in the buffer gas of a proportional counter. The photodissociation cross-section value was attained by measuring the photoionization signal as a function of the photodissociating laser pulse fluence and wavelength $\sigma_{\text{phd}}(\lambda)$.

The atoms formed by pulsed photodissociation in a strictly localized region begin participating in different processes. First, diffusion of atoms in the buffer gas and, second, chemical reactions with the impurities in the buffer gas take place. Both processes can be investigated by observing the variation of the Cs concentration in time and space.

The diffusion of Cs atoms was studied thus (Hurst *et al.*, 1978b). A narrow laser beam with λ_{phd} in the region of CsI photoabsorption set up a certain concentration of Cs atoms inside the cell at the instant $t = 0$. In parallel with this beam, a laser pulse was passed at the distance ρ with delay Δt. This pulse ionized the Cs atoms which diffused at the ρ distance. The diffusion of atoms from a linear source is described by the relation

$$N(\rho, t) = \frac{\lambda}{4\pi D_1 t} \exp\left(-\frac{\rho^2}{4D_1 t}\right) \exp(-\beta t), \tag{10.3}$$

where D_1 is diffusion coefficient of trace Cs atoms in a buffer gas, β is the rate of loss of reactive Cs atoms duc to their participation in the chemical reaction. Both coefficients D_1 and β can be determined by measuring the dependence $N(\rho, t)$.

It is essential that the D_1 quantity be measured accurately even in a highly reactive medium when the β coefficient is great. For this purpose, it is convenient to measure the concentrations of atoms at two different distances ρ_1 and ρ_2. Then the value

$$\ln[N(\rho_1, t)/N(\rho_2, t)] = (\rho_2^2 - \rho_1^2)/(4D_1 t) \tag{10.4}$$

does not depend on the β coefficient. The theoretical dependence (10.4) agrees well with the experimental one. The slope of the curve (10.4) enables the diffusion coefficient D_1 to be determined accurately.

The kinetics of damping of $N(\rho, t)$ due to the death of the atoms, i.e., the β coefficient, can be measured by varying the delay between the dissociating and ionizing laser pulses Δt. It is very convenient to take such measurements in the geometry of coaxial laser beams when the ionizing laser beam is a cylinder coaxial about the dissociating laser beam (Grossman *et al.*, 1977a). This method was applied to investigate the death of Cs atoms in their chemical reaction with the oxygen molecules added into the buffer gas. The presence of a buffer gas makes it more difficult, of course, to interpret the obtained results because the collisions affect the energy of the intermediate products of chemical reactions.

10.2.1.2. IR Multiphoton Dissociation. This type of monomolecular decay of a polyatomic molecule under collisionless conditions has been considered actively in recent years, because this process is ideal for checking the basic statements of the statistical theory of monomolecular reactions (RRKM theory) (Robinson and Holbrook, 1972) and because of its practical importance in isotope separation (Bagratashvili *et al.*, 1985c). When a molecule undergoes IR dissociation in the ground electronic state, the weakest bond is usually broken. In many polyatomic molecules absorbing 10-μm radiation of a CO_2 laser, bonds to halogen atoms (I, Br, Cl) are the weakest bonds, since they usually break away in the course of molecular dissociation. It is rather difficult to realize spectroscopic detection of halogen atoms because their resonant absorption lines lie in the vacuum UV region. This is also true for the absorption, fluorescence, and photoionization methods to an equal degree. However, in the case of photoionization, due to a high detection sensitivity of the excited atoms by their subsequent ionization, it is possible to use two- or three-photon resonant excitation of atoms by tunable laser radiation in the accessible UV region (the schemes in Figs. 3.18 and 3.19).

Such photoionization experiments with I atoms were carried out in the course of IR photodissociation of perfluoroalkyl iodiodes (CF_3I, C_2F_5I, C_3F_7I, and n-C_6F_7I) (Hackett *et al.*, 1983). As these molecules are irradiated by CO_2 laser pulses with an energy fluence of about 1 J/cm^2, their photodissociation occurs to yield an iodine atom in the ground electronic state. The gas pressure being about 0.1 Torr, the concentration of I atoms (10^{13}–10^{15} cm^{-3}) can be detected even in a simple ionization diode. For this purpose, a dye laser pulse with an energy of 2 mJ and a wavelength tuned in the range from 470 to 500 nm, was focused into the region between the cathode and the anode. The radiation intensity in the focus exceeded 10^9 W/cm^2, so the three-photon resonant excitation of iodine atoms at the transition $5s(^2P^0_{3/2}) \rightarrow 6s(^4P^0_{3/2})$ at $\lambda = 485.28$ nm became probable. Subsequent two-photon ionization of excited atoms at the transition to the continuum (Fig. 3.19b) came about under saturation. So the ion signal of such a five-photon process was proportional to the third-step intensity of laser radiation. The delay between the ionizing dye laser pulse and the photodissociating CO_2 laser pulse varied between 0 and 100 μs, which allowed the detection of iodine atoms at different times after the dissociation of the molecule.

This method was used to measure the dependence of the molecular IR photodissociation yield on the CO_2 laser pulse fluence. The absence of an ionization signal from electronically excited iodine atoms shows that, in the process of IR photodissociation, only iodine atoms in the ground state are formed. Although the sensitivity attained in the simple diode experiment was not high (3×10^{13} atoms/cm^3), it can be increased apparently by many orders with optimized parameters.

The next step in using this method of detection was taken by Bagratashvili *et al.* (1985a,b) with the CF_3I molecule. The essence of these experiments is explained by Fig. 10.3. Under the action of a CO_2 laser pulse, multiphoton excitation of CF_3I molecules to the vibrational levels near the dissociation limit D_0 took place. A fraction of the molecules β dissociated to form $I(^2P^0_{3/2})$. The concentration of these atoms, and hence the IR photodissociation yield, were measured through three-photon ionization by dye-laser second-harmonic radiation at $\lambda_1 = 3047$ Å. Figure 10.4 shows a scheme of quantum

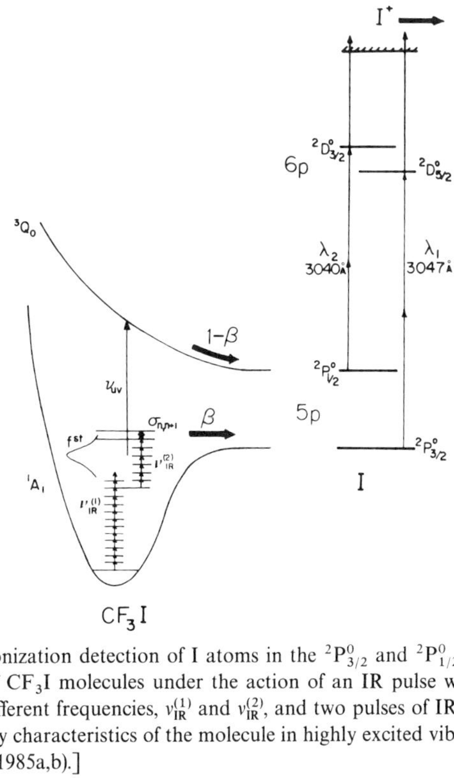

FIG. 10.3. Photoionization detection of I atoms in the $^2P^0_{3/2}$ and $^2P^0_{1/2}$ states formed by the photodissociation of CF_3I molecules under the action of an IR pulse with $\nu^{(1)}_{IR}$, two pulses of IR radiation with different frequencies, $\nu^{(1)}_{IR}$ and $\nu^{(2)}_{IR}$, and two pulses of IR and UV radiation. It allows studying many characteristics of the molecule in highly excited vibrational states. [From Bagratashvili *et al.* (1985a,b).]

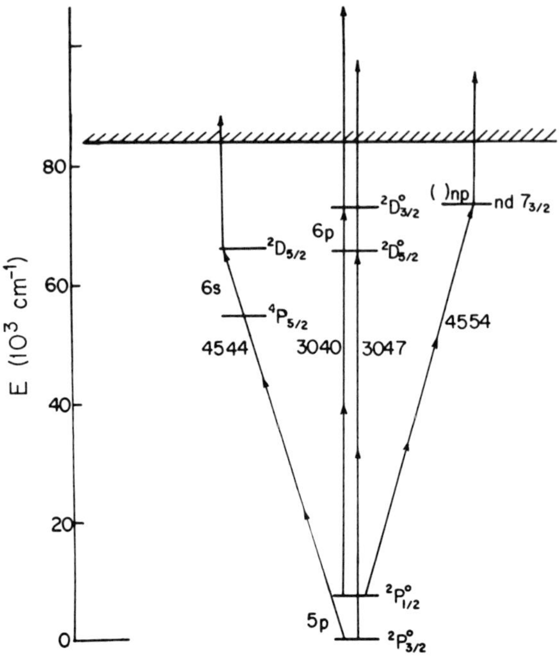

FIG. 10.4. Multistep photoionization of I atoms in the ground $^2P^0_{3/2}$ and excited $^2P^0_{1/2}$ states at two-photon and three-photon resonant excitation of intermediate high-lying states.

transitions bringing about the ionization of iodine atoms in the ground ($^2P^0_{3/2}$) and electronically excited ($^2P^0_{1/2}$) states. In the case of short-wave radiation, the ionization occurs during absorption of only three photons with two-photon resonant excitation. Such a scheme allows the use of less intense radiation with it is focused onto a spot with a diameter of just 0.25 mm.

The nondissociated CF_3I molecules, their fraction being $1 - \beta$, remain in highly excited vibrational states. Their absorption band is shifted from the UV region to the visible range. If the excited molecules are irradiated by an additional pulse of visible radiation at $\lambda_2 = 450$ nm, it is possible to induce their photodissociation through an excited electronic term yielding electronically excited iodine atoms ($^2P^0_{1/2}$). They are detected by means of three-photon ionization at $\lambda_1 = 3040$ Å (Fig. 10.4). By measuring the dependence of the concentration of excited I* atoms on the visible laser pulse fluence, it is possible to determine the photodissociation cross section of vibrationally excited CF_3I molecules ($\sigma = 5 \times 10^{-21}$ cm^2) and the fraction of molecules $(1 - \beta)$ not subjected to IR multiphoton dissociation. This method of double IR–visible photodissociation is of interest because it permits the measurement of small fractions of nondissociated molecules, not by measuring the value of β, but $(1-\beta)$ directly. Finally, the vibrationally excited molecules on the last discrete vibrational levels below the dissociation limit D_0 can be irradiated by the second pulse of the CO_2 laser. The frequency of the second pulse can be chosen in a more exact resonance to the IR absorption of highly excited molecules. If the second IR pulse energy is sufficient, the vibrationally excited molecules at the transitions to the vibrational continuum will dissociate. The resulting additional atoms $I(^2P^0_{2/2})$ can be detected again by means of three-photon ionization at $\lambda_1 = 3047$ Å. In this way, by varying the $\nu_{IR}^{(2)}$ frequency and energy fluence of the second IR pulse, it is possible to measure the cross section of the last vibrational transition σ_{n_D,n_D+1} ($\simeq 10^{-19}$ cm^2) and its dependence on the IR radiation frequency (Bagratashvili *et al.*, 1985b).

The method of photoionization detection of atoms with a time resolution of 10^{-8}–10^{-12} s has great potential for studying the dynamics of the monomolecular decay of molecules excited over the dissociation limit, as well as for direct investigation of the fluctuations in the probability of decay of single molecules (Letokhov and Makarov, 1981).

10.2.2. Detection of Radicals

Monomolecular decay of polyatomic molecules always results in one or two primary radicals. It is particularly interesting to detect the radicals after the molecule has decayed and to determine the store of their internal (electronic, vibrational, and rotational) and kinetic energies. The detection of simple radicals, like halogen atoms, usually calls for laser VUV or UV sources or multiphoton ionization by high-intensity visible radiation. Both approaches have been applied in a number of experiments.

10.2.2.1. "Soft" Photoionization by VUV Radiation. Vacuum VUV laser pulses can be used to realize one- or two-step photoionization of, perhaps, any simple radical whose absorption lines lie in the VUV range. It is advisable to use the mass-selective detection of the resulting molecular photoions for their unambiguous identification. This method was applied in successful experiments on the two-step photoionization detection of the CH_3 and CD_3 radicals formed by UV photoionization of the CH_3I and CD_3I molecules (Danon *et al.*, 1982). In the experiments, two laser sources were used at $\lambda_1 = 215$–216 nm with $I = 10^6$ W/cm^2 (for resonant excitation of radicals at the transition $\tilde{B}^2A_1' \leftarrow \tilde{X}^2A_2''$) and at $\lambda_2 = 266$ nm with $I = 10^7$ W/cm^2 (for ionization of excited radicals). It is essential that the $\tilde{B}$ state not fluoresce because of fast predissociation. Nevertheless, the good time resolution (6–7 ns) and high sensitivity of the photoionization method make it possible to detect such a fast decaying intermediate state.

Single-step photoionization by more rigid VUV radiation at $\lambda = 118.4$ nm was used to detect the primary CH_3CO radicals formed in the process of IR multiphoton dissociation of $(CH_3CO)_2O$ molecules by a CO_2 laser pulse

(Feldman *et al.*, 1983). To produce a VUV radiation pulse, the frequency of the third-harmonic radiation of a Nd:YAG laser at $\lambda = 335$ nm was tripled. Even a low-energy VUV pulse (10^{10} photon/pulse) was sufficient to produce a reliable photoion signal.

10.2.2.2. Multiphoton Ionization by Visible and UV Radiation. In a number of experiments, multiphoton ionization was used to detect the simple radicals formed after pulsed IR photolysis of polyatomic molecules. It is known, for example, that the CH_3NO_2 molecule dissociates under the action of a CO_2 laser pulse into the NO_2 molecule and the CH_3 radical (Ambartzumian *et al.*, 1975b). Multiphoton ionization was successfully used to detect both of these fragments (Rockney and Grant, 1981, 1982). Figure 10.5 shows the geometry of the irradiation region in these experiments. A pulsed supersonic molecular jet irradiated by a focused CO_2 laser pulse was a source of CH_3NO_2 molecules. The photodissociation of CH_3NO_2 gave rise to CH_3 radicals and NO_2 molecules, which were photoionized by a focused pulse of a dye laser tunable in the visible range (450–510 nm). The resultant molecular photoions were detected with a quadrupole mass spectrometer. The NO_2 molecules were detected through three-photon ionization by radiation at $\lambda = 511.0$ nm. The sensitivity was quite sufficient for detection the NO_2 molecule at a pressure of 5×10^{-4} Torr. By tuning the dye laser radiation frequency, it was possible to measure the three-photon ionization linewidth, which was abnormally broadened due to the formation of vibrationally "hot" NO_2 molecules in the process of IR photodissociation of CH_3NO_2 (Rockney and Grant, 1981).

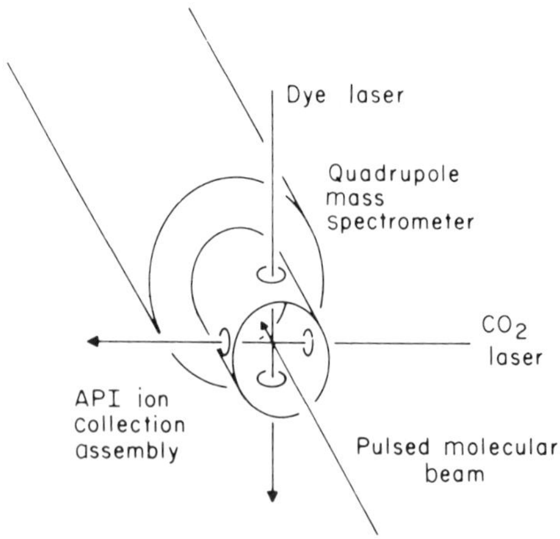

FIG. 10.5. Geometry of irradiating a pulsed supersonic jet of molecules by a CO_2 laser pulse inducing photodissociation of polyatomic molecules and by a dye laser pulse inducing multiphoton ionization of radical, i.e., IR photodissociation products. Molecular photoions are detected with a quadrupole mass spectrometer. [From Rockney and Grant (1981).]

The NO_2 molecule can also be detected when it is irradiated by a pulse at $\lambda = 450.0$ nm, which brings about two-photon dissociation of NO_2 to NO and $O(^1D)$. The same radiation ionizes NO through a resonant intermediate state A (Rockney and Grant, 1982). In much the same way, the CH_3 radical is detected by three-photon resonant excitation of the $\tilde{C}$ state at $\lambda = 450.9$ nm.

The photodissociation fragments can be detected with angular and time-of-flight resolution by scanning the focusing point of the photoionizing pulse around the region of molecular photoionization and varying the delay time between these pulses. This was demonstrated with success in experiments with the CH_3NO_2 molecule (Rockney and Grant, 1982). The measuring accuracy of the angular distributions of both fragments was sufficient to determine the excess energy transferred to the translational motion of fragments during photodissociation ($E_{tr} = 2800\ cm^{-1}$).

Multiphoton ionization detection of radicals can be realized using time-of-flight mass-selective detection of photoions. Such a technique was applied in experiments by Chou *et al.* (1983). The IR photodissociation of dimethyl methylphosphonate (DMMP) gave birth to the PO radical, which was detected through multiphoton ionization by either one-frequency radiation at $\lambda_1 = 324$–327 nm or additional radiation at $\lambda_2 = 266$ nm. The linewidth of the photoionization spectrum shows that the resultant PO radicals are cold to vibrations and rotations.

The ionization technique for detection of primary radicals is rather simple and very effective. According to many published papers, it is becoming routine for the study of chemical kinetics. Table 10.1 contains the results of some

TABLE 10.1. Laser Photoionization Detection of Some Radicals

Radical	Wavelengths of laser pulses (nm)	Ionization scheme	Reference
CO	300–310; 480	$2\omega_1, \omega_2, \omega_1$	Sha *et al.*, 1984;
	1060; 532		Huillier *et al.*, 1984
NO	383.3	$2\omega, \omega$	Radhakrishnan and Estler, 1983
PH	380–395	$2\omega, \omega$	Ashfold *et al.*, 1984
PO	324–327	2ω	Chou *et al.*, 1983
	266; 324–327	ω_1, ω_2	Chou *et al.*, 1983
NO_2	511	3ω	Rockney and Grant, 1981
CH_3	450.9	$3\omega, \omega$	Rockney and Grant, 1982
	216; 266	ω_1, ω_2	Danon *et al.*, 1982
CD_3	216; 266	ω_1, ω_2	Danon *et al.*, 1982
CH_2CO	118.4	ω	Feldman *et al.*, 1983
UF_5	433	$n\omega$	Chou *et al.*, 1982

experiments on the photoionization detection of radicals and denotes the wavelengths of laser radiation and the types of multiphoton processes. It is evident that in the next few years optimal schemes of photoionization will be found for most simple radicals, and their photoionization detection will be as reliable and convenient as for atoms.

10.3. Study of Statistical Effects

In experiments on the detection of one or a few particles (atoms, molecules), we must deal with statistical effects. The average amplitude of signal fluctuation ΔN_{fl}, as N particles are detected, is

$$\Delta N_{fl} \simeq \sqrt{N}. \tag{10.5}$$

In the measurement of classical particles, the fluctuation distribution obeys the Poisson distribution (10.1). The photoionization detection of single particles makes the direct measurement of these fluctuations and their distributions in space and time possible.

10.3.1. Fluctuations of Atomic and Molecular Processes

In the first experiments on photoionization detection of single atoms with saturation at each excitation step, the number of atoms in the irradiated volume fluctuated. In the experiments with Cs atoms in a buffer gas (Hurst *et al.*, 1977a,b), the Poisson distribution for the probability of atomic detection was observed when N was small. The form of this distribution is given in Fig. 4.8. In the subsequent experiments of the same group, this method made it possible to observe the fluctuations of gas gain of the proportional counter (Hurst *et al.*, 1978a). Similarly, in the experiments with low-intensity atomic beams, fluctuations in the number of atoms detected in the irradiated volume were observed. Figure 4.3 shows the Poisson distribution attained in experiments with a beam of Yb atoms.

The observation of the fluctuation in atomic and molecular processes was the subject of the special experiments carried out in Oak Ridge with a photoionization proportional atom counter (Iturbe *et al.*, 1982). The essence of this experiment is explained in Fig. 10.6. A small addition of LiI molecules is introduced into the chamber of the counter. A laser pulse at $\lambda = 290$ nm photodissociates these molecules, forming Li atoms in the ground state. The Li atoms were detected through two-step resonant photoionization in a two-frequency field (Kramer *et al.*, 1979). The photoelectrons formed by ionization drifted in the proportional counter with a gain of over 10^3. The photoelectron cell signal could be detected both with digital techniques for counting one

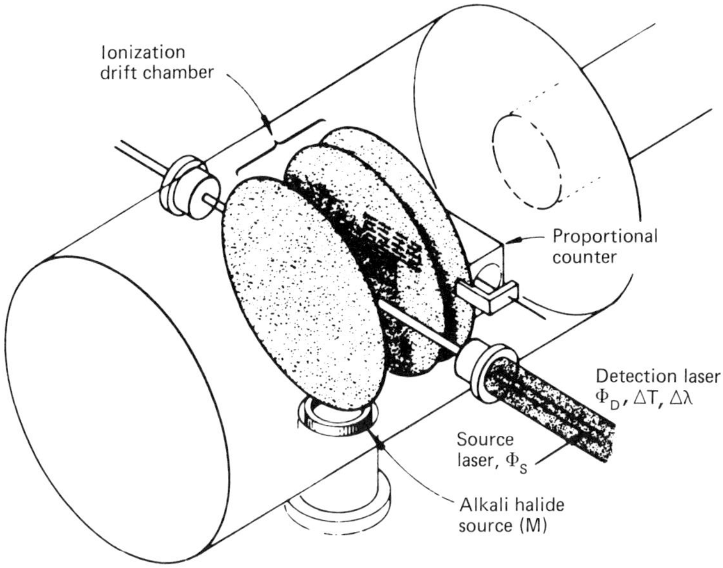

Physical phenomena that fluctuate	(M)	Φ_S	Φ_D	ΔT	$\Delta\lambda$
None-No observable fluctuation	Large	Large	Large	0	0
1. Molecular concentration	Small	Large	Large	0	0
2.Photodissociation yield	Large	Small	Large	0	0
3. Atomic concentration	Large	Large	Large	>0	0
4. Photoionization yield	Large	Large	Small	0	0
5. Collision frequency	Large	Large	Large	0	>0

FIG. 10.6. Scheme of an experiment and its modes for observing the fluctuations of different physical phenomena by the method of photoionization detection. [From Iturbe *et al.* (1982, 1984).]

photoelectron and with analog techniques for measuring the number of photoelectrons

Depending on the parameters of the processes of laser photodissociation of LiI molecules and laser photoionization of Li atoms, it was possible to study the fluctuations of five different phenomena, as shown in Fig. 10.6. With a small number of Li molecules in the irradiated volume and high yields of photodissociation (a high fluence of photodissociating radiation Φ_S) and photoionization (a high fluence of photoionizing radiation Φ_D), it was possible to investigate the fluctuations in the number of molecules. With high concentrations of molecules but with a low value of Φ_S, it was possible to study the fluctuations of the photodissociation yield. By varying the delay

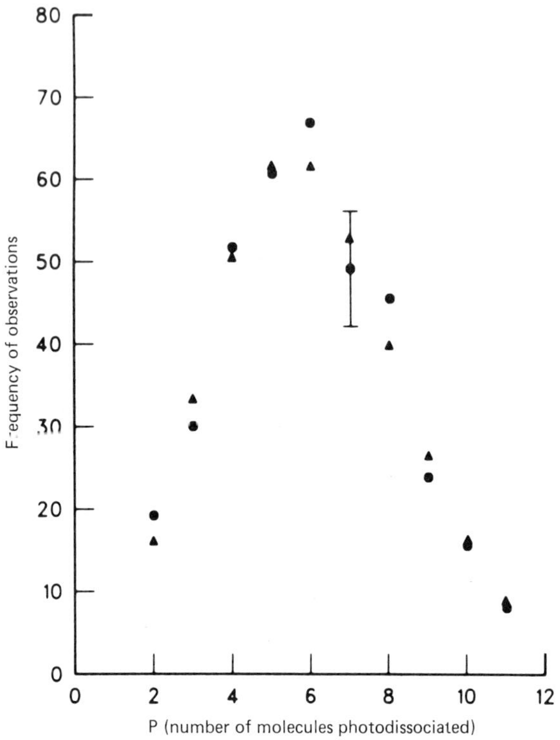

FIG. 10.7. Fluctuations in the photodissociation yield of the LiI molecule measured experimentally and the theoretical Poisson distribution. [From Iturbe *et al.* (1982, 1984).]

Δt between the dissociating and ionizing pulses, it was possible to detect a small number of remaining free Li atoms and measure the fluctuations of their number in the region of irradiation. Similarly, with a low photoionization yield, the fluctuations of this process could be measured, and by detuning the wavelength $\Delta\lambda$ of the radiation exciting the Li atoms, it was possible to investigate the fluctuations in the number of collisions responsible for the collisional wing of the resonant absorption line. The fluctuations of all five studied obey the Poisson distribution. For illustration, Fig. 10.7 shows the measured distribution of fluctuations in the photodissociation yield (Iturbe *et al.*, 1982, 1984). The experimental results agree well with the Poisson distribution.

10.3.2. Test of the Ergodic Hypothesis

Statistical physics is based on the ergodic hypothesis, according to which averaging over an ensemble can be replaced by averaging over time. This hypothesis is valid when, in the course of evolution, the system considered successively holds all the admissible volume of phase space. It is interesting, of course, to test this hypothesis in a direct experiment, considering, for example, the gas diffusion from one wall to another, as was suggested by Einstein (1956). Assume that at the instant of time $t = 0$, many atoms release in a plane and diffuse to a parallel plane spaced a distance ρ. It is possible to measure the atomic density $N(\rho, t)$ in this plane, which must obey Eq. (10.3). In an experiment, when in the plane $\rho = 0$ at the instant $t = 0$ many atoms release, summation over space or ensemble comes about. On the other hand, it is possible to release a few atoms and measure the time of their arrival t_j to the ρ plane. To measure the distribution function $N(\rho, t)$, we should repeat this experiment many times, that is, carry out time averaging.

The measurement of $N(\rho, t)$ in two such cases was realized with the photoionization method used for measuring concentrations of atoms (Iturbe *et al.*, 1983, 1984). Instead of a plane source of atoms, a linear source of Li atoms based on the pulsed photodissociation of LiI molecules by a laser beam was used. The photoionizing laser radiation detected the Li atoms in the plane parallel to the linear source at the instant of time t. The distribution function $N(\rho, t)$ could be measured by varying the delay between the pulses that photodissociated the LiI molecules and photoionized the Li atoms. Such distributions were attained for two cases: (a) release of several hundreds of Li atoms in a pulse and measurement of $N(\rho, t)$ with summation over several pulses; and (b) the release of a few atoms, measurement of the probability of arrival of an atom at the ρ line at the instant t, and repetition of such measurement several hundreds of times. Both measurements with summation over ensemble and time gave the same distributions.

Thus, the experiment with laser ionization detection of single atoms confirms the validity of the ergodic hypothesis, at least for the case of diffusive atomic motion.

11. Multiphoton Ionization Photoelectron Spectroscopy

Photoionization spectroscopy is based on a process in which a neutral particle (atom, molecule) is converted to an "ion–electron" pair by laser radiation. The foregoing chapters are concerned with (1) *photoionization resonant spectroscopy*, in which the dependence of the ionization yield on the radiation wavelength (or several wavelengths) is measured and (2) *photoionization mass spectrometry*, in which, in addition, the mass spectrum of the resulting ions and fragments is analyzed. It is, in principle, possible, to obtain additional information about the excited neutral and ionic states of the particle being ionized by analyzing the energy spectrum and the angular distribution of photoelectrons. This case underlies the third trend: *photoelectron spectroscopy* with multiphoton ionization (MPI) (see Fig. 1.7). It is this trend of laser ionization spectroscopy that is under consideration in this chapter.

11.1. Principles and History of the Method

Analysis of the energy spectrum of the photoelectrons generated as a sample is exposed to x-ray or VUV radiation forms the basis for a very efficient method developed by Siegbahn *et al.* (1967, 1969, 1974) to investigate the electron state spectrum of atoms, molecules, and solids. Figure 11.1 compares photoelectron spectroscopy based on VUV single-photon excitation with photoelectron spectroscopy based on multiphoton resonant ionization. In this first case, direct single-electron ionization occurs from the ground state, and it is possible to analyze only the electronic states that can be excited as one photon is absorbed and in the course of subsequent relaxation. This method is valuable since it enables measuring the energies of atomic and molecular states and not the energies of the transitions between different states measured in optical spectroscopy. In the second case, the resonant excitation of the electronic states, which then, after absorbing one or several more photons of the same or a different color, are subjected to photoionization. Here photoionization comes about from the excited state rather than

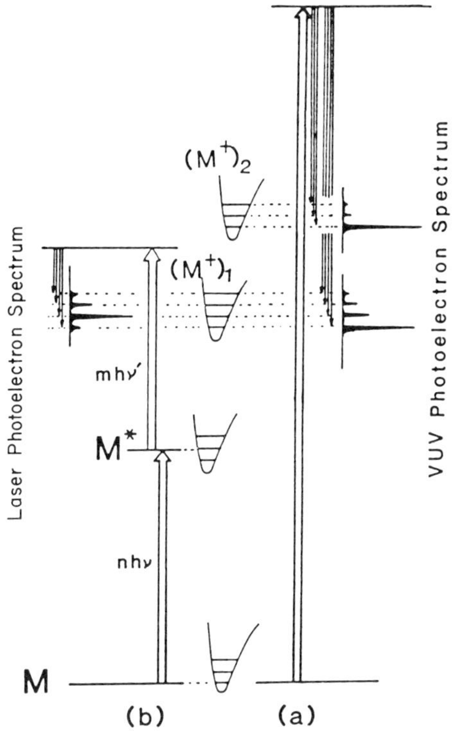

FIG. 11.1. Schematic energy level diagram relevant to (a) VUV single-photon ionization from the ground state and (b) laser two-step ionization through an excited state. Differences in photoelectron vibrational structure between (a) and (b) are illustrated schematically. [From Kimura (1985).]

from the ground state of the neutral particle. In essence, the ESCA method can be applied in this way to the excited states of neutral particles. Besides, the selection rules for single-photon and multiphoton or multistep excitation differ greatly. For example, the two-electron and high-lying Rydberg states of atoms and molecules can be excited by using the two-step method (Chapters 3 and 7). This factor extends considerably the class of ion states that can be analyzed by electron spectroscopy.

The method of photoelectronic spectroscopy with multiphoton ionization was developed by several independent groups in Japan (Achiba *et al.*, 1980; 1981) and the United States (Meek *et al.*, 1980; Compton *et al.*, 1980). At present, this method has been applied in several tens of experiments with atoms and molecules of various complexity, which are reviewed by Kimura in his review (1985).

11.1.1. Information from Photoelectron Spectroscopy

The first step of research is to analyze the multiphoton ionization spectrum, that is, the dependence of the ion yield on the radiation wavelength (or several wavelengths). The next step is to study the energy spectrum of photoelectrons at wavelengths corresponding to certain, most intense, photoion lines and certain excited intermediate states. The photoelectronic spectrum gives additional information on the properties of both excited and ionic states.

There are four classes of problems that can be successfully solved using photoelectron spectroscopy of excited states with resonant multiphoton ionization (Kimura, 1985). These are

(1) identification of highly excited molecular states and determination of their electronic structure, molecular structure, and dynamics;
(2) formation of specific ionic states and photoelectron study of their electrons and molecular structure and dynamics;
(3) selective excitation of autoionization states and their photoelectron study;
(4) study of short-lived products produced by dissociation from excited or ionic states.

This list may also include the study of the angular distribution of the photoelectrons produced by polarized-light photoionization of atoms (Lambropoulos, 1976; Leuchs *et al.*, 1979; Hansen *et al.*, 1980), which gives
(5) information on the symmetry of excited and ionic states of particles.

The photoelectron parameters that must be measured (energy spectrum, angular distribution, and spin) are associated with the structure and properties of the energy levels of the particle under ionization.

11.1.2. Kinetic Energy of Photoelectrons

The kinetic energy K of photoelectrons is related to the energy E_{res} of the resonant excited state from which photoionization occurs, the internal energy of ionic state E_{ion}, and the energy $\hbar\omega_{\text{i}}$ of the photon that photoionizes the excited state as

$$K = \hbar\omega_{\text{i}} - E_{\text{ion}} - (I_1 - E_{\text{res}}), \tag{11.1}$$

where I_1 is the first ionization potential, or the adiabatic ionization potential in the case of a molecule. Figure 11.2 shows a diagram of energy levels which explains this relation. If the ionization from an excited state is realized by means of an n-photon process, $n\hbar\omega_{\text{i}}$ should be used in (11.1) instead of $\hbar\omega_{\text{i}}$.

The probability of, say, single-photon ionization of an excited molecular state is described by the relation

$$P \sim \langle\psi_{\text{e}''}^{\text{res}}|M|\psi_{\text{e}'}^{\text{ion}}\rangle\langle\psi_{\text{v}''}^{\text{res}}|M|\psi_{\text{v}'}^{\text{ion}}\rangle, \tag{11.2}$$

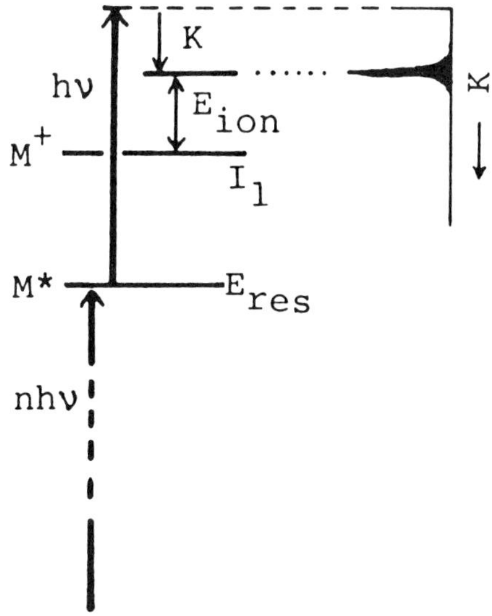

FIG. 11.2. Energy level diagram relevant to resonant MPI photoelectron spectroscopy. [From Kimura (1984).]

where $\psi_{e''}^{res}$ and $\psi_{v''}^{res}$ are the electron and vibrational wave-functions of excited state, respectively, and $\psi_{e'}^{ion}$ and $\psi_{v'}^{ion}$ are the wave functions for the final ionic state, respectively. The cross section of ionization from the excited state depends on both the selection rules for electronic transition and the Frank–Condon factors governed by the difference between the equilibrium geometries of the excited and ionic molecular states. It can be easily seen that at the multistep excitation of the ionic state, investigations can be carried out over a wider range of the potential curve (and, consequently, the vibrational structure of ionic molecular state) than at single-photon excitation from the ground state.

11.1.3. Angular Distribution of Photoelectrons

In the case of VUV photoelectron spectroscopy, when one VUV photon brings about the transition of a particle from the ground to the ionic state, the angular distribution of the photoelectrons emitted by the particle in a linearly polarized light field is expressed as (Cooper and Zare, 1968)

$$I(\theta) \sim 1 + (\beta/2)(3\cos^2\theta - 1), \tag{11.3}$$

where θ is the angle between the directions of polarization and photoelectron flight, and β the parameter of asymmetry varying between -1 and 2. In the case of single-photon ionization from an excited state, relation (11.3) can be also used.

More detailed calculations for atomic systems show that, besides the term $\cos^2\theta$, higher-order Legendre polynomials (such as $P_{2m}(\cos\theta)$, with $m > 1$) can make a material contribution (Leuchs *et al.*, 1979; Hansen *et al.*, 1980). According to theory, the angular distribution of the photoelectrons formed in the course of n-photon ionization of an anisotropic ensemble of molecules by linearly polarized radiation can be expressed as (Lambropoulos, 1976)

$$I(\theta) = 1 + \sum_{m=1}^{n} a_{2m} P_{2m}(\cos\theta). \tag{11.4}$$

The highest-order terms in (11.4) reflect the orientation of the total angular momentum of an excited particle. The a_{2m} coefficients are, generally speaking, functions of laser radiation intensity and spectral width (Dixit and Lambropoulos, 1983).

11.1.4. Photoelectron Spin

It is quite possible to perform, in principle, a complete quantum-mechanical experiment when the spin of a photoelectron emitted in a definite direction is measured. It was assumed some time ago that the degree of polarization of photoelectrons was proportional to $\hbar\omega_i/mc^2$, where m is the electron mass, that is, it was essentially only at relativistic photon energies $\hbar\omega_i \simeq mc^2$ (Sauter, 1931; Pratt *et al.*, 1964). Such a conclusion was drawn in studying the photoionization of the K shells of atoms in the first-order perturbation theory with respect to the αZ parameter, where α is the fine-structure constant and Z is the nuclear charge. Fano (1969) showed that within the Cooper minimum of the photoionization cross section of the outer shells of alkali metals, the degree of polarization of the total photoelectron flow may be about 100%. Later, Cherepkov (1973, 1979) showed that the photoelectrons coming out at the definite angle from any subshell of the atom by circularly polarized, linearly polarized, and unpolarized radiation turned out to be polarized. The degree of polarization of the electrons is about 100%, if it is possible to separate photoelectrons corresponding to ionization from a certain fine structure sublevel, and of the order of $(\alpha Z)^2$ in other cases. The total flow of photoelectrons is polarized only when circularly polarized light is absorbed.

The polarization effect of photoelectrons also takes place during photoionization of both nonoriented and oriented molecules (Cherepkov, 1981).

The measurement of photoelectron polarization in the course of photoionization from the ground or excited states makes it possible to identify autoionization resonances, because the degree of polarization depends on the total momenta of the initial and final states of the ionization transition (Heinzmann *et al.*, 1975).

11.2. Experimental Technique

The experimental technique of photoelectron spectroscopy with multiphoton ionization is similar to the technique applied in photoionization molecular spectroscopy (Chapters 5, 6). Another important feature is the use of a photoelectron energy analyzer. There are two types of analyzers: time-of-flight electron velocity analyzers and spherical electrostatic energy analyzers.

Time-of-flight photoelectron energy analyzers, which are very simply in design and have good resolution, are most generally employed. The kinetic energy K of photoelectrons is related to the velocity v by

$$v\,(\mathrm{cm/s}) = 5.93 \times 10^7 \sqrt{K}\,(\mathrm{eV}). \tag{11.5}$$

The typical time resolution of 10 ns corresponds to the energy resolution of 20 meV for $K = 0.5$ eV. Time-of-flight photoelectron analyzers are used in most experiments (Achiba *et al.*, 1980, 1981; Meek *et al.*, 1980; Kruit *et al.*, 1981).

In multiphoton ionization, a comparatively small number (10^2–10^4) of photoelectrons are usually formed during a laser pulse. They are distributed in a solid angle of 4π steradians and over an energy range from 0 to 10 eV. So, to ensure good resolution and a good signal-to-noise ratio, it is advisable to collect a considerable fraction of photoelectrons. For this purpose, a modification of a time-of-flight analyzer is proposed that is based on the ability of a divergent magnetic field to make the trajectrories of the electrons moving from the strong magnetic field to the weak magnetic field parallel (Kruit *et al.*, 1981; Kimman *et al.*, 1982).

In some experiments, spherical electrostatic analyzers were used (Compton *et al.*, 1980; Pratt *et al.*, 1983a). Figure 11.3 shows schematically a photoionization and photoelectron spectrometer with multiphoton ionization of molecules in a beam with simultaneous recording of the ion mass spectrum by a time-of-flight mass analyzer and measurement of the photoelectron energy spectrum in an analyzer with a 160° spherical section (Miller and Compton, 1981). The photoelectron analyzer operates in the energy range of 4 to 20 eV with a resolution of about 150 meV. The photoelectron spectrum observed in such an analyzer is greatly affected by the Earth's magnetic field, which must be compensated for with Helmholtz coils.

The photoelectron energy scale is usually calibrated by the Xe photoionization signal, which has ionization potentials of 12.130 eV ($^2P_{3/2}$) and 13.436 eV ($^2P_{1/2}$). The resonant excitation of the $6s[11/2]^0J = 1$ state of Xe can be achieved at three-photo absorption with $\lambda = 440.9$ nm and its

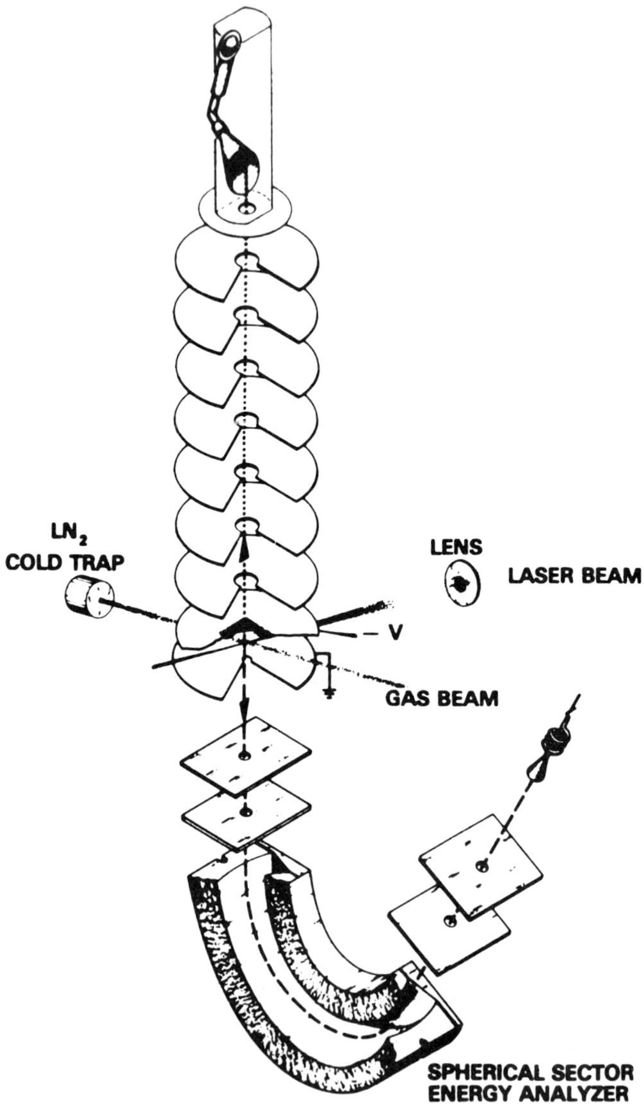

FIG. 11.3. Schematic diagram of the apparatus used for multiphoton ionization mass spectrometry and photoelectron energy analysis. [From Miller and Compton (1981).]

photoionization at two-photon absorption of radiation with the same wavelength, that is, on account of the (3 + 2)-photon process. As a result, the photoelectron spectrum develops two peaks with known energies: 1.93 and 0.63 eV (see Fig. 11.5).

The angular distribution of photoelectrons is measured by rotating the light wave polarization vector, the direction of photoelectron recording being fixed.

11.3. Atomic Photoelectron Spectroscopy with MPI

Various experiments have been carried out for atoms with the aim of analyzing the energy spectrum and the angular distribution of photoelectrons with different schemes of multistep and multiphoton ionization. The experiments may be classified according to the excitation and ionization schemes listed in Section 3.4 (Fig. 3.16–3.19).

11.3.1. $(\omega_1, \tilde{\omega}_2)$: Two-Step Photoionization

For Na atoms the angular distribution of photoelectrons was studied with ionization through the resonant intermediate $3p^2P_{3/2}$ state (Hansen *et al.*, 1980). Ionization occurs at two transitions to continuum: $3p \rightarrow ks$ and $3p \rightarrow kd$, with the contribution of each ionization channel depending on the phase shift of the final s and d states and the intermediate state coherence. This manifests itself in the dependence of the angular distribution of photoelectrons on the delay time between the exciting and the ionizing pulses. The results of the experiments are in good agreement with theory.

If the laser radiation produces excited atoms in considerable concentration, then the observation of a photoelectron signal from them, even in the case of nonlaser sources of radiation photoionizing the excited atoms, becomes possible. Such an experiment was performed with Na atoms in the $3p^2P_{3/2}$ state in a concentration of $2 \times 10^{12}\ \mathrm{cm}^{-3}$ with a synchrotronic source of VUV radiation with the quantum energy $\hbar\omega_2 \simeq 30$–75 eV (Bizau *et al.*, 1982).

11.3.2. $(2\omega, \tilde{\omega})$: Two-Step Photoionization

This scheme was applied in detailed experiments with the aim of studying the angular distribution of the photoelectrons formed when the atoms were photoionized through the intermediate 4^2D state excited by a two-photon process (Ohnesorge *et al.*, 1984). Figure 11.4a illustrates the geometry of measurement of the angular distribution of photoelectrons by rotating the radiation polarization vector. A distinct dependence of the angular distribution on the angular momentum of intermediate state ($^2D_{3/2}$ or $^2D_{5/2}$) was revealed. The angular distributions of photoelectrons are given for these two

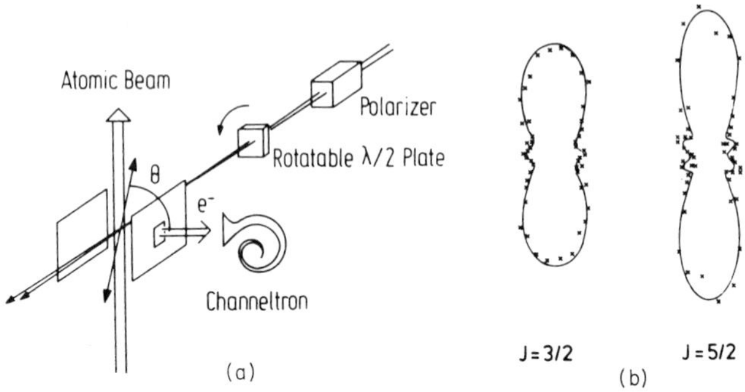

FIG. 11.4. Measurement of the angular distribution of photoelectrons. (a) schematic of the interaction region (the arrow shown under the angle θ indicates the direction of laser polarization); (b) polar diagrams of photoelectron angular distributions for ionization by way of the $4^2D_{3/2}$ and $4^2D_{5/2}$ levels of Na atoms at low laser intensity (i.e., no Stark effect). Solid curves are the theoretical distributions. [From Ohnesorge *et al.* (1984).]

cases. Here we can see the theoretical distributions obtained from Eq. (11.4) and calculated with the use of the a_{zm} coefficients given in Table 11.1. Taking into account, the Stark effect, this enables us to describe the angular distribution evolution with increasing laser radiation intensity.

The same ionization scheme was applied in the experiments aimed at studying the photoelectron energy spectrum for Fe atoms (Nagano *et al.*, 1982). Iron atoms were produced by photodissociating the $Fe(CO)_5$ molecule with the same laser pulse that then ionized the Fe atoms. There were strong ion peaks observed at $\lambda = 463.4$, 464.9, and 465.9 nm, which corresponds to the two-photon excitation of the $a^5D_J \rightarrow e^7D_J$ transition for $J = 2$, 3, and 4 with $\Delta J = 0$.

TABLE 11.1. Coefficients Calculated for the Fourier Expansion of the Angular Distribution of Photoelectrons in the Process of Na Ionization through the $4^2D_{3/2}$, $4^2D_{5/2}$, or 4D Levels with Neglect of Fine Structure[a]

Excitation pathway	a_2/a_0	a_4/a_0	a_6/a_0
$^2S_{1/2} \rightarrow {}^3D_{3/2} \rightarrow \varepsilon f$	4/5	1/3	0
$^2S_{1/2} \rightarrow {}^2D_{5/2} \rightarrow \varepsilon f$	387/410	23/41	25/82
$S \rightarrow D \rightarrow \varepsilon f$	39/34	15/17	25/34

[a] From Ohnesorge *et al.* (1984).

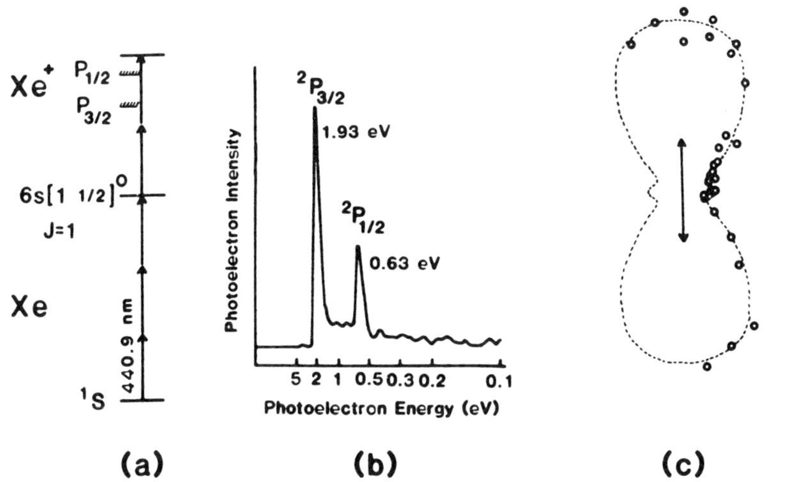

FIG. 11.5. The $(3\omega, 2\tilde{\omega})$ resonant MPI of Xe via the $6s[11/2]^0$ $J = 1$ state. (a) Energy level diagram; (b) photoelectron kinetic energy spectrum; and (c) photoelectron angular distribution. [From Sato *et al.* (1984a).]

11.3.3. $(3\omega, \tilde{\omega})$ and $(3\omega, 2\tilde{\omega})$: Two-Step Photoionization

This ionization scheme was used in analyzing the photoelectron spectra for Kr (Sato *et al.*, 1984a) and Xe (Compton *et al.*, 1980; Miller and Compton, 1982; Kruit *et al.*, 1983; Sato *et al.*, 1984a). For illustration, Fig. 11.5 schematically shows some transitions for $(3\omega, 2\tilde{\omega})$: the ionization of Xe and the shape of the spectrum and angular distribution of the photoelectrons. The ionization was performed under the action of a laser pulse at $\lambda = 440.9$ nm through the Rydberg state $6s[11/2]^0$ $J = 1$. Two arising photoelectron peaks correspond to the $^2P_{3/2}$ and $^2P_{1/2}$ ionic states. These peaks can be used for calibrating the photoelectron spectrometer. The angular distribution of photoelectrons can be approximately described by Eq. (11.4), even when $n = 2$ (Kruit *et al.*, 1983; Sato *et al.*, 1984a).

11.3.4. $(\omega_1, \omega_2, \tilde{\omega}_3)$: Three-Step Photoionization

Experiments were carried out for Na and Ba atoms to analyze the angular distribution of photoelectrons. By taking different intermediate quantum states, it was possible to observe a strong dependence of the angular distribution form on them (Leuchs, 1984). These studies demonstrate that the measurement of the angular distribution of photoelectrons can give information on the wave function of an intermediate resonant state. Such a method is of interest for studying and identifying autoionization states if ionization is realized on account of the $(\omega_1, \omega_2^{a.i.})$ or $(\omega_1, \omega, \omega_3^{a.i.})$ schemes (see Section 3.4).

11.4. Molecular Photoelectron Spectroscopy with MPI

Many successful experiments on photoelectron spectroscopy with multiphoton ionization of both simple and complex molecules listed by Kimura (1985) have been so far performed.

11.4.1. Experiments with Simple Molecules

The following simple molecules have been the subject of photoelectron spectroscopy with multiphoton ionization: H_2 (Pratt *et al.*, 1983b), NO (Miller and Compton, 1981, 1982; Kimman *et al.*, 1982; White *et al.*, 1982 Achiba *et al.*, 1983b), CO (Pratt *et al.*, 1983a), H_2S (Achiba *et al.*, 1982) and NH_3 (Achiba *et al.*, 1982, 1983b; Glownita *et al.*, 1982). In all cases, multiphoton ionization was realized by means of resonant two- or three-photon excitation of high-lying electronic states.

For illustration, we shall consider experiments with the NO molecule, the resonant ionization of which was realized with different intermediate electronic states and identification of vibrational levels. Figure 11.6 presents the photoelectron spectrum produced during multiphoton ionization through

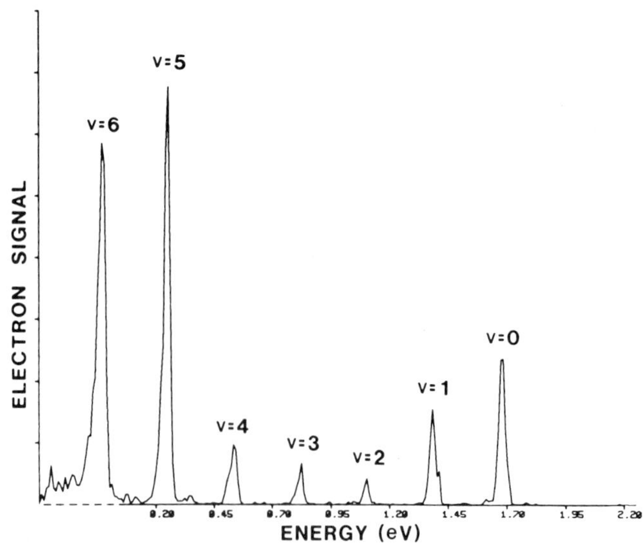

FIG. 11.6. Electron energy spectrum for four-photon $(2\omega, 2\tilde{\omega})$ ionization of NO via the intermediate $A^2\Sigma^+(v = 0, J = 6\frac{1}{2})$ state at $\lambda = 454.2$ nm. [From Kimman *et al.* (1982).]

the intermediate $A^2\Sigma^+$ state (Miller and Compton, 1981; Kimman *et al.*, 1982),

$$X^2\Pi_{3/2}(v = 0, J = 8\tfrac{1}{2}) \xrightarrow{2\hbar\omega} A^2\Sigma^+(v = 0, J = 6\tfrac{1}{2}) \xrightarrow{2\hbar\omega}$$
$$X^1\Sigma^+(v = 0, \ldots, 6; 4\tfrac{1}{2} \leqslant J \leqslant 8\tfrac{1}{2}),$$

under radiation at $\lambda = 454.2$ or other wavelengths in the range from 391 to 452 nm.

We can clearly see a vibrational structure in the photoelectron spectrum, since all seven energetically allowed vibrational levels of the ground state of NO^+ are populated. There are two maxima in the vibrational distribution: when $v = 0$ and $v = 5$. All the vibrational peaks correspond to direct ionization ($v = 0$) and electronic (vibrational) autoionization (for $v > 0$).

First, these results demonstrate the efficiency of photoelectron spectroscopy for studying the vibrational structure of molecular ions in the ground electronic state. With the intermediate state properly chosen, it is possible, in principle, to widen considerably the range of vibrational states of the molecular ion accessible for investigation.

Second, these results demonstrate the possibility of studying the effects of electronic and vibrational autoionization for molecular ions. When the spectroscopic constants of the molecular ion are known, it is possible to calculate beforehand the positions of vibrational peaks in the photoelectron spectrum, using relation (11.1). Such calculations for NO agree well with experiment (Achiba *et al.*, 1983b).

11.4.2. Experiments with Complex Molecules

The complex molecules benzene (Achiba *et al.*, 1981; Meek *et al.*, 1980), toluene (Meek *et al.*, 1982), chlorobenzene (Anderson *et al.*, 1982), and naphthalene (Hiraya *et al.*, 1985), as well as van der Waals complexes Ar–NO (Sato *et al.*, 1984b), H_2O–phenol, H_2O–7-azaindole (Fuke *et al.*, 1984), and others, have been the subject of photoelectron study with multiphoton ionization.

Since the electronic absorption bands of complex molecules lie in a near-UV range, in most cases photoionization can be realized using the simplest two-step scheme with single-photon absorption at each step. Two color radiation is often used for ionization. Cooled molecular jets are usually used to produce and identify well-allowed multiphoton ionization spectra and to study poorly bound van der Waals complexes.

To illustrate the potential of photoelectron spectroscopy for complex molecules, we consider here briefly the experiments that have revealed information on the vibrational spectrum of the molecular ion (Anderson *et al.*, 1982), the vertical and adiabatic ionization potential of the van der

TABLE 11.2. Comparison of the Vibrational Frequencies for the Chlorobenzene Cation and Neutral Molecules in the Ground and Excited States[a]

Vibrational mode (C_{2v} symmetry)	Neutral ground state (cm^{-1})	Neutral excited state (cm^{-1})	Ionic ground state (cm^{-1})
18b (b_2)	294.7	287.3	266 ± 15
6a (a_1)	416.8	378.2	419 ± 10
6b (b_2)	614.9	521.0	451 ± 10
12 (a_1)	706.5	671.3	661 ± 10
9a (a_1)	1153.0	980.6	1145 ± 15
9b (b_2)	1167.1	1149.4	

[a] From Anderson *et al.* (1982).

Waals complex (Fuke *et al.*, 1984), and the intramolecular conversion in benzene (Achiba *et al.*, 1983a, 1984).

(1) A new method based on a combination of multiphoton ionization and photoelectron spectroscopy of excited molecules was proposed to define the vibrational structure of molecular ions (Anderson *et al.*, 1982). In the experiment, the molecule of chlorobenzene (C_6H_5Cl) was ionized by a two-step scheme. It was possible to determine the vibrational frequencies of the first five vibrational modes of the chlorobenzene cation by ionizing through different vibronic levels of the excited $^1B_2$ state and analyzing the vibrational structure of the photoelectron spectrum. For comparison, Table 11.2 presents the vibrational frequencies of the chlorobenzene cation with the vibrational frequencies of neutral benzene in the ground and excited $^1B_2$ states measured in this experiment. It can be seen that the vibrational frequencies for the ionic state are somewhat smaller than for the neutral ground and excited states.

Such a technique is quite suitable for systematic studies of the vibrational structure of molecular ions. It was applied successfully, for example, for molecular ionic naphthalene (Hiraya *et al.*, 1985).

(2) Measurement of the vertical and adiabatic ionization potentials was successfully demonstrated in experiments with molecular complexes of phenol and 7-azaindole with water in a supersonic jet (Fuke *et al.*, 1984). The $(\omega_1, \tilde{\omega}_2)$, two-step ionization, was realized through the intermediate electronic S_1 state. Figure 11.7 shows the photoelectron spectrum for the phenol molecule and its complexes with H_2O. The peak for phenol (Fig. 11.7a) at 0.53 eV is associated with the level $v = 0$ of the molecular cation

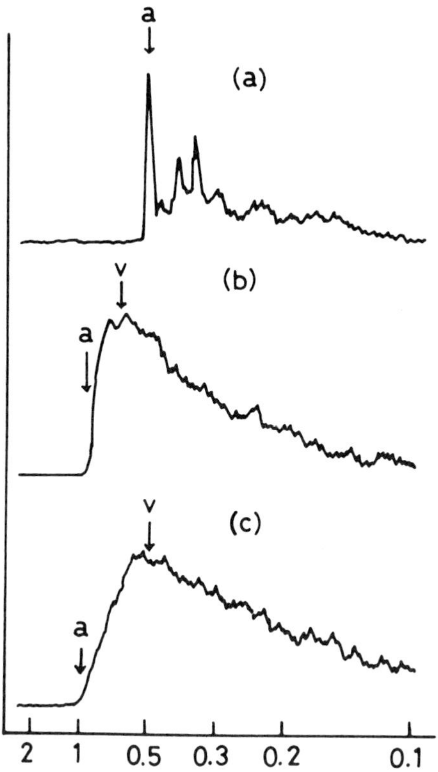

FIG. 11.7. Photoelectron spectra (in electron volts) of several hydrogen-bonded complexes of phenol (Ph), obtained by $(\omega_1, \tilde{\omega}_2)$ resonant ionization through the S_1 state origins, the letters *a* and *v* indicate the positions of I_a and I_v, respectively. (a) Ph, $\lambda_e = 36,352\ \text{cm}^{-1}$; (b) Ph–$H_2O$, $\lambda_e = 35{,}996\ \text{cm}^{-1}$; and (c) Ph–$(H_2O)_n$, $\lambda_e = 36{,}220\ \text{cm}^{-1}$. [From Fuke *et al.* (1984).]

in the ground state. The adiabatic ionization potential can be found from the energies of photoelectrons and exciting photons. Its value ($I_a = 8.49$ eV) is in good agreement with those measured by other methods. Figure 11.7b shows the photoelectron spectrum for the complex Ph–(H_2O) with excitation performed in the 0—0 band of the $S_0 \rightarrow S_1$ transition at a slightly shifted frequency. From this spectrum, the values of the adiabatic ionization potential $I_a = 8.09$ eV and the vertical ionization potential of Ph–(H_2O) $I_v = 8.30$ eV were determined. The value of I_a was smaller than that of I_v by the value of ion stabilization energy 0.40 eV. This value was found to be an order of magnitude higher than the stabilization energy of the S_1 state of Ph–(H_2O). The values of I_a and I_v for solvated phenol Ph–$(H_2O)_n$, $n \geqslant 4$, can be determined from the spectrum given in Fig. 11.7c ($I_a = 8.05$ eV and $I_v = 8.45$ eV). It can be seen that the stabilization energy of solvated phenol

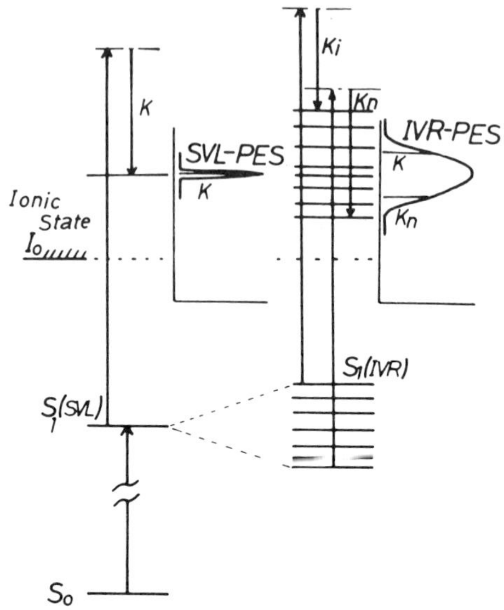

FIG. 11.8. Schematic drawing of the ionization band shapes expected for the photoelectron spectra through vibrationally unrelaxed and relaxed levels, where only $\Delta v = 0$ ionization transitions are considered. The differences in photoelectron kinetic energies represented by K_i and K_n correspond to the differences in vibrational frequency changes on ionization among many vibrational modes. [From Achiba *et al.* (1984).]

does not differ from that of monohydrated phenol. Thus the first molecule of water stabilizes the molecular ion of phenol by forming a direct hydrogen bond, and the other molecules of water act as a solvent and do not have a pronounced effect on the stabilization of the phenol ion. The values of I_a and I_v for phenol clusters were measured in a similar way.

The same method was applied to investigate the van der Waals complex Ar–NO, and to determine the adiabatic potential of Ar–NO ($I_a = 9.148 \pm 0.005$ eV, for NO $I_a = 9.266$ eV) and the dissociation energy of the $Ar–NO^+$ ($X^1\Sigma$) ion ($D_0 = 0.129 \pm 0.005$ eV) (Sato *et al.*, 1984b).

(3) If in the process of stepwise excitation a molecule is able to relax to other intermediate states, there are several ways for ionization, which clearly manifests itself in the photoelectron spectrum. This method was used to investigate the process of intramolecular vibrational relaxation in the excited electronic states of benzene (Achiba *et al.*, 1983a, 1984) and napthalene molecules (Hiraya *et al.*, 1985). Figure 11.8 explains how the photoelectron spectrum changes due to the process of intramolecular vibrational relaxation for the S_1 state of benzene. A laser pulse excites a definite vibronic level of the S_1 state. This process only involves optically active vibrational modes. As benzene transfers to the ionic state, we can observe a well-pronounced

vibrational structure in the photoelectron spectrum. However, along with the vibrational structure, a continuous background appears in the vicinity of the vibrational peak $\Delta v = 0$, its part growing as the excess of vibrational energy in the S_1 state increases over $\Delta E = 300\ \mathrm{cm}^{-1}$. This fact cannot be explained by intercombination conversion, because the ionization of the isoenergetic levels of a triplet state resulting in the same excess of vibrational energy in the molecular ion requires at least two photons, which is inconsistent with the experiment. So this can be explained by the mechanism of intramolecular vibrational relaxation within the S_1 state. Due to this process, the vibrational excitation level in optically active modes decreases, and thus the vibrational excitation of nonactive modes increases. As a result, ionization from a wider set of vibrational states becomes possible, which gives rise to a broad pedestal in photoelectron spectrum, as shown in Fig. 11.8. Given the ionization rate of excited levels K_i, it is possible to evaluate the intramolecular relaxation rate, which in the experiment was no higher than $10^{11}\mathrm{s}^{-1}$ even for vibronic levels with $\Delta E = 5000\ \mathrm{cm}^{-1}$.

These examples, of course, do not limit the possibility of obtaining information on complex molecules, their excited and ionic states. All the potential of laser ionization spectroscopy is to be used in studying complex molecules.

In conclusion, it should be emphasized that the channel of photoionization spectroscopy complements rather well the channels of resonant optical and mass spectrometry in photoionization laser spectroscopy. All these channels together can give more information about the atoms and molecules under analysis than other methods of laser spectroscopy.

12. Photoionization Detection of Molecules with Spatial Resolution

The photoionization of an atom or a molecule gives rise to charged particles at the point of electron detachment. In principle, it is quite possible to localize the point where a photoelectron is formed with an accuracy much better than the photoionizing radiation wavelength. This is the basis for an interesting possibility of application of photoionization spectroscopy to the detection of molecules and chromophores with a spatial resolution of several angstroms, that is, the development of photoionization microspectroscopy (Letokhov, 1975b,c, 1976a). Even though this problem is still far from solution, its importance justifies discussion of the ways of its solution in the last chapter of this monograph.

12.1. Laser Resonant Photoelectron and Photoion Microscopy

The idea of photoionization spatial localization of the molecular bonds or chromophore of a large molecule is based on a combination of spectrally selective photoionization of a chosen bond or chromophore of the molecule with field-ionization electron or ion microscopy (Letokhov, 1975b,c). This idea is based on the effect of photoelectron emission or laser field desorption of molecular ions from the surface. When this proposal was made in 1975, the predicted effect of laser-stimulated field desorption was not observed yet. Field desorption was observed by Müller in 1956 (Müller, 1956; Müller and Tsong, 1969, 1973). The physical principles of this effect were discussed in a number of papers and reviews (Gomer, 1959; Beckey, 1978; Panitz, 1978). After field-ionization-electron and then ion-microscopy were discovered, attempts were made to produce images of some organic molecules by virtue of their absorption on the emitter of electronic or ionic projectors. However, the resolution of field-ionization-electron microscopy is no better than 20 Å, which is not sufficient to resolve atomic details. An even easier goal, i.e., the determination of the configuration of molecular contours at an atomic level, has proven to be almost impossible. Field-ionization-ion microscopy demands such strong electric fields that macromolecules evaporate in them. Their microscopy also involves great difficulties that have not yet been overcome.

The direct spatial localization of atoms and bonds is a very important problem, since it is considered to be one of the most important potential applications of short-wavelength x-ray and γ-ray lasers (molecular holography). However, in our opinion, lasers in the optical spectral region can also be effectively used to study the spatial and chemical structure of biomolecules. In this section, in particular, some principal modifications of the methods of field-ionization-electron and -ion microscopy are discussed. They provide reasons to hope for an effective application of these methods, i.e., the observation of the structure of macromolecules.

12.1.1. Principle of a Laser Photoelectron (Photoion) Microscope

A diagram of a laser photoelectron microscope for molecular bonds is given in Fig. 12.1, according to the first proposal in 1975. A hemi-spherical tip of a needle with a radius of curvature r is used as a cathode, and a cofocal screen with a radius of curvature R is used as an anode. A macromolecule with a cross section $b \ll r$ is adsorbed on the surface of the cathode. Under a series of ultra short light pulses of different frequencies $\omega_1, \omega_2, \ldots$ the successive multistep excitation of the quantum levels of the chosen molecular bond AB takes place followed by subsequent ionization, i.e., the extraction of

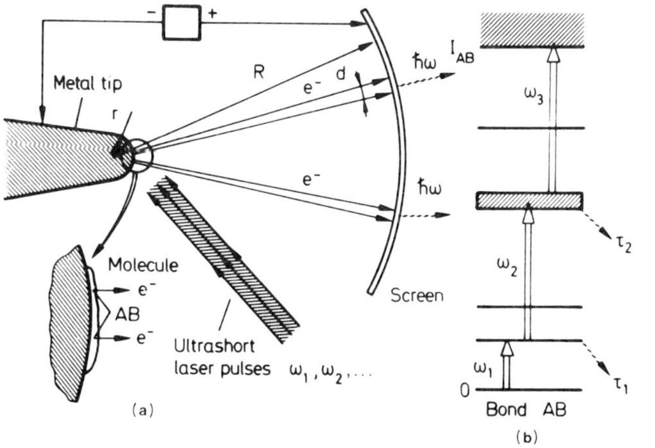

FIG. 12.1. (a) Simplified scheme of a laser photoelectron (photoion) microscope and (b) selective multistep photoionization scheme of the selected molecular bond. [From Letokhov (1975b,c).]

an electron from the bond AB. Under the influence of a strong electric field near the cathode, the photoelectrons move along radial lines toward the screen. As in the field-ionization-electron microscope (Müller's), the screen displays a magnified image of the section of the cathode that emits electrons. The difference lies in the fact that the electric field only serves to transfer the photoelectrons to the anode rather than to extract them out of the cathode. Such a "soft" action allows certain kinds of molecules or particular bonds (parts) of a molecule to be ionized on the surface of the cathode without involving the cathode electrons.

The idea of combining the selective laser generation of photoelectrons and the formation of the image of the emitting surface in an electric field is rather general. It may be applied to the generation of photoions.

The field-ionization-ion microscope is based on the display of the point surface by the ions formed in the field ionization of the atoms of the residual (sample) gas in the projector chamber, the instant they approach the point surface. For this purpose, an electric field of great strength (about 500 MV/cm) is produced at the surface of the tip. The image of the surface is displayed on the screen and is formed by positive ions which fly perpendicular to the surface. The magnification of field-ionization-ion microscope (M) can reach 10^6 while its resolution is 2–3 Å. An enhancement in the resolution of an ion microscope can be achieved by using a heavy ion. Attempts to use a field-ionization-ion microscope to produce an image of biological molecules have failed. The main obstacle to producing ion images of large organic molecules is that the necessary, strong electric fields decompose and evaporate the molecules. Also, different distortions take place (Müller and Tsong, 1973).

The ion microscope may possess two laser modifications (Letokhov, 1975b,c). First of all, the strength of the electric field required for atomic ionization near the tip can be substantially reduced if the atoms are excited to states near the ionization limit. The strength of the electric field can probably be reduced to values less than 10^7 V/cm. At such strengths, the electric field can completely eliminate molecular evaporation and decomposition. Second, attempts can be made to selectively detach certain functional groups of the macromolecules absorbed on the surface of the tip as ionic fragments using laser radiation. The detached, positive photoions can be directed to the screen using an electric field of moderate strength. Such a version of photoion microscopy has no analog in field-ionization-ion microscopy, since the photoions are directly emitted from the surface of the tip where molecules are absorbed in the absence of a sample (residual) gas. Such a laser photoion emission microscope is quite similar to a laser photoelectron microscope. It is interesting to estimate and compare the ultimate spatial resolution of emission photoelectron and photoion microscopes.

12.1.2. Ultimate Spatial Resolution

The resolution of a photoelectron (photoion) laser microscope is limited by two fundamental factors: the ratio of indeterminacy and the presence of a nonzero tangential component of the velocity of the emitted photoelectron (photoion). The same factors restrict the spatial resolution of the field-ionization-electron (ion) microscope (Müller and Tsong, 1973). It must be emphasized that one difference lies in the fact that for a photoemission microscope, there is no need for a strong (ionizing) electric field that distorts and desorbs the macromolecules. Also, the laser radiation allows photoelectrons (photoions) to be selectively extracted from certain parts of a macromolecule.

The indeterminacy principle causes a spread of the transversal velocity of the photoelectron (photoion) $\Delta v_1 = \hbar/2ma_1$, if its coordinate is accurately determined to be a_1, where m is the mass of the photoelectron (photoion). This spread, Δv_1, results in a circle of electron (ion) scattering on the screen with a diameter $d_1 = 2Tv_1$, where $T = R[m/(2eU)]^{1/2}$ is the flight time of the particle from the tip to the screen, and R is the radius of the screen, i.e., the distance between the tip and the screen (see Fig. 12.1). On the other hand, the circle of scattering on the screen, d_1, is also related to the indeterminacy of the coordinate at the point of extraction a_1 ($a_1 = d_1/M$), where M is the projector magnification coefficient. As a result, the spatial resolution is limited by the indeterminacy ratio at the level,

$$a_1 = (2\hbar T/mM)^{1/2} = [\chi\hbar r(meU/2)^{-1/2}]^{1/2}. \qquad (12.1)$$

In Eq. (12.1) the relationship between the magnification coefficient M and the radii of the needle tip r and the screen R, i.e., $M = R/(\chi r)$, was taken into account. Here χ is a numerical factor ranging from 1.5 to 2.0 (Dyke and Dolan, 1956). The maximum potential difference is limited by the value of the destructive field U near the tip of the needle (equal to 0.5 V/Å) (Müller and Tsong, 1973). For macromolecules, it may be assumed that $F_{max} = eU/r <$ 0.2 eV/Å. Thus, a simple estimate of the ultimate resolution can be made

$$a_1 = [\chi\hbar(2r/mF_{max})^{1/2}]^{1/2}. \qquad (12.2)$$

When the radius of the needle tip $r = 10^3$ Å, $a_1 \simeq 20$ Å for the photoelectron microscope. For the photodetachment of H^+, $a_1 \simeq 3$ Å. When a molecular ion with a molecular weight $A = 100$ photodetaches, the indeterminacy ratio limits the spatial resolution to 1 Å.

An emitted photoelectron (photoion) can have a nonzero transversal velocity, Δv_2, if a part of the excitation energy is changed to kinetic energy during the photodetachment from the surface. It is assumed that the average kinetic energy of the transverse motion of a photoelectron (photoion) equals $\mathscr{E}_{kin}$. Then, in a way similar to the above procedure, it is possible to estimate the spatial resolution due to this effect as

$$a_2 = 2\chi(r\mathscr{E}_{\rm kin}/F_{\rm max})^{1/2}. \tag{12.3}$$

Let the photoelectron (photoion) have thermal energy, i.e., under normal conditions the energy of transversal motion $\mathscr{E}_{\rm kin} = 0.01$ eV. Then, all other parameters being the same ($\chi = 1.5$–2, $r = 10^3$ Å, and $F_{\rm max} = 0.2$ eV/Å), $a_2 = 20$ Å. First, it is essential that the limit of the spatial resolution due to the nonzero transverse velocity of the emitted electrons (ions) be independent of the mass of the particle. Second, the temperature of the needle and the adsorbed molecules must be reduced to 4–10 K to reach a resolution of about several angstroms.

These simple estimates show that with a photoion microscope it is possible to reach resolutions of several angstroms, which may be sufficient for the visualization of macromolecules. For this purpose, however, photoions must be detached from the surface.

12.2. Laser Surface Desorption of Molecular Ions

The most crucial point in the creation of a laser photoionic projector with a molecular resolution is the effect of selective photodetachment of molecular photoions as the molecules adsorbed on the needle surface are exposed by laser light. Even though the period of study is ten years old, this problem is still unsolved; nevertheless, its solution seems to be quite possible.

12.2.1. First Experiments

Soon after the idea of a photoionic laser projector had been published, the first experiments were carried out (Tsong *et al.*, 1976; Niu *et al.*, 1977), in which the authors could observe an enhancement of the field ionization yield as they irradiated the needle point by laser radiation in the presence of residual gas. In the experiment (Tsong *et al.*, 1976) a five-fold increase of field-ionization yield was reported as the surface of a semiconductor was exposed to low-intensity laser radiation at $\lambda = 300$ nm. In another experiment (Niu *et al.*, 1977), the low-intensity radiation of a Hg lamp at $\lambda = 250$–390 nm was used. When the needle surface was irradiated, the field-ionization yield of residual gas ($C_{14}H_{10}$) increased. This fact was interpreted as a result of an electron distribution variation in metal that affected the field-ionization yield. The mechanism of such field ionization stimulated or provided by photons was discussed in the literature (Viswanathan *et al.*, 1979; Niu and Bryant 1979). In most cases, the laser stimulation of ion field desorption is explained by the heating of the surface with radiation (Drachsel *et al.*, 1980; Kellogg and Tsong, 1980). From the standpoint of photoionic microscopy, the quantum processes stimulating field desorption are of particular interest. The possibility of such stimulation of field desorption is pointed to in the experiment in which the increased flow of ethylene molecular ions under laser radiation in subcritical fields was explained by the photogeneration of vacancies in the conductivity zone of the material and by the tunneling of an electron from the ground level of ethylene molecule to the metal (Nishigaki *et al.*, 1979).

In another group of experiments aimed at realizing the idea of a photoionic microscope, the formation of molecular ions was observed as the surface of a molecular crystal was exposed to intense UV radiation (Antonov *et al.*, 1980c, 1981b). Later it was found out that the molecular ions were caused by two successive processes: thermal desorption of molecules from the surface under slight pulse heating and subsequent multiphoton resonant ionization of the free molecules near the surface by the same laser pulse (Egorov *et al.*, 1983, 1984a,b). This process is discussed in more detail in Section 6.3 because it is of particular interest for photoionization detection of molecules rather than for the photoionic microscope.

Thus, all the experiments showed that there is a serious difficulty in the photodetachment of molecular photoions when the molecules adsorbed on a surface are irradiated. It consists of inevitable thermal heating of the surface. This problem can be overcome in several ways.

First, as it was assumed in the first experiments (Letokhov, 1975b,c), it is possible to realize multistep selective excitation of chosen types of molecular bonds or groups by a sequence of picosecond laser pulses having a high intensity but a low energy, as shown in Fig. 12.1b. However, the extraction of molecular ions with angstrom spatial resolution is very difficult for the following fundamental reasons. Picosecond laser radiation is capable of depositing significant amounts of energy to localized parts of a molecule in short time, $\tau_{exc} = 10^{-12}$–10^{-13} s, which is determined by the intensity of the laser light. However, a molecular fragment with a thermal velocity $v_0 \simeq 10^4$ cm/s needs at least 10^{-12} s to travel the distance $\delta r = 1$ Å. However, during this time a fast transfer of energy from the excited parts of the large molecule to other parts may occur. For example, recent picosecond experiments have shown that the time required for a transfer of energy is less than 10^{-12} s (Kopainsky *et al.*, 1978). This is a serious obstacle to the selective photodetachment of ions, even when picosecond laser pulses with tunable frequencies are used. It seems to be necessary that femtosecond pulses should be used. In principle, it is quite possible since the absorption bands of the large molecules adsorbed on the surface are rather wide, and pulses with a negligible energy (10^{-9} J) are needed to irradiate a very small area of the needle (10^{-8} cm^2).

Second, to overcome this problem and provide photoion (photoelectron) detachment from certain parts of the molecule, an obvious way out was proposed in the first experiment (Letokhov, 1975b). It consisted of preliminary chemical labeling of certain bonds or groups of a macromolecule with chromophores. It is possible to select a chromophore whose electronic excitation energy is lower than that of the adjacent parts of the molecule. This can result in the highly selective laser excitation of the labeled part, prevent the transfer of energy to the adjacent parts of the molecule, and make the photodetachment of the excited chromophore at low kinetic energies easier. Thus, chemical labeling of macromolecules holds much promise for the development of a laser photoion microscope with angstrom spatial resolution. It should be emphasized that some methods of labeling large biomolecules with heavy metals (with dimensions of the order of angstroms) have been developed and have produced images with the electron microscope. For example, some parts of polynucleotides can be labeled with Os and other heavy atoms (Cole *et al.*, 1977). For a photoion microscope, there is no need for the heavy metals to have a high electron density. It suffices that the compact label have a long-wavelength electron absorption band.

Finally, the photodetachment of molecular ions must be studied in the presence of an electric field with a strength that is inadequate for field desorption but sufficient for a considerable shift of the photoabsorption limit of the molecule or the chromophore (Letokhov, 1975b). In this case it is probably possible to realize resonant action of laser light on the field-induced structure arising in the absorption spectrum in order to induce joint laser-field desorption of molecular ions. Some of these possibilities are already under experimental study. We consider them here briefly.

12.2.2. Photoionization of Large Molecules at Surfaces by Picosecond Pulses

In experiments with nanosecond and even picosecond pulses, the electron excitation energy relaxed to thermal energy and gave rise to nonselective thermal desorption and subsequent photoionization of molecules. The search for nonequilibrium processes of detachment of the molecular ion of an excitable chromophore of a large molecule on a surface is of great interest. Of course, nonequilibrium, chromophore-selective, detachment of a molecular ion can probably be observed only with irradiation of a large molecule by a subpicosecond laser pulse. Proceeding along this path, we performed experiments with radiation pulses whose duration varied from 10 ns to 5 ps (Antonov *et al.*, 1982a,c).

The experiments were performed with a fine crystalline powder of peptide, whose structure is a chain consisting of three amino acids (tryptophan, alanine, and glycine) with protective acetate and ester groups at the ends. One of the amino acids entering into the peptide, tryptophan, contains a

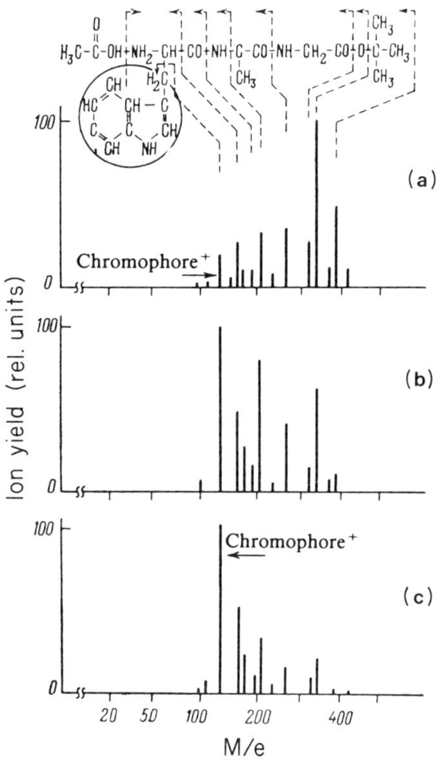

FIG. 12.2. Mass spectra forming upon irradiation of peptide powder by laser pulses of different durations. (a) 10 ns, (b) 30 ps, and (c) 5 ps. [From Antonov *et al.* (1982a).]

chromophoric group, an indole ring, whose system of conjugate bonds determines the absorption of the peptide in the near-UV region.

The experiment consisted of irradiating a peptide specimen with pulses of different duration: 10 ns, 30 ps (Nd:YAG laser, fourth harmonic), and 5 ps (Nd:phosphate glass laser, fourth harmonic). The experimental technique is quite similar to the one described in Section 6.3.

The mass spectra of the positive ions formed on irradiation are shown in Fig. 12.2. The typical mass spectrum consists of several intense peaks that correspond to different fragments of the peptide molecule. Each fragment contains a chromophoric group and forms as a result of fragmentation of the molecule primarily along the peptide bonds. One of the lightest fragment ions corresponds to the ion of the chromophore (it is circled in Fig. 12.2a).

The mass spectra obtained are not normalized to the sensitivity of the apparatus for ions of different masses. For this reason, the relationship between the ion peaks within a single mass spectrum is only of a qualitative character. However, this does not interfere with the comparison of different mass spectra with one another.

The dependence of the total ion yield on the energy flux of the laser pulses is very sharp, as in the case of simpler molecules. At the same time, the ratio between the different peaks in the mass spectrum remains practically constant in the range of energy fluxes investigated from the threshold of appearance ($\Phi_{th} = (2-4) \cdot 10^{-3}$ J/cm^2) to the maximum recorded magnitude (10^4 ions/pulse, $\Phi = 10^{-2}$ J/cm^2) with fixed pulse duration.

Comparison of mass spectra obtained with different radiation pulse durations shows the following: (1) a decrease in the duration leads to a significant increase in the fraction of chromophore ions and with $\tau_p = 5$ ps the magnitude of the corresponding peak in the mass spectrum becomes maximum; and (2) in this case, new fragment ions do not appear in the mass spectrum and this indicates the absence of additional fragmentation of molecules.

The sharp dependence of the ion yield on the energy flux of the laser radiation and simultaneously the constancy of the pattern of the mass spectrum with fixed pulse duration suggests that in this case the recorded ions are formed as a result of ionization and fragmentation, above the surface of the specimen, of neutral molecules evaporating during the course of the pulse.

The growth of the chromophore ion fraction in the case of picosecond pulses is probably related to the fact that the chromophoric group can absorb two and more photons and can become detached before total randomization of the absorbed energy in the molecule.

For picosecond pulse durations, together with intramolecular nonequilibrium processes, the surface can play an important role in the formation of the mass spectrum, because over the time of the pulse, molecules have time to fly away from the surface to distances comparable to the dimensions of the molecules themselves. For example, for $\tau_p = 5$ ps, the displacement of molecules over the time of the pulse is only 7–10 Å. In the case of subpicosecond pulses, detachment of the molecular ion of a chromophore molecule, which has not yet moved away from the surface, can be observed. In other words, with the use of subpicosecond laser pulses it is possible to go from photoionization spectroscopy of free molecules to photoionization spectroscopy of molecules on a surface.

12.2.3. Field Desorption of Molecular Ions Stimulated by Laser Radiation

The possibility of stimulating field desorption by resonant laser excitation of the adsorbate has been studied recently (Antonov and Letokhov, 1984; Antonov *et al.*, 1986). The experiments were performed with molecular layers of tetracene (naphthacene) deposited on a sharp razor blade, which was irradiated by nanosecond pulses of KrF laser ($\lambda = 248$ nm) in an electric field with a strength of about 10^7 V/cm. When the radiation intensity varied between 10^4 and 10^5 W/cm^2, it was possible to observe the photostimulation of the field desorption of tetracene molecular ions with a clear quadratic dependence of the ion yield on the laser radiation intensity. The results obtained are explained in terms of the mechanism of photostimulation consisting of two-step photoionization of molecules on a surface in an electric field, its intensity being less than critical, with subsequent desorption of the ions by the electric field.

In these experiments, a sharp razor blade with sprayed tetracene molecules (10–100 monolayers) was placed in the ion source of the time-of-flight mass spectrometer of the setup described in Section 6.3. The blade was cooled to 200 K. A potential of 10 kV allowed the electric field intensity at the pointed end to reach more than 10^7 V/cm (Beckey, 1978).

When the potential of the blade was 4–12 kV, the signal of tetracene molecular ions observed in this case exceeded the background level by 10^2–10^3 times. Figure 12.3a shows the dependence of this signal on the energy fluence of a laser pulse with a fixed length of 15 ns at different potentials of the blade. When the potential was 7–10 kV, this dependence was quadratic.

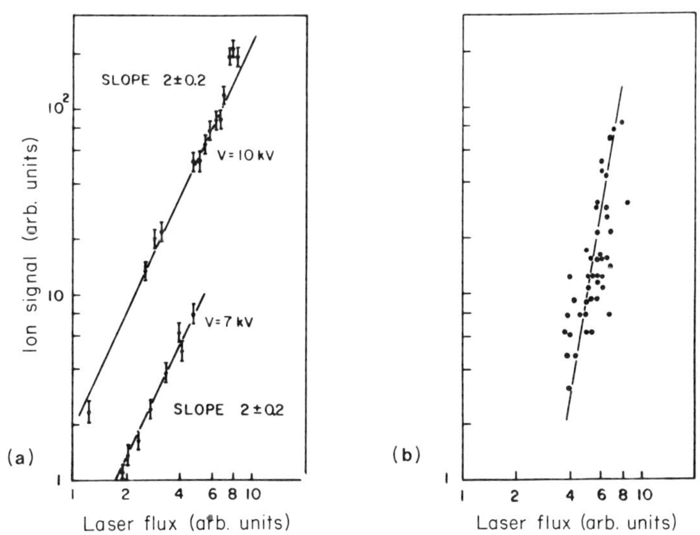

FIG. 12.3. Dependence of the desorption yield of tetracene molecular ions on laser radiation intensity. (a) Sharp razor blade at different potentials and (b) blunt blade at a potential of 9 kV. [From Antonov and Letokhov (1984); Antonov *et al.* (1986).]

To test the effect of electric fields on the process of tetracene desorption, similar measurements were performed for a uniformly blunted blade (the radius of curvature of the point is about 50 μm). Under these conditions, the laser radiation fluence was increased by several times so that the ion signal could be observed. Figure 12.3b shows the dependence of the ion signal from the point of such a blunt blade on the laser pulse fluence for a potential of 9 kV. This dependence can be approximated by a power function with its factor $n = 6 \pm 1$. Unlike the experiment with a sharp blade, the ion signal from a blunt blade slightly depends on electric field strength.

The results obtained for a sharp blade and a strong electric field (Fig. 12.3a) cannot be explained by the thermal stimulation of tetracene desorption. This is proved by a strong dependence of the ion signal on electric field strength. By contrast, for a blunt blade, lower electric field strengths, and higher laser pulse fluences, the steep curve of ion yield (Fig. 12.3b) is explained within the model of thermal desorption and subsequent photoionization of desorbed molecules near a surface (Egorov *et al.*, 1983, 1984a,b).

The observed quadratic dependence of the ion yield in the presence of a strong electric field seems to be connected with the resonant two-photon excitation and ionization of the molecules of the adsorbate. The KrF laser radiation at $\lambda = 248$ nm falls within the electronic absorption band of tetracene molecule in a crystal with the absorption cross section $\sigma = 8 \times 10^{-18}$ cm^2. The energy of two laser photons $2\hbar\omega = 10$ eV is much higher than the ionization potential of a free molecule, $I = 7$ eV, which provides a high efficiency of direct photoionization of molecules on a surface.

The formation of an ion on a surface, however, does not provide in the general case its desorption with the field, since the energy binding the ion with the surface increases, in comparison with a neutral molecule, by the value of Coulomb mirror image potential $E_k(\text{eV}) = 3.6/x$, where $x[\text{Å}]$ is the distance between the ion and the emitter surface. To understand the possibility of field desorption of ions, we should consider the joint action of electric field and laser radiation.

Figure 12.4 shows in a simplified way the potential curve of a neutral molecule and an ion in the absence (dashed curve) and in the presence of an electric field $-eFx$ for ordinary desorption and field desorption stimulated by the absorption of two photons. It also shows a vertical Frank–Condon transition to the saddle point of the ion potential curve in the presence of a field. When the electric field strength is high, the absorption of two quanta at once results in a transition to the repulsive potential curve "ion in the field + surface". Simple estimations give the following value for this critical strength of electric field F_{cr} (Antonov *et al.*, 1986),

$$eF_{er} = 3.6\left(\frac{\varepsilon - 1}{\varepsilon + 1}\right)\frac{1}{x_0^2}, \tag{12.4}$$

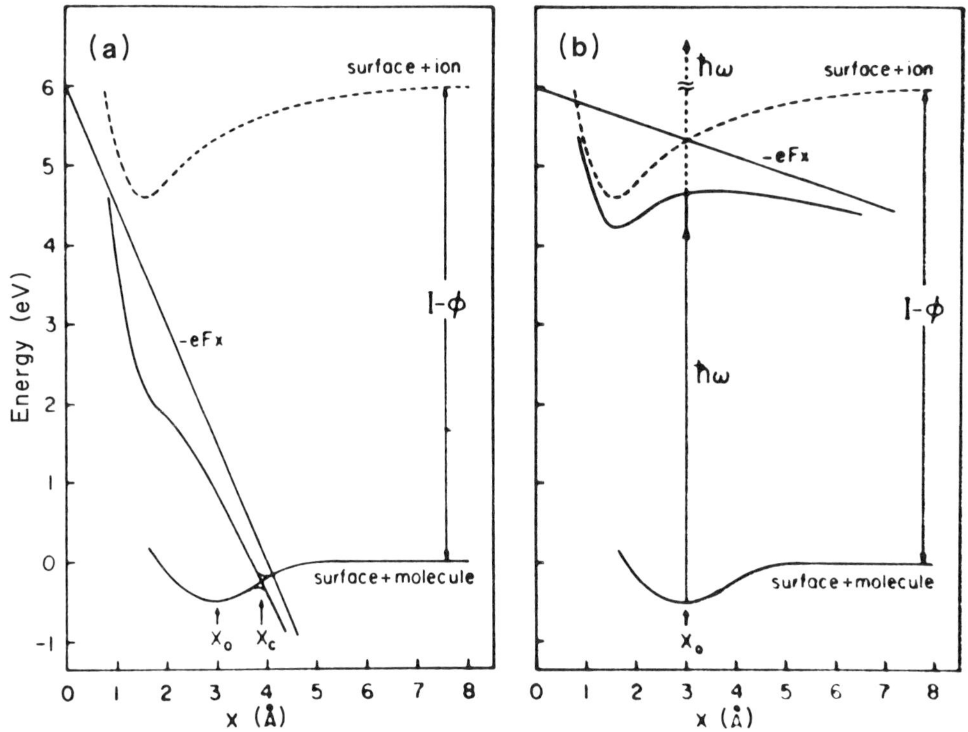

FIG. 12.4. Processes of ion desorption in the presence of a strong electric field. (a) Pure field desorption due to the intersection of the potential curve of a molecular ion in the presence of a field with the potential curve of a neutral molecule on the surface and (b) laser-stimulated field desorption due to the two-photon ionization of the adsorbate. [From Antonov and Letokhov (1984).]

where ε is the dielectric constant of tetracene layers, x_0 is the equilibrium position of a neutral molecule at the surface. Substituting $\varepsilon = 3$ and $x_0 = 4$ Å into Eq. (12.4) gives a numerical value for the critical strength $F_{cr} = 0.1$ V/Å $= 10^7$ V/cm. This value is in good qualitative agreement with experiment.

For comparison, it should be noted that the threshold strength during field desorption at low temperatures (Fig. 12.4a) is determined by an approximate expression

$$eFx_0 \simeq I - \varphi, \tag{12.5}$$

where I is the ionization potential of molecule, φ is the electronic work function of a medium (the Fermi energy for a metal), which equals about 1 V/Å.

Thus, an important condition for stimulating field desorption during photoionization of the adsorbate is a decrease of the energy of Coulomb interaction between the ion and the metal surface of the emitter due to the presence of a dielectric layer [the factor $(\varepsilon - 1)/(\varepsilon + 1)$ in Eq. (12.4)] as well as a sufficient distance between the surface and the part of the molecule with a positive charge (determined by the quantity x_0). The difficulties in observing the photostimulated desorption of He and H_2 from the surface of a tungsten emitter under VUV synchrotron radiation (Wiegmann *et al.*, 1984) are probably connected with the absence of a contribution of these two effects.

The observation of laser stimulation of field desorption opens the way for investigations aimed at the observation of a picture of the spatial distribution of different chromophore groups of biomolecules selectively ionized by laser radiation on the ion field microscope screen. The image can be formed by field desorption of an adsorbate transparent to radiation. This field desorption is stimulated by tunneling an electron of the adsorbate molecule to the ionized chromophore of the biomolecule. It is electron stimulation of the desorption of the imaging molecule of the adsorbate that enables us to hope to obtain a minimum kinetic energy of desorbed molecules at cryogenic temperatures, which is essential for attaining a high spatial resolution in electric fields of about 10^7 V/cm.

12.3. Wave–Corpuscular (Photoion or Photoelectron) Microscopy

Photoion microscopy combines the merits of two types of microscopy: the high spatial resolution of ion microscopy and the high spectral (energy) resolution of optical microscopy. From this point of view, photoion microscopy is an interesting example of wave–corpuscular microscopy. This peculiarity is considered in more detail below (Letokhov, 1981c).

There are two well-known types of microscopy: wave (optical) and corpuscular (electron, ion). In optical microscopy, good spectral resolution can be obtained since light reflection (transmission) is responsive to small changes in the photon energy ($\Delta E = 0.001$–0.01 eV). This is quite understandable since the photon energy is usually commensurate with the energy of the molecular bonds. However, the spatial resolution of optical microscopy is low ($\Delta x = 10^3$–10^4 Å). It depends on the wavelength λ of the light and the aperture of the beam 2α.

In corpuscular microscopy, high spatial resolution can be achieved. For example, with an electron energy of 1 MeV, a de Broglie wavelength $\lambda = 10^{-2}$ Å, and an angular aperture $2\alpha = 10^{-2}$ rad, the spatial resolution of an electron microscope may be as high as $\Delta x = 1$ Å. However, the spectral resolution of an electron microscope is not high since a 1-MeV electron is almost insensitive to any type of chemical bond. The fact that the electron interaction cross section depends on the electron density of the atom makes it possible to form an image with a low contrast. Therefore, to attain angstrom resolution with a high contrast, it is necessary that heavy atoms be incorporated into the molecules under study (Wall *et al.*, 1974).

Field-ion microscopy features extremely high spatial resolution but it is not very sensitive to the type of ion being produced. In combination with mass spectrometry, it can also provide information on the mass of the ion being produced (Beckey, 1978).

Thus, wave–corpuscular photoion microscopy can simultaneously provide high spatial and high spectral resolution. The new possibilities of this approach can be useful in solving the problem of picosecond time resolution.

Let us imagine that a device exists with which the picosecond laser pulses are able to photoionize certain parts of a molecule, and the points of their photodetachment from the photoelectron (photoion) projector can be visualized using the method of multistep excitation. It is not difficult to imagine a modification (Fig. 12.5) that allows us, for example, to directly observe the transfer of energy along a chain ABC of the molecule. For this purpose, it is necessary to excite the electronic state S_1 of the selected part of the molecule (A, for example, Fig. 12.5) using a picosecond laser pulse ω_1 whose duration τ_p is shorter than τ_{tr}, the time required for electronic excitation transfer. Now it is possible to excite the molecule by using another laser pulse of frequency ω_2. This corresponds to a stepwise photoionization of another part of the molecule (C, for example, Fig. 12.5). The appearance of photoelectrons (photoions) can be observed as the delay time between pulses, $\Delta\tau$, changes. In this way we can obtain information on the time it takes for energy to be transferred along a molecular chain. Of course, such an approach calls for diagrams of the successive quantum transitions for the selective stepwise photoionization and photodetachment of the ions. If this is impossible to realize, it is advisable to attain information on the origin of photoion detachment, not from the known irradiation frequencies $\omega_1, \omega_2, \ldots$ but from a mass analysis of the molecular ions to be detached.

Thus, the schemes of photoelectron and photoion microscopes can be modified to yield a device that allows molecular structures and molecular dynamics to be studied with high spatial, spectral (energy), and time resolution, all at the same time.

To end this discussion of the problems of laser visualization of molecules and the localization of molecular bonds using laser radiation, we should note that the problems are actually more complicated. It is sufficient to say that complex questions concerning the manipulation (adsorption, etc.) of macromolecules of surfaces, their complex three-dimensional structure, multiple measurements of the irreversible photodestruction of molecules after the very

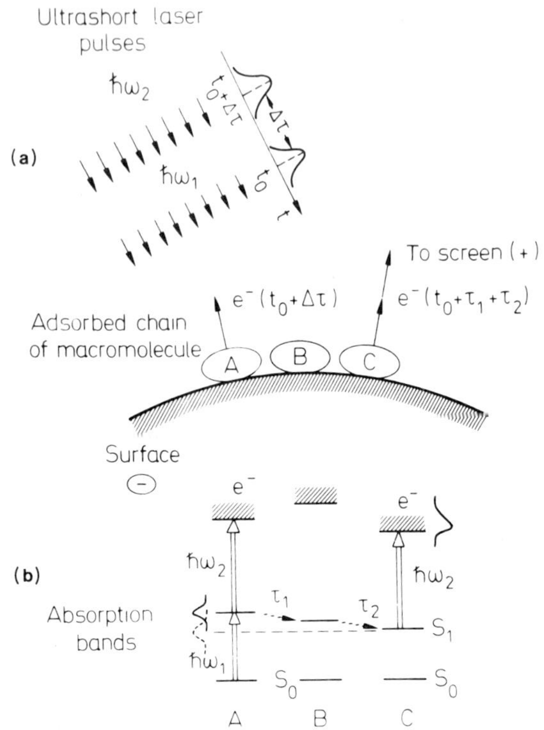

FIG. 12.5. Laser photoelectron microscopy of molecular bonds with picosecond time resolution. (a) geometry of irradiation and (b) simplified scheme of quantum transitions of chain molecule at the surface. Two subsequent laser ultrashort pulses with variable delay $\Delta\tau$ can induce the two-step ionization of some part (A, B, C) of chain molecules that depend on the delay $\Delta\tau$. Point of photoelectron coming out can be observed on the screen. [From Letokhov (1981c).]

first laser pulse, etc., have remained beyond discussion. On the other hand, new, interesting experimental results in this new field of laser photophysics and photochemistry are to be expected.

Thus, it is possible to combine wave (optical) and corpuscular (electron) microscopy. This kind of microscopy is based on the photodetachment of electrons or ions from certain parts of macromolecules using laser radiation. With it, it is possible, in principle, to realize high spatial (several angstroms), spectral (0.001–0.01 eV), and time (picoseconds) resolution, all at the same time.

Appendix. Transition Wavelengths for Ionization Schemes of Various Atoms

TABLE A1. Elements of Group I

Atom	E_i (eV)	Type of scheme[a]	λ_1 (Å)	λ_2 (Å)	λ_i(exper) (Å)	Reference	λ_{Lim}	λ_{Rydb}
Alkali elements								
Li	5.35	$\omega_1, \tilde{\omega}_2$	6708		3371	Kramer *et al.*, 1979	3497	3558
		$\omega_1, \omega_2, \tilde{\omega}_1(\tilde{\omega}_2)$	6708	6104	6104, 6708	Karlov *et al.*, 1977a	8192	8532
Na	5.14	ω_1, ω_2^R	5896		4210–4140	Ambartzumian *et al.*, 1975a	4082	4164
		ω_1, ω_2^R	5890		4196–4100	Ducas *et al.*, 1975	4084	4167
		$\omega_1, \omega_2, \tilde{\omega}_3$	5896	6154	10,600	Smith *et al.*, 1980	12,120	12,881
K	4.34	$\omega_1 \tilde{\omega}$	4047		4047	Beekman *et al.*, 1980	9698	10,180
		ω_1, ω_2^R	7665		4654	Kudriavtzev *et al.*, 1985	4551	4654
		$\omega_1, \omega_2, \omega_3^R$	7665	6939			13,226	14,137
Rb	4.18	$\omega_1, \tilde{\omega}_2$	7948		3471	Ambartzumian and Letokhov, 1972;	4375	4847
		$\omega_1, \tilde{\omega}_2$	4202		6943, 3472	Ambartzumian *et al.*, 1971	10,100	10,623
		$\omega_1, \omega_2, \omega_R$	7800	7758		Ambartzumian *et al.*, 1976a	12,516	13,330
Cs	3.89	$\omega_1 \tilde{\omega}$	4593		4593	Hurst *et al.*, 1977a,b	10,639	10,922
		ω_1, ω_2^R	8521				5081	5210
		$\omega_1, \omega_2, \omega_3^R$	8943	8761			11,325	19,987
d-shell elements								
Cu	7.72	$\omega_1, \tilde{\omega}_2(\omega_2^R)$	3274				3146	3195
		$\omega_1, \omega_2, \tilde{\omega}_3(\omega_3^R)$	3274	7933			5212	5348
Ag	7.57	$\omega_1, \tilde{\omega}_2(\omega_2^R)$	3283				3263	3316
		$\omega_1, \omega_2, \tilde{\omega}_3(\omega_3^R)$	3283	7698			5389	5535
Au	9.22	$\omega_1, \tilde{\omega}_2(\omega_2^R)$	2676				2698	2734
		$\omega_1, \omega_2, \tilde{\omega}_3(\omega_3^R)$	2676	5837			5017	5143

[a] See Section 3.4.
[b] Except H and Fr.

TABLE A2. Elements of Group II

Atom	E_i (eV)	Type of scheme[a]	λ_1 (Å)	λ_2 (Å)	λ_i(exper) (Å)	Reference	λ_{Lim}	λ_{Rydb}
Alkaline-earth elements								
Be	9.32	$\omega_1, \tilde{\omega}_2(\omega_2^R)$	2349				3064	3111
		$\omega_1, \omega_2, \tilde{\omega}_3(\omega_3^R)$	2349	8254			4873	4992
		$\omega_1, \omega_2, \tilde{\omega}_3(\omega_3^R)$	2349	4573			9288	9728
Mg	7.64	$\omega_1, \omega_2^{a.i.}$	2852		3009	Bradley *et al.*, 1976	3756	3826
		$\omega_1, \tilde{\omega}_2(\omega_2^R)$	2852				3756	3826
		$\omega_1, \omega_2, \tilde{\omega}_3(\omega_3^R)$	2852	8807			6548	6764
		$\omega_1, \omega_2, \tilde{\omega}_3(\omega_3^R)$	2852	5528			11,712	12,421
Ca	6.11	$\omega_1, \tilde{\omega}_2(\omega_2^R)$	4227				3897	3972
		$\omega_1, \tilde{\omega}_2$	6162		4880	Brinkman *et al.*, 1974	5627	5786
		$\omega_1, \omega_2, \tilde{\omega}_3(\omega_3^R)$	4227	7326			8326	8678
		$\omega_1, \omega_2, \tilde{\omega}_3(\omega_3^R)$	4227	5857			11,644	12,345
Sr	5.69	$\omega_1, \tilde{\omega}_2(\omega_2^R)$	4607				4125	4210
		$\omega_1, \omega_2, \tilde{\omega}_3(\omega_3^R)$	4607	6550			11,144	11,784
		$\omega_1, \omega_2, \tilde{\omega}_3(\omega_3^R)$	6893	6878			5918	6094
Ba	5.21	$\omega_1, \omega_2^{a.i.}$	5535		4000	Zimmerman *et al.*, 1978	4170	4257
		ω_1, ω_2^R	5535		4248	Cooke and Gallagher, 1978	4170	4257
		ω_1, ω_2^R	5535		4200	Aymar *et al.*, 1981	4170	4257
		$\omega_1, \omega_2, \tilde{\omega}_3(\omega_3^R)$	5535	5160			21,745	24,325
		$\omega_1, \omega_2, \tilde{\omega}_3(\omega_3^R)$	7911	7329			6299	6498

(continued)

TABLE A2. *(Continued)*

Atom	E_i (eV)	Type of scheme[a]	λ_1 (Å)	λ_2 (Å)	λ_i(exper) (Å)	Reference	λ_{Lim}	λ_{Rydb}
Ra	5.28	$\omega_1, \tilde{\omega}_2(\omega_2^R)$	4826				4573	4677
		$\omega_1, \omega_2, \tilde{\omega}_3(\omega_3^R)$	7141	7836			6318	6519
		$\omega_1, \omega_2, \tilde{\omega}_3(\omega_3^R)$	7141	5556			9446	9902
d-shell elements								
Zn	9.39	$\omega_1, \tilde{\omega}_2(\omega_2^R)$	2139				3445	3503
		$\omega_1, \omega_2, \tilde{\omega}_3(\omega_3^R)$	2139	6362			7511	7797
		$\omega_1, \omega_2, \tilde{\omega}_3(\omega_3^R)$	3076	4722			4524	4627
Cd	8.99	$\omega_1, \tilde{\omega}_2(\omega_3^R)$	2288				3466	3525
		$\omega_1, \omega_2, \tilde{\omega}_3(\omega_3^R)$	2288	6438			7505	7790
		$\omega_1, \omega_2, \tilde{\omega}_3(\omega_3^R)$	3261	4800			4748	4860
Hg	10.4	$\omega_1\tilde{\omega}$	3128		3128	Mizolek, 1981	4935	5057
		$\omega_1, \tilde{\omega}_2(\omega_2^R)$	3128				4935	5057
		$\omega_1, \omega_2, \tilde{\omega}_3$	2536	2857	5320	Dyer *et al.*, 1983	10,223	10,759
		$\omega_1, \omega_2, \tilde{\omega}_3(\omega_3^R)$	2536	4358			4579	4683

[a] See Section 3.4.

TABLE A3. Elements of Group III

Atom	E_i (eV)	Type of scheme[a]	λ_1 (Å)	λ_2 (Å)	λ_i(exper) (Å)	Reference	λ_{Lim}	λ_{Rydb}
Main subgroup								
B	8.30	$\omega_1, \tilde{\omega}_2(\omega_2^R)$	2089				5242	5380
		ω_1, ω_2^R	2497		3784	Bekov *et al.*, 1984b	3718	3787
Al	5.98	ω_1, ω_2^R	3082		6300–6700	Zherikhin *et al.*, 1984b	6310	6510
		ω_1, ω_2^R	3962		4474	Akilov *et al.*, 1982	4360	4454
Ga	5.99	$\omega_1, \tilde{\omega}_2$	2874		5782	Zherikhin *et al.*, 1983	7351	7625
		$\omega_1, \tilde{\omega}_2$	2874		6942	Wunderlich and Kirsten, 1984	7351	7625
		ω_1, ω_2^R	4172		4240–4380	Tursunov *et al.*, 1984	4236	4326
In	5.78	$\omega_1, \tilde{\omega}_2$	3039		5782	Muchnik *et al.*, 1983	7256	7522
		ω_1, ω_2^R	4102		4520–4490	Dönzelmann and Neijzen, 1983	4484	4584
Tl	6.11	$\omega_1, \tilde{\omega}_2(\omega_2^R)$	2768				7603	7896
		$\omega_1, \omega_2(\omega_2^R)$	3776				4387	4483
d-shell elements								
Sc	6.56	$\omega_1, \tilde{\omega}_2(\omega_2^R)$	2711				6295	6494
		$\omega_1, \tilde{\omega}_2(\omega_2^R)$	3019				5089	5219
		$\omega_1, \omega_2, \tilde{\omega}_3(\omega_3^R)$	6211	5455			5404	5550
Y	6.2	$\omega_1, \tilde{\omega}_2(\omega_2^R)$	2984				6204	6398
		$\omega_1, \tilde{\omega}_2(\omega_2^R)$	4102				3960	4038
		$\omega_1, \omega_2, \tilde{\omega}_3(\omega_2^R)$	6192	5880			5882	6056
Rare earth elements								
La	5.58	$\omega_1, \tilde{\omega}_2(\omega_2^R)$	4187				4737	4849
		$\omega_1, \tilde{\omega}_2(\omega_2^R)$	3574				5876	6050
		$\omega_1, \omega_2, \tilde{\omega}_3(\omega_2^3)$	5506	6793			8257	8603

[a] See Section 3.4.

(continued)

TABLE A3. (Continued)

Atom	E_i (eV)	Type of scheme[a]	λ_1 (Å)	λ_2 (Å)	λ_i(exper) (Å)	Reference	λ_{Lim}	λ_{Rydb}
Ce	5.54	ω_1, ω_2^R	4583		~4373	Worden *et al.*, 1978b	4373	4469
		$\omega_1, \omega_2, \omega_3^R$	6310	6697	~6874	Worden *et al.*, 1978b	6874	7112
Pr	5.46	$\omega_1, \tilde{\omega}_2(\omega_2^R)$	4951				4187	4274
		ω_1, ω_2^R	3945		~5761	Worden *et al.*, 1978b	5761	5927
		$\omega_1, \tilde{\omega}_2(\omega_2^R)$	5687	7065			8101	8434
Nd	5.52	ω_1, ω_2^R	4634		~4349	Worden *et al.*, 1978b	4349	4443
		$\omega, \tilde{\omega}$	4258		4258	Donohue *et al.*, 1982	5010	5135
		$\omega_1, \omega_2, \tilde{\omega}_3(\omega_3^R)$	6754	7256			6255	6452
Sm	5.64	$\omega, \tilde{\omega}$	4299		4299	Donohue *et al.*, 1982	4550	4653
		ω_1, ω_2^R	4226		~4749	Worden *et al.*, 1978b	4749	4861
		$\omega_1, \tilde{\omega}_2$	5916		3371	Karlov *et al.*, 1977b, 1978	3529	3591
		$\omega_1, \omega_2, \tilde{\omega}_3$	5874	5833	6200–5800	Dobrynin *et al.*, 1983	9475	9934
Gd	6.15	$\omega_1, \tilde{\omega}_2(\omega_2^R)$	4058				4039	4198
		$\omega_1, \tilde{\omega}_2$	5856		<3074	Karlov *et al.*, 1977b, 1978	3170	3220
		$\omega_1, \omega_2, \omega_3^{a.i.}$	5696	5976	6135	Fedoseyev *et al.*, 1985	6759	6989
		$\omega_1, \omega_2, \omega_3^R$	5618	6352	6222	Worden *et al.*, 1978b	6222	6416
		$\omega_1, \omega_2, \omega_3^{a.i.}$	5618	6352	6133	Bekov *et al.*, 1978b, 1979b	6222	6416
Tb	5.86	ω_1, ω_2^R	4139		~4374	Worden *et al.*, 1978b	4374	4469
		$\omega_1, \omega_2, \omega_3^{a.i.}$	6019	6139	6187	Fedoseyev *et al.*, 1985	7196	7458
		$\omega_1, \omega_2, \tilde{\omega}_3(\omega_3^R)$	6019	6139			7196	7458
Dy	5.94	ω_1, ω_2^R	4212		~4137	Worden *et al.*, 1978b	4137	4223
		$\omega_1, \tilde{\omega}_2$	5989		<3204	Karlov *et al.*, 1977c	3204	3254
		$\omega_1, \omega_2, \omega_3^R$	6259	6770	~5826	Worden *et al.*, 1978b	5826	5996

Ho	6.02	ω_1, ω_2^R	4164		~4130	Worden *et al.*, 1978b	4130	4215
		$\omega_1, \omega_2, \omega_3^R$	6305	6577	~5709	Worden *et al.*, 1978b	5709	5873
		$\omega_1, \omega_2, \tilde{\omega}_3(\omega_3^R)$	5983	6135	<6424(6632)		6424	6632
Er	6.11	$\omega_1, \tilde{\omega}_2$	5827		<3114	Karlov *et al.*, 1977c	3144	3152
		ω_1, ω_2^R	4088		~4030	Worden *et al.*, 1978b	4030	4111
		$\omega_1, \omega_2, \omega_3^R$	6221	6452	~5649	Worden *et al.*, 1978b	5649	5809
Tm	6.18	$\omega_1, \tilde{\omega}_2(\omega_2^R)$	4094				3926	4003
		$\omega_1, \omega_2, \omega_3^{a.i.}$	5631	6188	6280–5900	Vidolova-Angelova *et al.*, 1983, 1984	6280	6479
Yb	6.25	$\omega_1, \tilde{\omega}_2$	5556		3080	Ischenko *et al.*, 1977	3081	3128
		$\omega_1, \omega_2, \omega_3^R$	5556	6800	5843	Bekov *et al.*, 1978c,d	5633	5793
		$\omega_1, \omega_2, \omega_3^R$	5556	6800	5640–5730	Bekov *et al.*, 1980b	5633	5793
		$\omega_1, \omega_2, \tilde{\omega}_3(\omega_3^R)$	5556	6111	6050–6500	Blondel *et al.*, 1983	6213	6407
		$\omega_1, \omega_2, \omega_3^{a.i.}$	5556	4530	7100–4100	Bekov *et al.*, 1980a, 1981a	9632	10,106
		$\omega_1, \omega_2, \omega_3^{a.i.}$	5556	4530	4530	Bekov *et al.*, 1983b	9632	10,106
Lu	5.42	$\omega_1, \omega_2^{a.i.}$	4519		4607	Bekov and Vidolova-Angelova, 1981	4620	4727
		$\omega_1, \tilde{\omega}_2(\omega_2^R)$	4519		4380–4665	Miller and Nogar, 1982	4620	4727
		$\omega, \tilde{\omega}$	4519		4519	Miller *et al.*, 1982; Miller and Nogar, 1983	4620	4727
		$\omega_1, \tilde{\omega}_2(\omega_2^R)$	3376				7063	7315
		$\omega_1, \omega_2, \tilde{\omega}_3(\omega_3^R)$	5737	5134			14,568	15,682

TABLE A4. Elements of Group VII

Atom	E_i (eV)	Type of scheme[a]	λ_1 (Å)	λ_2 (Å)	λ_i(exper) (Å)	Reference	λ_{Lim}	λ_{Rydb}
F	17.42	$3\omega, 2\tilde{\omega}$	2921				2642	2676
		$3\omega_1, \tilde{\omega}_2(\omega_2^R)$	2921				2642	2676
		$3\omega_1, \omega_2, \tilde{\omega}_3$	2921	6349			4525	4627
Cl	12.97	$3\omega, \tilde{\omega}$	2411				4612	4718
		$3\omega_1, \tilde{\omega}_2(\omega_2^R)$	2411				4612	4718
		$3\omega, 2\tilde{\omega}$	4168				3064	3110
		$3\omega_1, \tilde{\omega}_2(\omega_2^R)$	4168				3064	3110
		$3\omega_1, \omega_2, \tilde{\omega}_1$	4168	7257			5301	5441
Br	11.8	$2\omega, \tilde{\omega}$	2678				4848	4966
		$2\omega_1, \tilde{\omega}_2(\omega_2^R)$	2678				4848	4966
		$3\omega, 2\tilde{\omega}$	4728				3138	3187
		$3\omega_1, \tilde{\omega}_2(\omega_2^R)$	4728				3138	3187
		$3\omega_1, \omega_2, \tilde{\omega}_1$	4728	7513			5390	5535
I	10.45	$2\omega, \tilde{\omega}$	3047		3047	Bagratashvili *et al.*, 1985a	5362	5506
		$2\omega_1, \tilde{\omega}_2(\omega_2^R)$	3047				5362	5506
		$3\omega, 2\tilde{\omega}$	4926		4926	Hackett *et al.*, 1983	4269	4360
		$3\omega, \tilde{\omega}$	4544		4544	Bagratashvili *et al.*, 1985b	5475	5626

[a] See Section 3.4.

TABLE A5. Elements of Group VIII

Atom	E_i (eV)	Type of scheme[a]	λ_1 (Å)	λ_2 (Å)	λ_i (exper) (Å)	Reference	λ_{Lim}	λ_{Rydb}
Rare gas atoms								
He*(2^1S)	3.97[b]	$\omega, \tilde{\omega}$	5016		5016	Hurst *et al.*, 1975	8262	8608
		$\omega_1, \tilde{\omega}_2^R$	2945		~26,000	Bloomfield *et al.*, 1982	22,168	24,855
Ne*	4.95[b]	$\omega_1, \tilde{\omega}_2$	6402		<4119	Ganz *et al.*, 1982; Siegel *et al.*, 1983	4119	4203
		$\omega_1, \tilde{\omega}_2$	6143		<4234	Ganz *et al.*, 1983, 1984	4234	4323
Ar*	4.21[b]	$\omega_1, \tilde{\omega}_2$	6965				5097	5227
Ar	15.76	$2\omega_1, \omega_2, \tilde{\omega}_2$	1993	1807	1807	Payne *et al.*, 1981; Hurst *et al.*, 1981	4985	5109
		$\omega_1, \omega_2, \tilde{\omega}_3$	1067	3660	10,640	Kramer, 1984	16,549	18,001
Kr*	4.08[b]	$\omega_1, \tilde{\omega}_2(\omega_2^R)$	5570		6600–6800	Delsart *et al.*, 1981	6678	6903
Kr	14.00	$2\omega_1, \omega_2, \tilde{\omega}_2$	1575	2482	2482	Payne *et al.*, 1981; Hurst *et al.*, 1981	9970	10,479
		$\omega_1, \omega_2, \tilde{\omega}_3$	1165	5581	10,640	Kramer *et al.*, 1983; Chen *et al.*, 1984	10,922	11,537
Xe*	3.82[b]	$\omega_1, \tilde{\omega}_2$	4671				10,676	11,262
Xe	12.13	$2\omega, \tilde{\omega}$	2524		2524	Chen *et al.*, 1980	5369	5513
		$2\omega, \tilde{\omega}$	2560,2		2560,2	Bushaw and Whitaker, 1981	5643	5803
		$\omega_1, \tilde{\omega}_2$	1250		5320	Kramer, 1984	5602	5759
Rn*	3.98[b]	$\omega_1, \tilde{\omega}_2$	7450				5360	5503
Rn	10.75	$2\omega, \tilde{\omega}$	2867				5900	6073

[a] See Section 3.4.

[b] Metastable level of atom is excited by a nonlaser method.

References

Achiba, Y., Shobotake, K., and Kimura, K. (1980). Annual Review (Institute of Molecular Science, Okazaki), p. 100.

Achiba, Y., Sato, K., Shindo, Y., and Kimura, K. (1981). Annual Review (Institute of Molecular Science, Okazaki), p. 105.

Achiba, Y., Sato, K., Shobotake, K., and Kimura, K. (1982). *J. Chem. Phys.*, **77**, 2709.

Achiba, Y., Sato, K., Shobotake, K., and Kimura, K. (1983a). *J. Chem. Phys.* **79**, 5213.

Achiba, Y., Sato, K., Shobotake, K., and Kimura, K. (1983b). *J. Chem. Phys.* **78**, 5474.

Achiba, Y., Sato, K., and Kimura, K. (1984). *J. Chem. Phys.* **80**, 6047.

Ackerhalt, J. R. and Eberly, J. H. (1976). *Phys. Rev. A* **14**, 1705.

Ackermann, R. J., Rauh, E. G., and Thorn, R. J. (1976). *J. Chem. Phys.* **65**, 1027.

Akilov, R. and Bekov, G. I. (1982). *Pis'ma Zh. Techn. Fiz.* **8**, 517 (in Russian).

Akilov, R., Bekov, G. I., Letokhov, V. S., Maximov, G. A., Mishin, V. I., Radaev, V. N., and Shishov, V. N. (1982). *Kvantovaya Elektronika* **9**, 1859 (in Russian).

Akilov, R., Bekov, G. I., Deviatikh, G. G., Letokhov, V. S., Maximov, G. A., Radaev, V. N., Shishov, V. N., and Shchepliagin, E. M. (1984). *Zh. Anal. Khim.* **39**, 31 (in Russian).

Akopian, M. E. (1977). "Advances in Photonics," p. 3 (in Russian). Leningrad State University, Leningrad.

Akopian, M. E. and Loginov, Yu. V. (1967). *High Energy Chemistry* **1**, 79 (in Russian).

Alkemade, C. Th. J. (1981). Appl. *Spectrosc.* **35**, 1.

Alkhazov, G. D., Barzakh, A. E., Berlovich, E. E., Denisov, V. P., Dernyatin, A. G., Ivanov, V. S., Zherikhin, A. N., Kompanetz, O. N., Letokhov, V. S., Mishin, V. I., and Fedoseyev, V. N. (1983a). *Pis'ma Zh. Eksp. Teor. Fiz.* **37**, 231 (in Russian).

Alkhazov, G. D., Barzakh, A. E., Berlovich, E. E., Denisov, V. P., Dernyatin, A. G., Invanov, V. S., Zherikhin, A. N., Kompanetz, O. N., Letokhov, V. S., Mishin, V. I., and Fedoseyev, V. N. (1983b). Measurements of Optical Isotopical Shifts for Radioactive Atoms on Laser–Nuclear Facility of IRIS Mass-Separator. Preprint N908 (Nuclear Physics Institute Leningrad, USSR.)

Alkhazov, G. D., Barzakh, A. E., Berlovich, E. E., Denisov, V. P., Dernyatin, A. G., Ivanov, V. S., Letokhov, V. S., Mishin, V. I., and Fedoseyev, V. N. (1984). *Pis'ma Zh. Eksp. Teor. Fiz.* **40**, 95 (in Russian).

Alkhazov, G. D., Barzakh, A. E., Byanov, N. B., Denisov, V. P., Dernyatin, A. G., Ivanov, V. S., Letokhov, V. S., Mishin, V. I., Sekatzkii, S. K., and Fedoseyev, V. N., Chubukov, I. Ya. (1985). Study of electromagnetic moments and charge radii of Tm isotopic nuclei. Preprint N1145. (Nuclear Physics Institute) Leningrad, USSR.

Allen, L. and Eberly, J. H. (1975). "Optical Resonance and Two-Level Atoms." Wiley-Interscience, New York.

Ambartzumian, R. V. and Letokhov, V. S. (1972). Appl. Opt. **11**, 354.

Ambartzumian, R. V., Kalinin, V. P., and Letokhov, V. S. (1971). *Pis'ma Zh. Eksp. Teor. Fiz.* **13**, 305 (in Russian).

Ambartzumian, R. V., Letokhov, V. S., Ryabov, E. A., and Chekalin, N. V. (1974). *Pis'ma Zh. Eksp. Teor. Fiz.* **20**, 597 (in Russian).
Ambartzumian, R. V., Bekov, G. I., Letokhov, V. S., and Mishin, V. I. (1975a). *Pis'ma Zh. Eksp. Teor. Fiz.* **31**, 595 (in Russian).
Ambartzumian, R. V., Chekalin, N. V., Letokhov, V. S., and Ryabov, E. A. (1975b). *Chem. Phys. Lett.* **36**, 301.
Ambartzumian, R. V., Apatin, V. M., Letokhov, V. S., Makarov, A. A., Mishin, V. I., Puretzky, A. A., and Furzikov, N. P. (1976a). *Zh. Eksp. Teor. Fiz.* **70**, 1660 (in Russian).
Ambartzumian, R. V., Furzikov, N. P., Letokhov, V. S., and Puretzky, A. A. (1976b). *Appl. Phys.* **9**, 335.
Anderson, S. L., Rider, D. M., and Zare, R. N. (1982). *Chem. Phys. Lett.* **93**, 11.
Andlauder, B. and Ottinger, Ch. (1971). *J. Chem. Phys.* **55**, 293.
Andreyev, S. V., Antonov, V. S., Knyazev, I. N., Letokhov, V. S., and Movshev, V. G. (1975). *Phys. Lett. A* **54**, 91.
Andreyev, S. V., Antonov, V. S., Knyazev, I. N., and Letokhov, V. S. (1977). *Chem. Phys. Lett.* **45**, 166.
Andreyev, S. V., Mishin, V. I., and Sekatzkii, S. K. (1985). *Kvantovaya Elektron.* **12**, 611 (in Russian).
Andreyev, S. V., Mishin, V. I., and Letokhov, V. S. (1986). *Opt. Comm.* **51**, 317.
Anton, K.-R., Kaufman, W., Moruzzi, G., Neugart, R., Otten, E.-W., and Schinzler, E. (1978). *Phys. Rev. Lett.* **40**, 642.
Antonov, V. S. (1978). *Kvantovaya Elektron.* **5**, 915 (in Russian).
Antonov, V. S. (1982). *Opt. Spektosk.* **52**, 10 (in Russian).
Antonov, V. S. and Letokhov, V. S. (1981). *Appl. Phys.* **24**, 89.
Antonov, V. S. and Letokhov, V. S. (1984). *In* "Multiphoton Processes" (P. Lambropoulos and S. J. Smith, eds.) Vol. 2, p. 182. Springer Series on Atoms and Plasmas. Springer, Berlin.
Antonov, V. S. and Shibanov, A. N. (1982). *Opt. Spektrosk.* **52**, 390 (in Russian).
Antonov, V. S., Knyazev, I. N., Letokhov, V. S., Matiuk, V. M., Movshev, V. G., and Potapov, V. K. (1977a). *Pis'ma Zh. Tekhn. Fiz.* **3**, 1287 (in Russian).
Antonov, V. S., Knyazev, I. N., Letokhov, V. S., and Movshev, V. G. (1977b). *Zh. Eksp. Teor. Fiz.* **73**, 1325 (in Russian).
Antonov, V. S., Knyazev, I. N., Letokhov, V. S., Matiuk, V. M., Movshev, V. G., and Potapov, V. K. (1978). Opt. Lett. **3**, 37.
Antonov, V. S., Letokhov, V. S., and Shibanov, A. N. (1980a). *Appl. Phys.* **22**, 293.
Antonov, V. S., Letokhov, V. S., and Shibanov, A. N. (1980b). *Zh. Eksp. Fiz.* **78**, 2222 (in Russian).
Antonov, V. S., Letokhov, V. S., and Shibanov, A. N. (1980c). *Pis'ma Zh. Eksp. Teor. Fiz.* **31**, 471 (in Russian).
Antonov, V. S., Letokhov, V. S., and Shibanov, A. N. (1981a). *Opt. Comm.* **38**, 182.
Antonov, V. S., Letokhov, V. S., and Shibanov, A. N. (1981b). *Appl. Phys.* **25**, 71.
Antonov, V. S., Egorov, S. E., Letokhov, V. S., and Shibanov, A. N. (1982a). *Pis'ma Zh. Eksp. Teor. Fiz.* **36**, 29 (in Russian).
Antonov, V. S., Letokhov, V. S., Matveetz, Yu. A., and Shibanov, A. N. (1982b). *Laser Chem.* **1**, 37.
Antonov, V. S., Khoroshilova, E. V., Kuzmina, N. P., Letokhov, V. S., Matveetz, Yu. A., Shibanov, A. N., and Egorov, S. E. (1982c). *In* "Picosecond Phenomena III," (K. N. Eisenthal, R. M. Hochstrasser, W. Kaiser, and A. Laubereau, eds.), Vol. 23, p. 31. Springer Series in Chemical Physics, Springer, Berlin.
Antonov, V. S., Egorov, S. E., Letokhov, V. S., and Shibanov, A. N. (1983). *Pis'ma Zh. Eksp. Teor. Fiz.* **38**, 185 (in Russian).

Antonov, V. S., Letokhov, V. S., and Shibanov, A. N. (1984). *Usp. Fiz. Nauk* **142**, 177 (in Russian).
Antonov, V. S., Letokhov, V. S., and Moskovetz, E. V. (1986). *Izv. USSR Acad. Sci.* **50**, 690 (in Russian).
Aoki, T., Morikawa, T., and Sakurai, K. (1973). *J. Chem. Phys.* **59**, 1543.
Apatin, V. M., Letokhov, V. S., and Mishin, V. I. (1978). *Kvantovaya Elektron.* **5**, 632 (in Russian).
Armstrong, J. A., Esherick, P., and Wynne, J. J. (1977). *Phys. Rev. A* **15**, 180.
Armstrong, J. A., Wynne, J. J., and Esherick, P. (1979). *J. Opt. Soc. Am.* **69**, 211.
Ashfold, M. N. R., Dixon, R. N., and Stickland, R. J. (1984). *Chem. Phys. Lett.* **111**, 226.
Autler, S. H. and Townes, C. H. (1950). *Phys. Rev.* **78**, 340A.
Aymar, M., Champen, R.-J., Deslart, C., and Keller, J.-C. (1981). *J. Phys. B* **14**, 4489.
Bagratashvili, V. N., Kuzmin, M. V., Letokhov, V. S., and Shibanov, A. N. (1983). *Pis'ma Zh. Eksp. Teor. Fiz.* **37**, 92 (in Russian).
Bagratashvili, V. N., Ionov, S. I., Mishakov, G. V., Semchishen, V. A., and Masalov, A. V. (1985a). *Chem. Phys. Lett.* **115**, 144.
Bagratashvili, V. N., Ionov, S. I., Kuzmin, M. V., and Mishakov, G. V. (1985b). *Chem. Phys. Lett.* **115**, 149.
Bagratashvili, V. N., Letokhov, V. S., Makarov, A. A., and Ryabov, E. A. (1985c). "Multiple Photon Infrared Laser Photophysics and Photochemistry." Harwood Academic Publishers, London and New York.
Bahcall, J. N. (1982). *Rev. Mod. Phys.* **50**, 881.
Bahcall, J. N., Cleveland, B. T., Davis, R., Dostrovsky, I., Evans, J. C., Frati, W., Friedlander, G., Zande, K., Rowley, J. K., Stoenner, R. W., and Weneser, J. (1978). *Phys. Rev. Lett.* **40**, 1351.
Bahcall, J. N., Hubner, W. F., Lubow, S. H., Parker, P. D., and Ulrich, R. K. (1982). *Rev. Mod. Phys.* **54**, 767.
Bakulina, I. N. and Ionov, N. I. (1959). *Zh. Eksp. Teor. Fiz.* **35**, 1001 (in Russian).
Balykin, V. I., Letokhov, V. S., Mishin, V. I., and Semchishen, V. A. (1977). *Pis'ma Zh. Eksp. Teor. Fiz.* **26**, 492 (in Russian).
Balykin, V. I., Bekov, G. I., Letokhov, V. S., and Mishin, V. I. (1978). *Proc. Int. Conf. At. Phys. 6th, Riga USSR* p. 565. Plenum Press, New York and Zinatne, Riga.
Balykin, V. I., Letokhov, V. S., and Mishin, V. I. (1980a). *Appl. Phys.* **22**, 245.
Balykin, V. I., Bekov, G. I., Letokhov, V. S., and Mishin, V. I. (1980b). *Usp. Fiz. Nauk* **132**, 293 (in Russian). *Sov. Phys. Uspekhi* **23**(10), 651.
Balykin, V. I., Letokhov, V. S., and Minogin, V. G. (1984). *Appl. Phys. B* **33**, 247.
Beckey, H. D. (1978). "Principles of Field Ionization and Field Desorption Mass Spectrometry." Pergamon, Oxford.
Beekman, D. W. and Calcott, T. A. (1984). *In* "Resonance Ionization Spectroscopy," 1984 (G. S. Hurst and M. G. Payne, eds.), Publ. N71. Conf. Ser. The Institute of Physics, Bristol and Boston. p. 143.
Beekman, D. W., Calcott, T. A., Kramer, S. D., Arakawa, F. T., Hurst, G. S., and Nussbaum, E. (1980). *Int. J. Mass Spectr. Ion Phys.* **34**, 89.
Beer, J., Andree, M., Oeschger, H., Stauffer, B., Balzer, R., Bonani, G., Stoller, G., Suter, M., Wolfli, W., and Finkel, R. C. (1983). *Radiocarbon* **25**, 269.
Behrisch, R., ed. (1981). "Sputtering by Particle Bombardment. I. Physical Sputtering of Single-Element Solids," Vol. 47, Springer Series "Topics in Applied Physics." Springer-Verlag, Berlin, Heidelberg and New York.
Beigang, R. and Timmermann, A. (1982a). *Phys. Rev. A* **25**, 1496.
Beigang, R. and Timmermann, A. (1982b). *Phys. Rev. A* **26**, 2990.

Beigang, R. and Timmermann, A. (1983a). *In* "Laser Spectroscopy VI" (H. P. Weber and W. Lüthy, eds.), Vol. 40, p. 233, Springer Series in Optical Sciences. Springer-Verlag, Berlin, Heidelberg and New York.

Beigang, R. and Timmermann, A. (1983b). *In* "Laser Spectroscopy VI," (H. P. Weber and W. Lüthy, eds.), Vol. 40, p. 235, Springer Series in Optical Sciences. Springer-Verlag, Berlin, Heidelberg, and New York.

Beigang, R., Makat, W., and Timmermann A. (1984). Opt. Comm. **49**, 253.

Bekov, G. I. and Letokhov, V. S. (1983a). *Appl. Phys. B* **30**, 161.

Bekov, G. I. and Letokhov, V. S. (1983b). *Trends Anal. Chem.* **2**, 252.

Bekov, G. I. and Vidolova-Angelova, E. P. (1981). *Kvantovaya Elektron.* **8**, 227 (in Russian).

Bekov, G. I., Letokhov, V. S., and Mishin, V. I. (1977a). *Zh. Eksp. Teor. Fiz.* **73**, 157 (in Russian).

Bekov, G. I., Letokhov, V. S., and Mishin, V. I. (1977b). *Opt. Comm.* **23**, 85.

Bekov, G. I., Letokhov, V. S., and Mishin, V. I. (1978a). *Pis'ma Zh. Eksp. Teor. Fiz.* **27**, 52 (in Russian).

Bekov, G. I., Letokhov, V. S., Matveev, O. I., and Mishin, V. I. (1978b). *Pis'ma Zh. Eksp. Teor. Fiz.* **28**, 308 (in Russian).

Bekov, G. I., Letokhov, V. S., Matveev, O. I., and Mishin, V. I. (1978c). *Opt. Lett.* **3**, 159.

Bekov, G. I., Letokhov, V. S., Matveev, O. I., and Mishin, V. I. (1978d). *Zh. Eksp. Teor. Fiz.* **75**, 2092 (in Russian).

Bekov, G. I., Vidolova-Angelova, E. P., Letokhov, V. S., and Mishin, V. I. (1979). *In* "Laser Spectroscopy IV," (H. Walther and K. W. Rothe, eds.) Vol. 21, p. 283, Springer Series in Optical Sciences. Springer-Verlag, Berlin, Heidelberg and New York.

Bekov, G. I., Vidolova-Angelova, E. P., Ivanov, L. N., Letokhov, V. S., and Mishin, V. I. (1980a). *Opt. Comm.* **35**, 194.

Bekov, G. I., Vidolova-Angelova, E. P., Letokhov, V. S., and Mishin, V. I. (1980b). *Opt. Spektrosk.* **48**, 435 (in Russian).

Bekov, G. I., Vidolova-Angelova, E. P., Ivanov, L. N., Letokhov, V. S., and Mishin, V. I. (1981a). *Zh. Eksp. Teor. Fiz.* **80**, 866 (in Russian).

Bekov, G. I., Zherikhin, A. N., Letokhov, V. S., Mishin, V. I., and Fedoseyev, V. N. (1981b). *Pis'ma Zh. Eksp. Teor. Fiz.* **33**, 467 (in Russian).

Bekov, G. I., Yegorov, A. S., Letokhov, V. S., and Radayev, V. N. (1983a). *Nature* (*London*) **301**, 410.

Bekov, G. I., Yegorov, A.S., Letokhov, V. S., and Mishin, V. I. (1983b). *Zh. Anal. Khim.* **38**, 429 (in Russian).

Bekov, G. I., Yegorov, A. S., Letokhov, V. S., and Radayev, V. N. (1983c). *Oceanology* **23**, 171 (in Russian).

Bekov, G. I., Letokhov, V. S., Radayev, V. N., Egorov, A. S., Baturin, G. N., Kursky, A. N., and Narseyev, V. A. (1984a). *Nature* (*London*) **312**, 748.

Bekov, G. I., Maximov, G. A., Nikogosyan, D. N., and Radayev, V. N. (1984b). *Kvantovaya* Elektron. **11**, 1262 (in Russian).

Bekov, G. I., Letokhov, V. S., and Radayev, V. N. (1984c). *Laser Chem.* **5**, 11.

Bekov, G. I., Letokhov, V. S., and Radayev, V. N. (1985). *J. Opt. Soc. Am. B* **2**, 1554.

Berkowitz, J. (1979). "Photoabsorption, Photoionization, and Photoelectron Spectroscopy." Academic Press, New York.

Berman, E. (1979). *In* "Toxic Metals and Their Analysis," (L. C. Thomas, ed.), p. 6, International Topics in Science. Heiden, London, Philadelphia and Rheine.

Bernhardt, A. F., Duerre, D. E., Simpson, J. R., and Wood, L. L. (1974). *Appl. Phys. Lett.* **25**, 617.

Bernhardt, A. F., Duerre, D. E., Simpson, J. R., and Wood, L. L. (1976). *Opt. Comm.* **16**, 169.

Bernheim, R. A., Gold, L. P., and Tipton, T. (1982). *Chem. Phys. Lett.* **92**, 13.

Beterov, I. M., Fateev, N. V., and Chebotayev, V. P. (1982). *Pis'ma Zh. Eksp. Teor. Fiz.* **36**, 251 (in Russian).

Bethe, H. A. and Salpeter, E. A. (1957). "Quantum Mechanics of One- and Two-Electron Atoms." Academic Press, New York.

Bizau, J. M., Wuillenmier, F., Dhez, P., Ederer, D. L., Picque, J. L., LeGouët, J. C., and Koch, P. (1982). *In* "Laser Techniques for Extreme Ultraviolet Spectroscopy," (T. J. McIlrath and R. F. Freeman, eds.), Publ. 90, p. 331. American Institute of Physics, New York.

Bjorkholm, J. E. and Liao, P. F. (1974). *Phys. Rev. Lett.* **33**, 128.

Bjorkholm, J. E. and Liao, P. F. (1975). *In* "Laser Spectroscopy," Proceedings of 2nd Int. Conf., p. 176, June 23–27, 1975, Megeve, France. Springer-Verlag, Berlin, Heidelberg, and New York.

Bloch, F. (1946). *Phys. Rev.* **70**, 460.

Bloembergen, N. (1965). "Nonlinear Optics." Benjamin, New York.

Bloembergen, N. and Levenson, M. D. (1976). *In* "High Resolution Laser Spectroscopy," (K. Shimoda, ed.) Vol. 13, p. 315, Topics in Applied Physics, Springer-Verlag, Berlin, Heidelberg, and New York.

Blondel, C., Champean, R.-J., and Delsart, C. (1983). *Phys. Rev. A* **27**, 583.

Bloomfield, L. A., Gerhardt, H., and Hänsch, T. W. (1982). *Appl. Phys. B* **29**, 162.

Boesl, U., Neusser, H. J., and Schlag, E. W. (1978). *Z. Naturforsch.*, **33A**, 1546.

Boesl, U., Neusser, H. J., and Schlag, E. W. (1980). *J. Chem. Phys.* **72**, 4327.

Boesl, U., Neusser, H. J., and Schlag, E. W. (1981). *Chem. Phys.* **55**, 193.

Boesl, U., Neusser, H. J., and Schlag, E. W. (1982a). *Chem. Phys. Lett.* **87**, 1.

Boesl, U., Neusser, H. J., Weinkauf, R., and Schlag, E. W. (1982b). *J. Chem. Phys.* **86**, 4857.

Bonch-Bruevich, A. M. and Khodovoi, V. A. (1967). *Usp. Fiz. Nauk* **93**, 1967; *Sov. Phys. Usp.* **10**, 637 (1968).

Bradley, D. J., Dudan, C. H., Ewart, P., and Purdie, A. F. (1976). *Phys. Rev. A* **13**, 1416.

Brinkman, U., Hartwig, W., Telle, H., and Walther, H. (1974). *Appl. Phys.* **5**, 109.

Brophy, J. H. and Rettner, C. T. (1979a). *Chem. Phys. Lett.* **67**, 351.

Brophy, J. H. and Rettner, C. T. (1979b). *Opt. Lett.* **4**, 337.

Burgess, A. and Seaton, M. J. (1960). *Mon. Nat. R. Astron. Soc.* **120**, 121.

Burmakin, V. A., Evtunin, A. N., Lesnoi, M. A., and Bilkin, V. I. (1978). *Kvantovaya Elektron.* **5**, 1000.

Burmakin, V. A., Evtunin, A. N., and Lesnoi, M. A. (1979). *Kvantovaya Elektron.* **6**, 1589.

Bushaw, B. A. and Whitaker, T. J. (1981). *J. Chem. Phys.* **74**, 6519.

Cagnac, B., Crynberg, G., and Biraben, F. (1973). *J. Phys.* (*Paris*) **34**, 56.

Camus, P. and Tomkins, F. (1969). *J. Phys.* (*Paris*), **30**, 545.

Camus, P., Débarre, A., and Morillon, C. (1980). *J. Phys. B* **13**, 1073.

Cantrell, C. D., Letokhov, V. S., and Makarov, A. A. (1980). *In* "Coherent Nonlinear Optics: Recent Advances," (M. S. Feld and V. S. Letokhov, eds.) Vol. 21, p. 165, Springer Series "Topics in Current Physics". Springer-Verlag, Berlin, Heidelberg, and New York.

Capelle, G. A., Jessup, D. A., Borella, H. M., and Franks, L. A. (1984). *Appl. Phys. Lett.* *44*(2), 177.

Carlson, L. R., Paisner, J. A., Worden, E. W., Johnson, S. A., May, C. A., and Solarz, R. W. (1976). *J. Opt. Soc. Am.* **66**, 849.

Carlson, N. W., Taylor, A. J., and Schawlow, A. L. (1980). *Phys. Rev. Lett.* **45**, 18.

Carlson, N. W., Taylor, A. J., Jones, K. M., and Schawlow, A. L. (1981). *Phys. Rev. A* **24**, 822.

Caschetto, S. and Wollast, R. (1979a). *Geochim. Cosmochim. Acta* **43**, 3.

Caschetto, S. and Wollast, R. (1979b). *Marine Chem.* **7**, 2.

Chang, C. S. (1980). *Phys. Rev. A* **21**, 1480.

Chen, C. H., Hurst, G. S., and Payne, M. G. (1980). *Chem. Phys. Lett.* **75**, 473.

Chen, C. H., Kramer, S. D., Allman, S. L., and Hurst, G. S. (1984). *Appl. Phys. Lett.* **44**, 640.
Cherepkov, N. A. (1973). *Zh. Eksp. Teor. Fiz.* **65**, 933 (in Russian).
Cherepkov, N. A. (1979). *J. Phys. B* **12**, 1278.
Cherepkov, N. A. (1981). *J. Phys. B* **14**, 2165.
Chien, R.-L., Millins, O. C., and Berg, R. S. (1983). *Phys. Rev. A* **28**, 2078.
Chou, J. S., Sumida, D., Stuke, M., and Wittig, C. (1982). *Laser Chem.* **1**, 1.
Chou, J. S., Sumida, D., and Wittig, C. (1983). *Chem. Phys. Lett.* **100**, 397.
Chu, S. and Mills, A. P. (1982). *Phys. Rev. Lett.* **48**, 1333.
Chu, S., Mills, A. P., Jr., and Hall, H. L. (1983). *In* "Laser Spectroscopy VI," (H. P. Weber and W. Lüthy, eds.), Vol. 40, p. 28, Springer Series in Optical Sciences. Springer-Verlag, Berlin, Heidelberg, and New York.
Cole, M. D., Wiggins, J. W., and Beer, M. (1977). *J. Mol. Biol.* **117**, 387.
Collins, C. B., Johnson, B. W., Mirza, M. Y., Popescu, D., and Popescu, I. (1974). *Phys. Rev. A* **10**, 813.
Compton, R. N., Miller, J. C., Carter, A. E., and Kruit, P. (1980). *Chem. Phys. Lett.* **71**, 87.
Conzemius, R. J. and Capellen, J. M. (1980). *Int. J. Mass. Spectr. Ion Phys.* **34**, 197.
Cooke, W. E. and Gallagher, T. F. (1978). *Phys. Rev. Lett.* **41**, 1648.
Cooke, W. E., Gallagher, T. F., Edelstein, S. A., and Hill, R. M. (1978). *Phys. Rev. Lett.* **40**, 178.
Cooke, W. E., Bhatti, S. A., and Cromer, C. L. (1982). *Opt. Lett.* **7**, 69.
Cooper, J. and Zare, R. N. (1968). *J. Chem. Phys.* **48**, 942.
Corliss, C. H. and Bozman, R. W. (1962). "Experimental Transition Probabilities for Spectral Lines of Seventy Elements," NBS Monograph 53. U.S. Government Printing Office, Washington, D.C.
Cowan, G. A. and Haxton, W. C. (1982). *Science* **216**, 51.
Cowan, G. A., Daniels, K. S., Haxton, W. C., Rokop. D. V., Treher, E. N., and Wolfsberg, K. (1984). *In* "Resonance Ionization Spectroscopy, 1984" (G. S. Hurst and M. G. Payne, eds.), p. 263, Publ. Conf. Ser. N71 (The Institute of Physics, Bristol and Boston).
Crocket, H. (1972). *Geochim. Cosmochim. Acta* **36**, 517.
Curry, S. M., Collins, C. B., Mirza, M. Y., and Popescu, D. (1976). *Opt. Comm.* **16**, 251.
Damburg, R. and Kolosov, V. (1976). *J. Phys. B* **9**, 3149.
Danon, J., Zacharias, H., Rottke, H., and Welge, K. S. (1982). *J. Chem. Phys.* **76**, 2399.
Darznek, S. A., Zverev, M. M., and Kopyt, S. N. (1983). *Kvantovaya Elektron.* **10**, 1270 (in Russian).
Davis, R., Harmer, D. S., and Hoffman, K. C. (1968). *Phys. Rev. Lett.* **20**, 1205.
Davis, R., Jr. (1978). *In* "Proceedings of an Informal Conference on the Status and Future of Solar Neutrino Research," (G. Friedlander, ed.), BNL Report 50879. Brookhaven National Laboratory, Upton, New York.
Delone, N. B. (1975). *Usp. Fiz. Nauk* **115**, 361 (in Russian).
Delone, N. B. and Krainov, V. P. (1984). "Atoms in Strong Light Field," Vol. 28, Springer Series in Chemical Physics. Springer-Verlag, Berlin, Heidelberg, and New York.
Delsart, C., Keller, J. C., and Thomas, C. (1981). *J. Phys. B* **14**, 4241.
Demtröder, W., Eisel, D., Hemmerlong, B., and Rai, S. B. (1983). *In* "Laser Spectroscopy VI," (H. P. Weber and W. Lüthy, eds.), p. 250. Springer-Verlag, Berlin, Heidelberg, and New York.
Devdariani, A. Z., Klucharev, A. N., Lazarenko, A. V., and Scheverev, V. A. (1978). *Pis'ma Zh. Techn. Fiz.* **4**, 1013 (in Russian).
Diels, J.-C. (1976). *Phys. Rev. A* **13**, 1520.
Dietz, T. G., Duncan, M. A., Liverman, M. G., and Smalley, R. E. (1980). *J. Chem. Phys.* **73**, 4816.
Dietz, T. G., Duncan, M. A., and Smalley, R. E. (1982a). *J. Chem. Phys.* **76**, 1227.
Dietz, W., Neusser, H. J., Boesl, V., Schlag, E. W. (1982b). *Chem. Phys.* **66**, 105.

Dirac, P. A. M. (1930). "The Principles of Quantum Mechanics, 1st edition." Clarendon Press, Oxford.

Dirac, P. A. M. (1958). "The Principles of Quantum Mechanics, 2nd edition." Clarendon Press, Oxford.

Dixit, S. N. and Lambropoulos, P. (1983). *Phys. Rev. A* **27**, 861.

Dobrynin, V. E., Karpov, N. A., Kotochigova, S. A., Krynetzkii, B. B., Mishin, V. A., Stel'makh, O. M., and Shustriakov, V. M. (1983). *Opt. Spektrosk.* **54**, 415 (in Russian).

Donnelly, V. M. and Kaufman, F. (1978). *J. Chem. Phys.* **69**, 1456.

Donohue, D. L., Young, J. P., and Smith, D. H. (1982). *Int. J. Mass. Spectr. Ion Phys.*, **43**, 293.

Dönzelmann, A. and Neijzen, J. H. M. (1983). *Acta Phys. Pol. A* **63**, 201.

Drachsel, W., Nishigaki, S., and Block, J. H. (1980). *Int. J. Mass Spectr. Ion Phys.* **32**, 333.

Ducas, T. W., Littman, M. G., Freeman, R. R., and Kleppner, D. (1975). *Phys. Rev. Lett.* **35**, 366.

Duncan, M. A., Dietz, T. G., and Smalley, R. E. (1981). *J. Chem. Phys.* **75**, 2118.

Duong, H. T. and Vialle, J.-L. (1974). *Opt. Comm.* **12**, 71.

Duong, H. T., Jacquinot, P., Liberman, S., Pinard, J., and Vialle, J.-L. (1973). *Compt. Rend. (Paris) 276B*, 909.

Dyer, P., Baldwin, G. C., Kittrel, C., Imre, D. G., and Abramson, E. (1983). *Appl. Phys. Lett.* **42**, 311.

Dygdala, R. S. and Stefanski, K. (1980). *Chem. Phys.* **53**, 51.

Dyke, W. P. and Dolan, W. W. (1956). *Adv. Electron. Electron Phys.* **8**, 90.

Dylla, H. F. and Jarell, J. A. (1976). *Rev. Sci. Instrum.* **47**, 331.

Egorov, S. E., Letokhov, V. S., and Shibanov, A. N. (1983). *In* "Surface Studies with Lasers," Vol. 33, p. 156, Springer Series in Chem. Phys. (F. Aussenegg, A. Leither, and M. E. Lippitsch, eds.), Springer-Verlag, Berlin.

Egorov, S. E., Letokhov, V. S., and Shibanov, A. N. (1984a). *Chem. Phys.* **85**, 349.

Egorov, S. E., Letokhov, V. S., and Shibanov, A. N. (1984b). *Kvantovaya Elektron.* **11**, 1393 (in Russian).

Einstein, A. (1956). *In* "Investigation on the Theory of the Brownian Movement," (R. Fürth, ed.), Dover, New York.

Eisel, D. and Demtröder, W. (1982). *Chem. Phys. Lett.* **88**, 481.

Esherick, P. and Anderson, R. J.-M. (1980). *Chem. Phys. Lett.* **70**, 621.

Esherick, P. and Owyoung, A. (1983). *Chem. Phys. Lett.* **103**, 235.

Esherick, P., Armstrong, J. A., Dreufus, R. W., and Wynne, J. J. (1976). *Phys. Rev. Lett.* **36**, 1296.

Esherick, P., Wynne, J. J., and Armstrong, J. A. (1977). *In* "Laser Spectroscopy III," ed. (J. L. Hall and J. L. Carlsten, eds.), Vol. 7, p. 170. Springer Series in Optical Sciences. Springer-Verlag, Berlin, Heidelberg, and New York.

Fairbank, W. M., Jr. and She, C. Y. (1979). *Opt. News*, Spring Issue, **4**.

Fano, U. (1961). *Phys. Rev.* **124A**, 1866.

Fano, U. (1969). *Phys. Rev.* **178**, 131.

Fano, U. (1970). *Phys. Rev. A* **2**, 353.

Fano, U. and Cooper, J. W. (1965). *Phys. Rev. 137A*, 1964.

Fedoseyev, V. N. and Yusupov, A. K. (1984). *Opt. Spektrosk.* **57**, 905 (in Russian).

Fedoseyev, V. N., Letokhov, V. S., Mishin, V. I., Alkhazov, A. D., Barzakh, A. E., Denisov, V. P., Dernyatin, A. G., and Ivanov, V. S. (1984). *Opt. Comm.* **52**, 24.

Feld, M. S. and Letokhov, V. S., eds. (1980). "Coherent Nonlinear Optics: Recent Advances," Vol. 21, Springer Series "Topics in Current Physics". Springer-Verlag, Berlin, Heidelberg, and New York.

Feldman, D., Lauremper, J., and Welge, K. M. (1983). *J. Chem. Phys.* **79**, 278.

Feynman, R. P., Vernon, F. L., and Hellwarth, R. W. (1957). *J. Appl. Phys.* **28**, 49.

Fitch, P. S. H., Wharton, L., and Levy, D. H. (1978). *J. Chem. Phys.* **69**, 3424.
Foote, P. D. and Mohler, E. L. (1925). *Phys. Rev.* **26**, 195.
Foster, P. J., Leckenby, R. E., and Robbins, E. J. (1969). *J. Phys. B* **2**, 478.
Franzen, W. F. (1959). *Phys. Rev.* **115**, 850.
Freeman, R. R., Jobson, R. M., Boxor, J., and Cooke, W. E. (1983). *In* "Laser Spectroscopy VI," (H. P. Weber and W. Lüthy, eds.), Vol. 40, p. 220, Springer Series in Optical Sciences. Springer-Verlag, Berlin, Heidelberg, and New York.
Freuholz, R., Wessel, J., and Wheatley, E. (1980). *Anal. Chem.* **52**, 281.
Fuke, K., Yoshinchi, H., Kaya, K., Achiba, Y., Sato, K., and Kimura, K. (1984). *Chem. Phys. Lett.* **108**, 179.
Gallagher, T. F. (1982). *In* "Laser Techniques for Extreme Ultraviolet Spectroscopy," (T. J. McIlrath and R. R. Freeman, eds.), N90 Publ. p. 358. American Institute of Physics, New York.
Gallagher, T. F., Edelstein, S. A., and Hill, R. M. (1975). *Phys. Rev.* **11**, 1504.
Gallas, J. A. C., Walther, H., and Werner, E. (1982). *Phys. Rev. Lett.* **49**, 867.
Ganz, J., Lewandowski, B., Siegel, A., Bussert, W., Waibel, H., Raf, M.-W., and Hotop, H. (1982). *J. Phys. B* **15**, 1485.
Ganz, J., Siegel, A., Bussert, W., Harth, K., Ruf, M.-W., Hotop, H., Geiger, J., and Fink, M. (1983). *J. Phys. B* **16**, 1569.
Ganz, J., Raab, M., Hotop, H., and Geiger, J. (1984). *Phys. Rev. Lett.* **53**, 1547.
Geller, Yu. I. and Popov, A. K. (1981). "Laser-Induced Nonlinear Resonances in Continuous Spectra." Publication of Sibirian Branch USSR Acad. Sci., Novosibirsk (in Russian).
Georges, A. T. and Lambropoulos, P. (1978). *Phys. Rev. A* **18**, 1072.
Glownita, J. H., Riley, S. J., Colson, S. D., Miller, J. C., and Compton, R. N. (1982). *J. Chem. Phys.* **77**, 68.
Goeppert-Mayer, M. (1931). *Ann. Phys.* **9**, 273.
Gomer, R. (1959). *J. Chem. Phys.* **31**, 341.
Gonohe, N., Shimizu, A., Abe, H., Mikami, N., and Ito, M. (1984). *Chem. Phys. Lett.* **107**, 22.
Green, R. B., Keller, R. A., Luther, G. G., Schenck, P. K., and Travis, J. C. (1976). *J. Am. Chem. Soc.* **98**, 8517.
Greene, B. I. (1984). *In* "Ultrafast Phenomena IV," (D. H. Auston and K. B. Eisenthal, eds.), Vol. 38, p. 308, Springer Series in Chemical Physics. Springer-Verlag, Berlin, Heidelberg, and New York.
Greene, B. I. and Farrow, R. C. (1983). *J. Chem. Phys.* **78** (6), part 1, 3336.
Greenlees, G. W., Clark, D. L., Kaufman, S. L., Lewis, D. A., Tonn, J. F., and Broadhurst, J. H. (1977). *Opt. Comm.* **23**, 236.
Grossman, L. W., Hurst, G. S., Kramer, S. D., Payne, M. G., and Young, J. P. (1977a). *Chem. Phys. Lett.* **50**, 207.
Grossman, L. W., Payne, M. G., and Allman, S. L. (1977b). *Chem. Phys. Lett.* **50**, 70.
Grubb, S. G., Whetten, R. L., Albrecht, A. C., and Grant, E. R. (1984). *Chem. Phys. Lett.* **108**, 420.
Hackett, P. A., John, P., Myhew, M., and Rayner, D. M. (1983). *Chem. Phys. Lett.* **96**, 139.
Haffty, J., Riley, L. B., and Goss, W. D. (1977). *Geol. Sur. Bull.* (*US*) **1445**.
Hagen, D. F. (1979). *Anal. Chem.* **51**, 870.
Hänsch, T. W. (1972). *Appl. Opt.* **11**, 895.
Hansen, J. C., Duncanson, J. A., Jr., Chien, R. L., and Berry, R. S. (1980). *Phys. Rev. A* **21**, 222.
Harris, S. E. and Lidow, D. B. (1976). *Appl. Phys. Lett.* **26**, 104.
Harvey, K. C. (1981). *Rev. Sci. Instrum.* **52**, 204.
Harvey, K. C. and Stoicheff, B. P. (1977). *Phys. Rev. Lett.* **38**, 537.
Heinzmann, U., Heuer, H., and Kessler, J. (1975). *Phys. Rev. Lett.* **34** (8), 441.

Heinzmann, U., Schinkowski, D., and Zeman, H. D. (1977). *Appl. Phys.* **12**, 113.
Hering, P., Maaswinkel, A. G. M., and Kompa, K. (1981). *Chem. Phys. Lett.* **83**, 222.
Hermann, A., Schumacher, E., and Wöste, L. (1978). *J. Chem. Phys.* **68**, 2327.
Hertz, G. (1923). *Z. Phys.* **18**, 307.
Herzberg, G. (1950). "Nonlinear Spectra and Molecular Structure: I. Spectra of Diatomic Molecules." Van Nostrand-Reinhold, Princeton, New Jersey.
Hieftje, G. M., Travis, J. C., and Lyttle, F. G., eds. (1981). "Lasers in Chemical Analytics." Humana Press, Clifton, New Jersey.
Hilbig, R. and Wallenstein, R. (1982). *Appl. Opt.* **21**, 913.
Hillenkamp, F. (1982). *In* "Proceedings of the Second Workshop on Ion Formation from Organic Solids." Springer-Verlag, Berlin, Heidelberg, and New York.
Hiraya, A., Achiba, Y., Mikami, N., and Kimura, K. (1985). *J. Chem. Phys.* **82** (4), 1810.
Hotop, H. and Niehaus, A. (1968). *Z. Phys.* **215**, 395.
Huber, G., Touchar, F., Büttgenbach, S., Thibault, C., Klapisch, R., Duong, H. T., Liberman, S., Pinard, J., Vialle, J.-L., Juncar, P., and Jacquinot, P. (1978). *Phys. Rev. C* **18**, 2342.
Huillier, A. L., Mainfray, G., and Johnson, P. M. (1984). *Chem. Phys. Lett.* **103**, 447.
Hunt, D. F. and Crow, F. W. (1978). *Anal. Chem.* **50**, 1781.
Hurst, G. S. and Payne, M. G., eds. (1984). "Resonance Ionization Spectroscopy, 1984," Publ. Conf. Ser. N71. The Institute of Physics, Bristol and Boston.
Hurst, G. S., Payne, M. G., Nayfeh, M. H., Judish, J. P., and Wagner, E. B. (1975). *Phys. Rev. Lett.* **35**, 82.
Hurst, G. S., Nayfeh, M. H., and Young, J. P. (1977a). *Appl. Phys. Lett.* **30**, 229.
Hurst, G. S., Nayfeh, M. H., and Young, J. P. (1977b). *Phys. Rev. A* **15**, 2283.
Hurst, G. S., Nayfeh, M. H., Young, J. P., Payne, M. G., and Grossman, L. W. (1977c). *In* "Laser Spectroscopy III," (J. L. Hall and J. L. Carlsten, eds.), Vol. 7, p. 44, Springer Series in Optical Sciences. Springer-Verlag, Berlin, Heidelberg, and New York.
Hurst, G. S., Allman, S. L., Payne, M. G., Marshall, K. A., and Soohoo, K. L. (1978a). *Nuclear Instrum. Methods* **155**, 203.
Hurst, G. S., Allman, S. L., Payne, M. G., and Whitaker, T. J. (1978b). *Chem. Phys. Lett.* **60**, 150.
Hurst, G. S., Payne, M. G., Kramer, S. D., and Young, J. P. (1979a). *Rev. Mod. Phys.* **51**, 767.
Hurst, G. S., Payne, M. G., Kramer, S. D., and Young, J. P. (1979b) *Chem. Phys. Lett.* **63**, 1.
Hurst, G. S., Payne, M. G., Kramer, S. D., and Chen, C. H. (1980). *Phys. Today* **33**, N9, 24.
Hurst, G. S., Payne, M. G., Chen, C. H., Willis, R. D., Lehmann, B. E., and Kramer, S. D. (1981). *In* "Laser Spectroscopy V," (A. R. W. McKeller, T. Oka, and B. P. Stoicheff, eds.), Vol. 30, p. 59, Springer Series in Optical Sciences. Springer-Verlag, Berlin, Heidelberg, and New York.
Hurst, G. S., Chen, C. H., Kramer, S. D., Cleveland, B. T., Davis, R., Jr., Rowley, R. K., Gabbard, F., and Schima, F. J. (1984a). *Phys. Rev. Lett.* **53**, 1116.
Hurst, G. S., Payne, M. G., Phillips, R. C., Dabbs, J. W. T., and Lehmann, B. E. (1984b). *J. Appl. Phys.* **55**, 1278.
Ischenko, V. N., Karlov, N. V., Krynetskii, B. B., Lisitzin, V. N., Mishin, V. A., and Razjev, A. M. (1977). *Pis'ma Zh. Techn. Fiz.* **3**, 1044 (in Russian).
Iturbe, J., Allman, S. L., Hurst, G. S., and Payne, M. G. (1982). *Chem. Phys. Lett.* **93**, 460.
Iturbe, J., Allman, S. L., Hurst, G. S., and Payne, M. G. (1983). *Chem. Phys. Lett.* **94**, 505.
Iturbe, J., Allman, S. L., Hurst, G. S., and Payne, M. G. (1984). *In* "Resonance Ionization Spectroscopy, 1984," Publ. Conf. Ser. N71, p. 49. The Institute of Physics, Bristol and Boston.
Ivanov, L. N. and Letokhov, V. S. (1975). *Kvantovaya Elektron.* **2**, 585; Erratum:(1977), **5**, 877 (in Russian).
Ivanov, L. N. and Letokhov, V. S. (1985). *Comm. Atom. Mol. Phys.* **16**, 169.

Jäger, H. (1978). *Appl. Atom. Spectr.* **2**, 1.
Janes, G. S., Itzkan, I., Pike, C. T. R., Levy, R. H., and Levin, L. (1976). *IEEE J. Quant. Electr.* **QE-12**, 111.
Johnson, P. M. (1976). *J. Chem. Phys.* **64**, 4143.
Johnson, P. M. (1980). *Acc. Chem. Res.* **13**, 20.
Johnson, P. M., Berman, M. R., and Zakheim, D. (1975). *J. Chem. Phys.* **62**, 2500.
Kapitza, P. L. (1966). *In* "The Science of Science," (M. Goldsmith and A. Mackay, eds). Souvenier Press, New York.
Karasek, F. W., Kim, S. H., and Hill, H. H., Jr. (1976). *Anal. Chem.* **48**, 1133.
Karlov, N. V., Krynetskii, B. B., Mishin, V. A., Prokhorov, A. M., Savel'ev, A. D., and Smirnov, V. V. (1976). *Pis'ma Zh. Techn. Fiz.* **2**, 961.
Karlov, N. V., Krynetzkii, B. B., and Stel'makh, O. M. (1977a). *Kvantovaya Elecktron.* **4**, 2275.
Karlov, N. V., Krynetzkii, B. B., Mishin, V. A., Prokhorov, A. M., Savel'ev, A. D., and Smirnov, V. V. (1977b). *Opt. Comm.* **21**, 384.
Karlov, N. V., Krynetzkii, B. B., Mishin, V. A., and Prokhorov, A. M. (1977c). *Pis'ma Zh. Eksp. Teor. Fiz.* **25**, 318.
Karlov, N. V., Krynetzkii, B. B., Mishin, V. A., and Prokhorov, A. M. (1978). *Appl. Opt.* **17**, 856.
Karplus, R. and Schwinger, J. (1948). *Phys. Rev.* **73**, 1020.
Kaufman, S. L. (1976). *Opt. Comm.* **17**, 309.
Keldysh, L. V. (1964). *Zh. Eksp. Teor. Fiz.* **47**, 1945.
Keller, R. A., Bomse, D. S., and Cremers, D. A. (1981). *Laser Focus*, **17**, N10, 75.
Kellogg, G. L. and Tsong, T. T. (1980). *J. Appl. Phys.* **51**, 1184.
Kimman, J., Kruit, P., and van der Wiel, M. J. (1982). *Chem. Phys. Lett.* **88**, 576.
Kimock, F. M., Baxter, J. P., and Winograd, N. (1983). *Surf. Sci.* **129**, L41.
Kimura, K. (1984). *In* "Multiphoton Processes," (P. Lambropoulos and S. J. Smith, eds.), Vol. 2, p. 168, Springer Series on Atoms and Plasmas. Springer-Verlag, Berlin and New York.
Kimura, K. (1985). *Adv. Chem. Phys.* **60**, 161.
Kingdon, K. H. (1923). *Phys. Rev.* **21**, 408.
Kingston, R. H. (1978). "Detection of Optical and Infrared Radiation," Vol. 10, Springer Series in Optical Sciences. Springer-Verlag, Berlin, Heidelberg, and New York.
Kirillov, A. E., Kukharev, V. N., and Soldatov, A. N. (1979). *Kvantovaya Elektron.* **6**, 473 (in Russian).
Kirsten, T. (1984). *In* "Resonance Ionization Spectroscopy, 1984," (G. S. Hurst and M. G. Payne, eds.), p. 251, Publ. Conf. Ser. N71. The Institute of Physics, Bristol and Boston.
Kliger, D., ed. (1983). "Ultrasensitive Laser Spectroscopy," Academic Press, New York.
Kligler, D. J. and Rhodes, C. K. (1978). *Phys. Rev. Lett.* **40**, 309.
Kligler, D. J., Bokor, J., and Rhodes, C. K. (1980). *Phys. Rev. A* **21**, 607.
Klimcak, C. M. and Wessel, J. E. (1980a). *Anal. Chem.* **52**, 1233.
Klimcak, C. M. and Wessel, J. E. (1989b). *Appl. Phys. Lett.* **37**, 138.
Kluge, H.-J., Neugart, R., and Otten, E. W. (1979). *In* "Laser Spectroscopy IV," (H. Walther and K. W. Rothe, eds.), Vol. 21, p. 517, Springer Series in Optical Sciences. Springer-Verlag, Berlin, Heidelberg, and New York.
Knox, B. E. (1972). *In* "Trace Analysis by Mass Spectrometry," (A. J. Ahearn, ed.) Academic Press, New York.
Knyazev, I. N., Letokhov, V. S., and Movshev, V. G. (1975). *IEEE J. Quant. Electron.* **QE-11**, 805.
Kolosova, L. P. (1982). *Zavod. Lab.* **48**, 8 (in Russian).
Kopainsky, B., Kaiser, W., and Schäfer, F. P. (1978). *Chem. Phys. Lett.* **56**, 458.
Kramer, S. D. (1984). *In* "Resonance Ionization Spectroscopy, 1984," (G. S. Hurst and M. G. Payne, eds.), Publ. Conf. Ser. N71, p. 205. The Institute of Physics, Bristol and Boston.
Kramer, S. D., Bemis, C. E., Jr., Young, J. P., and Hurst, G. S. (1978). *Opt. Lett.* **3**, 16.

Kramer, S. D., Young, J. P., Hurst, G. S., and Payne, M. G. (1979). *Opt. Comm.* **30**, 47.

Kramer, S. D., Chen, C. H., Payne, M. G., Hurst, G. S., and Lehmann, B. E. (1983). *Appl. Opt.*, **22**, 3271.

Krönert, U., Kluge, H. J., Kremmling, H., Ruster, W., Wallmeroth, K., Wolf, V., Peuser, P., and Trautmann, N. (1984). 16th EGAS Conference, London.

Kruit, P., Kimman, J., and van der Wiel, M. J. (1981). *J. Phys., B* **14**, 597.

Kruit, P., Kimman, J., Miller, H. C., and van der Wiel, M. J. (1983). *J. Phys. B* **16**, 937.

Kudriavtzev, Yu. A. and Letokhov, V. S. (1982). *Appl. Phys., B* **29**, 219.

Kudriavtzev, Yu. A., Letokhov, V. S., and Petrunin, V. V. (1985). *Pis'ma Zh. Eksp. Teor. Fiz.* **42**, 23 (in Russian).

Kuprianov, S. E. (1968). *Zh. Eksp. Teor. Fiz.* **55**, 460 (in Russian).

Kuz'min, A. F., Rafa'son, A. E., Kholin, N. A., and Tzigel'man, G. I. (1970). *Zh. Techn. Fiz.* **40**, 1531 (in Russian).

Kuz'min, V. A. (1965). *Zh. Eksp. Teor. Fiz.* **49**, 1532 (in Russian); *Sov. Phys.* JETP (1966), **22**, 1051.

Lamb, W. L., Jr., Retherford, R. C. (1947). *Phys. Rev.* **72**, 241.

Lambropoulos, P. (1976). *Adv. At. Mol. Phys.*, **12**, 87.

Lawrence, E. O. and Edlefson, N. E. (1929). *Phys. Rev.* **34**, 233.

Lee, D. S. (1983). *Nature (London)* **305**, 47.

Lee, K. T. (1971). *Phys. Rev., A* **4**, 579.

Lee, K. T. (1974). *J. Opt. Soc. Am.* **64**, 706.

Lehmann, B. E., Hurst, G. S., Chen, C. H., Kramer, S. D., Payne, M. G., and Willis, R. D. (1983). *In* "Laser Spectroscopy VI," (H. P. Weber and W. Lüthy, eds.), Vol. 40, p. 100, Springer Series in Optical Sciences. Springer-Verlag, Berlin, Heidelberg, and New York.

Letokhov, V. S., (1966). *Opt. Spektrosk.* **20**, 349 (in Russian).

Letokhov, V. S. (1969). Report of the P. N. Lebedev Physical Institute. "On the possibilities of isotope separating by the methods of the resonant photoionization of atoms and photodissociation of molecules by laser radiation." Published by the Institute of Spectroscopy (1979).

Letokhov, V. S. (1970). USSR Patent N784679 Appl. 30.03.1970. Published in Bulletin of Inventions (1982), N8, p. 308.

Letokhov, V. S. (1975a). *Science*, **190**, 344.

Letokhov, V. S. (1975b). *Kvantovaya Elektron.* **2**, 930 (in Russian).

Letokhov, V. S. (1975c). *Phys. Lett. A* **51**, 231.

Letokhov, V. S. (1976a). *In* "Tunable Lasers and Applications, (A. Mooradian, I. Jaeger, and P. Stokseth, eds.), Vol. 3, p. 122, Springer Series in Optical Sciences. Springer-Verlag, Berlin, Heidelberg, and New York.

Letokhov, V. S. (1976b). *Usp. Fiz. Nauk* **118**, 199 (in Russian).

Letokhov, V. S. (1977a). *Comm. Atom. Molec. Phys.* **7**, 93.

Letokhov, V. S. (1977b). *In* "Frontiers in Laser Spectroscopy," (R. Balian, S. Haroche, and S. Liberman, eds.) Vol. 2, p. 771. North-Holland Publ., Amsterdam.

Letokhov, V. S. (1978). *Optics and Laser Technology*, August, p. 175.

Letokhov, V. S. (1980). *In* "Chemical and Biochemical Applications of Lasers," (C. B. Moore, ed.) Vol. 5, p. 1. Academic Press, New York.

Letokhov, V. S. (1981a). *Comm. Atom. Molec. Phys.* **10**, 257.

Letokhov, V. S. (1981b). Invited Talk on First European Conference on Atomic Physics, Heidelberg, April, 1981.

Letokhov, V. S. (1981c). *Comm. Atom. Molec. Phys.* **11**, 1.

Letokhov, V. S. (1983a). "Nonlinear Laser Chemistry. Multiple-Photon Excitation," Vol. 22, p. 417, Springer Series in Chemical Physics. Springer-Verlag, Berlin, Heidelberg, and New York.

Letokhov, V. S. (1983b). *In* "Advances in Laser Spectroscopy," (F. T. Arecchi, F. Strumia, and H. Walther, eds.), p. 417. Plenum Press, New York.

Letokhov, V. S., ed. (1986). "Laser Analytical Spectrochemistry." A. Hilger, London.

Letokhov, V. S. and Ambartzumian, R. V. (1971). *IEEE J. Quant. Electr.* **QE-7**, 305.

Letokhov, V. S. and Chebotayev, V. P. (1977). "Nonlinear Laser Spectroscopy," Vol. 4, Springer Series in Optical Sciences. Springer-Verlag, Berlin, Heidelberg, and New York.

Letokhov, V. S. and Makarov, A. A. (1978). *Appl. Phys.* **16**, 47.

Letokhov, V. S. and Makarov, A. A. (1981). *Usp. Fiz. Nauk* **134**, 45 (in Russian).

Letokhov, V. S. and Minogin, V. G. (1976). *Zh. Eksp. Teor. Fiz.* **71**, 135 (in Russian).

Letokhov, V. S. and Mishin, V. I. (1979). *Opt. Comm.* **29**, 168.

Letokhov, V. S., Mishin, V. I., and Puretzky, A. A. (1977). *Prog. Quant. Electr.* **5**, 139.

Letokhov, V. S., Mishin, V. I., Eshkobilov, N., and Tursunov, A. T. (1982). *Opt. Comm.* **41**, 331.

Leuchs, G. (1984). *In* "Multiphoton Processes," (P. Lambropoulos and S. J. Smith, eds.), Vol. 2, p. 48, Springer Series on Atoms and Molecules. Springer-Verlag, Berlin and New York.

Leuchs, G. and Walther, H. (1979). *Z. Phys. A* **293**, 93.

Leuchs, G., Smith, S. J., Khawaja, E., and Walther, H. (1979). *Opt. Comm.* **31**, 313.

Leutwyler, S., Herrmann, A., Wöste, L., and Schumacher, E. (1980). *Chem. Phys.* **48**, 253.

Leutwyler, S., Hofmann, M., Harri, H.-P., and Schumacher, E. (1981). *Chem. Phys. Lett.* **77**, 257.

Leutwyler, S., Heinis, T., Jungen, M., Harri, H.-P., and Schumacher, E. (1982). *J. Chem. Phys.* **76**, 4290.

Liao, P. F. and Bjorkholm, J. E. (1975). *Phys. Rev. Lett.* **34**, 1.

Libby, W. F. (1980). "Collected Papers," (R. Berger and L. M. Libby, eds.), Vol. 1: Tritium and Radiocarbon. Geo Science Analytical, Santa Monica, California.

Liberman, S., Pinard, J., Duong, H. T., Juncar, P., Vialle, J. L., Pillet, P., Jacquinot, P., Huber, G., Touchard, F., Büttgenbach, S., Thibault, C., Klapisch, R., and Penselle, A. (1979). *In* "Laser Spectroscopy IV," (H. Walther and K. W. Rothe, eds.), Vol. 21, p. 275, Springer Series in Optical Sciences. Springer-Verlag, Berlin, Heidelberg, and New York.

Lichtin, D. A., Datta-Ghosh, S., Newton, K. R., and Bernstein, R. B. (1980). *Chem. Phys. Lett.* **75**, 214.

Ligon, W. V. (1979). *Science* **205**, 151.

Linder, G. A. and Möller, P. (1982). *Phys. Lett.* **110B**, 17.

Litherland, A. E., Buekens, R. P., Kilias, L. R., Rucklidge, J. C., Gove, H. E., Elmore, D., and Purser, K. H. (1981). *Nucl. Instrum. Methods* **18C**, 463.

Littman, M. G., Zimmerman, M. L., Ducas, T. W., Freeman, R. R., and Kleppner, D. (1976a). *Phys. Rev. Lett.* **36**, 788.

Littman, M. G., Zimmerman, M. L., and Kleppner, D. (1976b). *Phys. Rev. Lett.* **37**, 486.

Liverman, M. G., Berk, S. M., Monts, D. L., and Smalley, R. E. (1979). *J. Chem. Phys.* **70**, 192

Lompre, L. A., Mainfray, G., Manus, C., Thibault, J. (1978). *J. Phys. (Paris)* **39**, 610.

Lorenzen, C.-J. and Niemax, K. (1982). *J. Phys. B* **15**, L139.

Lorenzen, C.-J., Niemax, K., and Pendrill, L. R. (1983). *Phys. Rev. A* **28**, 2051.

Lorenzen, C.-J., Niemax, K., and Weber, K.-H. (1984). *Opt. Comm.* **52**, 178.

Lu, K. T., Fano, U. (1970). *Phys. Rev. A* **2**, 81.

Lubman, D. M. and Kronick, M. N. (1982). *Anal. Chem.* **54**, 1546.

Lubman, D. M., Naaman, R., and Zare, R. N. (1980). *J. Chem. Phys.* **72**, 3034.

Lucatorto, T. B., Clark, C. W., and Moore, L. J. (1984). *Opt. Comm.* **48**, 406.

McCorkell, R., Fireman, E. L., and Langway, C. C. (1967). *Science* **158**, 1690.

Mackenzie, T. F., Wollast, R., and Stoffyn, M. (1978). *Science* **199**, 4329.

McLaren, A. D. and Shugar, D. (1964). "Photochemistry of Proteins and Nucleic Acids," p. 97. Pergamon Press, Oxford.

Majewsky, U., Neukammer, J., and Rinueberg, H. (1982). *Phys. Rev. Lett.* **51**, 1340.

Makarov, A. A. (1983). *Zh. Eksp. Teor. Fiz.* **85**, 1192 (in Russian).

Mamyrin, B. A., Karataev, V. I., Shmikk, D. V., and Zagulin, V. A. (1973). *Zh. Eksp. Teor. Fiz.* **64**, 82 (in Russian).
Mann, J. B. (1964). *J. Chem. Phys.* **40**, 1632.
Marinero, E. E., Rettner C. T., and Zare, R. N. (1982). *Phys. Rev. Lett.* **48**, 1323.
Marinero, E. E., Vasudev, R., and Zare, R. N. (1983). *J. Chem. Phys.* **78**, 692.
Markova, S. V., Petrash, G. G., and Cheredov, V. M. (1978). *Kvantovaya Elektron.* **5**, 1585.
Marr, G. V. (1967). "Photoionization Processes in Gases," Academic Press, New York.
Martin, S., Chevaleyre F., Valiquant, S., Perrot, J. P., Broyer, M., Caband, B., and Hoarean, A. (1982). *Chem. Phys. Lett.* **87**, 235.
Martin, W. C., Hagan, L., Reader, J., and Sugar, J. (1974). *J. Phys. Chem. Ref. Data* **3**, 771
Martin, W. C., Zalubas, R., and Hagan, L. (1978). *Nat. Stand. Data Ser. NBS* **60**.
Matsuzawa, N. (1971). *J. Chem. Phys.* **55**, 2685.
Mayo, S., Lucarto, T. B., and Luther, G. G. (1982). *Anal. Chem.* **54**, 553.
Meek, J. T., Jones, R. K., and Reilly, J. P. (1980). *J. Chem. Phys.* **73**, 3503.
Meek, J. T., Long, S. R., and Reilly, J. P. (1982). *J. Phys. Chem.* **86**, 2809.
Mero, J. L. (1965). "The Mineral Resources of the Sea," Elsevier, Amsterdam, London, and New York.
Meuzelaar, H. L. C., Kistermaker, P. G., and Posthumus, M. A. (1974). *Biom. Mass. Spectr.* **1**, 312.
Miller, C. M. and Nogar, N. S. (1982). *In* "Laser Techniques for Extreme Ultraviolet Spectroscopy," (T. J. McIlrath and R. R. Freeman, eds.), p. 90, Publ. 90 American Institute of Physics, New York.
Miller, C. M. and Nogar, N. S. (1983). *Anal. Chem.* **55**, 1606.
Miller, C. M., Nogar, N. S., Gancarz, A. J., and Shields, W. K. (1982). *Anal. Chem.* **54**, 2377.
Miller, J. C. and Compton, R. N. (1981). *J. Chem. Phys.* **75**, 22.
Miller, J. C. and Compton, R. N. (1982). *Chem. Phys. Lett.* **93**, 453.
Mirza, M. Y. and Duley, W. W. (1978a). *Opt. Comm.* **25**, 185.
Mirza, M. Y. and Duley, W. W. (1978b). *Proc. R. Soc. London A* **364**, 255.
Mizolek, A. W. (1981). *Anal. Chem.* **53**, 118.
Mizushima, M. (1964). *Phys. Rev. A* **133**, 414.
Moore, C. B. (1973). *Acc. Chem. Res.* **6**, 323.
Muchnik, M. L., Orlov, Yu. V., Parshin, G. D., Cherniak, E. Ya., Letokhov, V. S., and Mishin, V. I. (1983). *Kvantovaya Electron.* **10**, 2331 (in Russian).
Müller, E. W. (1956). *Phys. Rev.* **102**, 618.
Müller, E. W. and Tsong, T. T. (1969). "Field Ion Microscopy: Principles and Applications," Elsevier Publ., Amsterdam.
Müller, E. W. and Tsong, T. T. (1973). *Prog. Surf. Sci.*, **4**.
Nagano, Y., Adeiba, Y., Sato, K., and Kimura, K. (1982). *Chem. Phys. Lett.* **93**, 510.
Nakayama, T., Kitamura, M. Y., and Watanabe, K. (1959). *J. Chem. Phys.* **30**, 1180.
Nayfeh, M. H., Hurst, G. S., Payne, M. G., and Young, J. P. (1977). *Phys. Rev. Lett.* **39**, 604.
Nayfeh, M. H., Hurst, G. S., Payne, M. G., and Young, J. P. (1978). *Phys. Rev. Lett.* **41**, 302.
Ng, C. Y., Mahan, B. H., and Lee, Y. T. (1976). *J. Chem. Phys.* **65**, 1956.
Niemax, K. (1982). *Acta Phys. Pol., A* **61**, 517.
Niemax, K. (1983). *Appl. Phys.* **32**, 59.
Niemax, K. and Pendrill, L. R. (1980). *J. Phys. B* **13**, L461.
Nishigaki, S., Drachsel, W., and Block, J. H. (1979). *Surf. Sci.* **87**, 389.
Niu, B. H. C. and Bryant, P. J. (1979). *J. Chem. Phys.* **70**, 2584.
Niu, B. H. C., Beacham, J. R., and Bryant, P. J. (1977). *J. Chem. Phys.* **67**, 2039.
Norcross, D. W., and Stone, P. W. (1966). *J. Quant. Spectrosc. Radiat. Transfer* **6**, 2779.
Nygaard, K. J., Hebner, R. E., Jones, J. D., and Corbin, R. J. (1975). *Phys. Rev. A* **12**, 1440.

Ohnesorge, W., Diedrich, F., Lenchs, G., Elliott, D. S., and Walther, H. (1984). *Phys. Rev. A* **29**, 1181.
Omenetto, N., ed. (1979). "Analytical Laser Spectroscopy." Wiley-Interscience, New York.
Opsal, R. B. and Reilly, J. P. (1983). *Chem. Phys. Lett.* **99**, 461.
Otten, E.-W. (1981). *Nucl. Phys. A* **354**, 471.
Paisner, J. A., Solarz, R. W., Worden, E. F., Conway, J. G. (1977). *In* "Laser Spectroscopy III," (J. L. Hall J. L. Carlsten, eds.), Vol. 7, p. 160, Springer Series in Optical Sciences. Springer-Verlag, Berlin, Heidelberg, and New York.
Pandolfi, R. S., Gobeli, D. A., and El-Sayed, M. A. (1981). *J. Chem. Phys.* **85**, 1779.
Panitz, J. A. (1978). *Prog. Surf. Sci.* **8**, 219.
Parks, J. E., Schmitt, H. W., Hurst, G. S., and Fairbank, W. M., Jr. (1983). *Thin Solid Films* **8**, 69.
Payne, M. G., Chen, C. H., Hurst, G. S., Kramer, S. D., Garrett, W. R., and Pindzola, M. (1981). *Chem. Phys. Lett.* **79**, 142.
Pendrill, L. R. and Niemax, K. (1982). *J. Phys. B* **15**, 147.
Persival, I. C. (1977). *Proc. R. Soc. Lond. A* **353**, 289.
Petty, G., Tai, C., Dalby, F. W. (1975). *Phys. Rev. Lett.* **34**, 1207.
Popescu, D., Collins, C. B., Johnson, B. W., and Popescu, I. (1974). *Phys. Rev.* **9A**, 1182.
Posthumus, M. A., Kistemaker, P. G., Menzelaar, H. L. C., and Ten Noever de Braw, M. C. (1978). *Anal. Chem.* **50**, 985.
Pratt, R. H., Levee, R. D., Pexton, R. L., and Aron, W. (1964). *Phys. Rev.* **134A**, 916.
Pratt, S. T., Poliakoff, E. D., Dehmer, P. M., and Dehmer, J. L. (1983a). *J. Chem. Phys.* **78**, 65.
Pratt, S. T., Dehmer, P. M., and Dehmer, J. L. (1983b). *J. Chem. Phys.* **78**, 4315.
Rabi, I. I. (1937). *Phys. Rev.* **51**, 652.
Radhakrishnan, G. and Estler, R. C. (1983). *Chem. Phys. Lett.* **100**, 403.
Ramsey, N. F., (1966). "Molecular Beams," Clarendon Press, Oxford.
Rauch, E. F. and Ackermann, R. J. (1974). *J. Chem. Phys.* **50**, 1396.
Reader, J. and Sugar, J. (1966). *J. Opt. Soc. Am.* **56**, 1189.
Ready, J. P. (1971). "Effects of High-Power Laser Radiation." Academic Press, New York.
Rebentrost, F. and Ben-Shaul, A. (1981). *J. Chem. Phys.* **74**, 3255.
Rebentrost, F., Kompa, K. L., and Ben-Shaul, A. (1981). *Chem. Phys. Lett.* **77**, 394.
Reid, N. W. (1971). *Int. J. Mass Spectr. Ion Phys.* **6**, 1.
Reilly, J. P. and Kompa, K. L. (1979). *Adv. Mass Spectr.* **8**, 1800.
Reilly, J. P. and Kompa, K. L. (1980). *J. Chem. Phys.* **73**, 5468.
Reisler, H., Mangir, M., and Wittig, C. (1980). *In* "Chemical and Biochemical Application of Lasers, ed. by (C. B. Moore, ed.), Vol. 5, p. 139. Academic Press, New York.
Rhodes, Ch. K., ed. (1979). "Excimer Lasers," Vol. 30, Springer Series in Applied Physics. Springer-Verlag, Berlin and New York.
Robbins, E. F., Leckenby, R. E., and Willis, P. (1967). *Advan. Phys.* **16**, 739.
Robieux, J. and Auclair, J. M. (1963). French Patent No. 1391738. Appl. 2110 1963. Publ. 1.02.1965.
Robin, M. B. (1980). *Appl. Opt.* **19**, 3941.
Robinson, P. I. and Holbrook, K. A. (1972). "Unimolecular Reactors." Wiley, London.
Rockney, B. H. and Grant, E. R. (1981). *Chem. Phys. Lett.* **79**, 15.
Rockney, B. H. and Grant, E. R. (1982). *J. Chem. Phys.* **77**(a) 4257.
Rockwood, S., Reilly, J. P., Hohla, K., and Kompa, K. L. (1979). *Opt. Comm.* **28**, 175.
Rothberg, L. J., Gerrit, D. P., and Vaida, V. (1981). *J. Chem. Phys.* **75**, 4403.
Rottke, H. and Welge, K. H. (1983). *Chem. Phys. Lett.* **99**, 456.
Rzazewski, K. (1983). *Phys. Rev. A* **28**, 2565.
Sakurai, K., Capelle, C., Broida, H. P. (1971). *J. Chem. Phys.* **54**, 1412.
Salomaa, R. and Stenholm, S. (1978). *Appl. Phys.* **17**, 309.

Sargent, M., III, Scully, M. O., and Lamb, W. E., Jr. (1974). "Laser Physics." Addison-Wesley, Reading, Massachusetts.

Sato, K., Achiba, Y., and Kimura, K. (1984a). *J. Chem. Phys.* **80**, 57.

Sato, K., Achiba, Y., and Kimura, K. (1984b). *J. Chem. Phys.* **81**, 57.

Sauter, F. (1931). *Ann. Phys.*, **11**, 454.

Schade, U., Stoll, R., and Röllgen, F. W. (1981). *Org. Mass. Spectr.* **16**, 441.

Schäfer, F. P., ed. (1977). "Dye Lasers," 2nd edition, Vol. 1, Springer Series "Topics in Applied Physics." Springer-Verlag, Berlin, Heidelberg, and New York.

Schlag, E. W. and Neusser, H. F. (1983). *Acc. Chem. Res.* **16**, 355.

Schueler, B. and Krueger, F. R. (1980). *Org. Mass Spectr.* **15**, 295.

Schulz, P. A., Sudbo, Aa. S., Krainovich, D. J., Kwok, H. S., Shen, Y. R., and Lee, Y. T. (1979). *Ann. Rev. Phys. Chem.* p. 30.

Scott, R. D. (1976). *Nature* (*London*) **264**, 729.

Seaver, M., Hudgens, F. W., and Decorpo, J. J. (1980). *Int. J. Mass. Spectr. Ion Phys.* **34**, 159.

Sha, G., Zhong, X., Zhao, S., and Zhang, C. (1984). *Chem. Phys. Lett.* **110**, 410.

Shimoda, K., ed. (1976). "High-Resolution Laser Spectroscopy," Vol. 13, Springer Series "Topics in Applied Physics." Springer-Verlag, Berlin, Heidelberg, and New York.

Siegbahn, K., Nordling, C., Fahlman, A., Nordberg, R., Hamrin, K., Hedman, J., Johansson, G., Bergmark, I., Karlsson, S. E., Lindgren, I., and Lindberg, B. (1967). ESCA Atomic, Molecular and Solid State Structure Studied by Means of Electron Spectroscopy (Uppsala).

Siegbahn, K., Nordling, C., Johansson, G., Hedman, J., Heden, P. F., Hamrin, K., Gelius, U., Bergmark, T., Werme, L. O., Manne, R., and Baer, Y. (1969). "ESCA Applied to Free Molecules." North-Holland Publ., Amsterdam.

Siegbahn, K., Allison, D. A., Allison, J. H. (1974). "ESCA-Photoelectron Spectroscopy" *In* "Handbook of Spectroscopy," (J. W. Robinson, ed.), Vol. 1, p. 257. CRC Press, Boca Raton, Florida.

Siegel, A., Ganz, J., Busert, W., and Hotop, M. (1983). *J. Phys. B* **16**, 2945.

Silberstein, J. and Levine, R. D. (1980). *Chem. Phys. Lett.* **74**, 6.

Silberstein, J. and Levine, R. D. (1981). *J. Chem. Phys.* **75**, 5735.

Smirnov, B. M. (1974). "Ions and Excited Atoms in Plasma." Fizmatgiz, Moscow.

Smith, A. V., Goldsmith, J. E. M., Nitz, D. B., and Smith, S. J. (1980). *Phys. Rev. A* **22**, 577.

Smith, D. H. and Hertel, G. R. (1969). *J. Chem. Phys.* **51**, 3105.

Smith, G. and Tomkins, F. S. (1975). *Proc. R. Soc. London, A* **342**, 149.

Smith, G. and Tomkins, F. S. (1976). *Philos. Trans. R. Soc. London* **283**, 345.

Smyth, K. C., Schiavone, J. A., and Freund, R. S. (1974). *J. Chem. Phys.* **61**, 1789.

Snyder, J. J., Lucatorto, T. B., Debenham, P. H., and Geltman, S. (1985). *J. Opt. Soc. Am. B* **2**, 1497.

Sobel'man, I. I. (1979). "Atomic Spectra and Radiative Transitions, Vol. 1, Springer Series in Chemical Physics. Springer-Verlag, Berlin, Heidelberg, and New York.

Solarz, R. W., May, C. A., Carlson, L. R., Worden, E. F., Johnson, S. A., Paisner, J. A., and Radziemski, L. J., Jr. (1976). *Phys. Rev. A* **14**, 1129.

Stebbings, K. F. (1976). *Science* **193**, 537.

Stebbings, R. F., Dunning, F. B., eds. (1983). "Rydberg States of Atoms and Molecules." Cambridge Univ. Press, London and New York.

Stebbings, R. F., Dunning, F. B., Tittel, F. K., and Rundel, R. D. (1973). *Phys. Rev. Lett.* **10**, 815.

Stebbings, R. F., Dunning, F. B., and Rundel, R. D. (1975a). "Atomic Physics," Vol. 4. Plenum, New York.

Stebbings, R. F., Latimer, C. F., West, W. P., Dunning, F. B., and Cook, T. B. (1975b). *Phys. Rev. A* **12**, 1453.

Stoicheff, B. P. (1981). *In* "Laser Spectroscopy V," (A. R. W. McKellar, T. Oka and B. P. Stoicheff, eds.), Vol. 30, p. 299, Springer Series in Optical Sciences. Springer-Verlag, Berlin, Heidelberg, and New York.

Stoicheff, B. P. and Weinberger, E. (1980). *Phys. Rev. Lett.* **44**, 733.

Stoll, R. and Röllgen, F. W. (1979). *Org. Mass. Spectr.* **14**, 642.

Sudbo, A. S. and Loy, M. M. T. (1981). *Chem. Phys. Lett.* **82**, 135.

Sudbo, Aa., Schulz, P. A., Krainovich, D. J., Lee, Y. T., and Shen, Y. R. (1979a). *Opt. Lett.* **4**, 219.

Sudbo, Aa., Schulz, P. A., Grant, E. R., Shen, Y. R., and Lee, Y. T. (1979b). *J. Chem. Phys.*, **70**, 912.

Sugar, J. (1974). *J. Chem. Phys.* **60**, 4103.

Svanberg, S. (1977). *In* "Laser Spectroscopy III," (J. L. Hall and J. L. Carlsten, eds.), Vol. 7, p. 183, Springer Series in Optical Sciences. Springer-Verlag, Berlin, Heidelberg, and New York.

Svanberg, S., Tsekeris, P. and Happer, W. (1973). *Phys. Rev. Lett.* **30**, 817.

Tehsenfeld, F. C. (1970). *J. Chem. Phys.*, **53**, 2000.

Thompson, D. C. and Stoicheff, B. P. (1982). *Rev. Sci. Instr.* **53**, 822.

Tjossen, P. J. H. and Cool, T. A. (1983). *Chem. Phys. Lett.* **100**, 479.

Tsong, T. T., Block, J. H., Nagasaka, M., and Vis Wandthan, B. (1976). *J. Chem. Phys.* **65**, 2469.

Tuccio, S. A., Dubrin, J. W., Peterson, O. G., and Snavely, B. B. (1974). *IEEE J. Quant. Electron.* **QE-10**, 790.

Tursunov, A. T. and Eshkobilov, N. B. (1982). *Kvantovaya Elektron.* **9**, 2097 (in Russian).

Tursunov, A. T., Eshkobilov, N. B., Kornienko, B. B. (1984). *Zh. Prikl. Spectr.* **41**, 653 (in Russian).

Tzigel'man, G. E. (1972). *Zh. Techn. Fiz.* **42**, 667 (in Russian).

Unsold, E., Hillenkamp, F., and Nitsche, R. (1976). *Analysis* **4**, 115.

Van der Pegl, G. J. Q., Haverkamp, J., and Kistemaker, P. G. (1981). *Org. Mass. Spectr.* **16**, 416.

Vasilenko, L. S., Chebotayev, V. P., and Shishaev, A. V. (1970). *Pis'ma Zh. Eksp. Teor. Fiz.*, **12**, 161.

Vidolova-Angelova, E. N., Ivanov, L. N., Ivanova, E. P., and Letokhov, V. S. (1981a) *Izv. USSR Acad. Sci. Phys. Ser.* **45**, 2300 (in Russian).

Vidolova-Angelova, E. P., Ivanov, L. N., and Letokhov, V. S. (1981b). *J. Opt. Soc. Am.*, **71**, 699.

Vidolova-Angelova, E. P., Ivanov, L. N., Letokhov, V. S. (1982). *J. Phys. B* **15**, 981.

Vidolova-Angelova, E. P., Angelov, D., Bekov, G. I., Fedoseyev, V. N., and Ivanov, L. N. (1983). *In* "Laser Spectroscopy VI," (H. P. Weber and W. Lüthy, eds.), Vol. 40, p. 217, Springer Series in Optical Sciences. Springer-Verlag, Berlin, Heidelberg, and New York.

Vidolova-Angelova, E. P., Bekov, G. I., Ivanov, L. N., Fedoseyev, V., and Atakhadjaev, A. (1984). *J. Phys. B* **17**, 953.

Viswanathan, B., Drachgel, W., Block, J. H., and Tsong, T. T. (1979). *J. Chem. Phys.* **70**, 2582.

Wall, J., Langmore, J., Isaakson, M., and Crewe, A. V. (1974). *Proc. Nat. Acad. Sci. USA* **71**, 1.

Weber, K.-H. and Niemax, K. (1979). *Opt. Comm.* **28**, 317.

Wedepohl, K. H. (1968). *In* "Origin and Distribution of the Elements," (L. H. Ahreus, ed.), Vol. 30, p. 999. Intern. Ser. Monograph Earth's Sciences, p. 999.

Weisheit, J. C. (1972). *J. Quant. Spectr. Radiat. Transfer* **12**, 1241.

Weiskopf, V. (1933). *Z. Phys.* **85**, 451.

West, W. P., Foltz, G. W., Dunning, F. B., Latimer, C. J., and Stebbings, R. F. (1976). *Phys. Rev. Lett.* **36**, 854.

Whetten, R. L., Fu, K. J., and Grant, E. R. (1983). *J. Chem. Phys.* **79**, 2626.

White, M. G., Seaver, M., Chupka, W. A., and Colson, S. D. (1982). *Phys. Rev. Lett.* **49**, 28.
Wiegmann, W., Drachsel, W., Jaenicke, S., and Block, J. H. (1984). *J. Phys.* (*Paris*) **45**, C9–105.
Wiesenfeld, J. M. and Greene, B. I. (1983). *Phys. Rev. Lett.* **51**, 1745.
Wiley, W. C. and McLaren, I. H. (1955). *Rev. Sci. Instrum.* **26**, 1150.
Williams, R. A. and Sara, R. I. (1962). *Int. J. Appl. Radiat. Isot.* **13**, 229.
Williams, R. A. and Naiditch, S. J. (1967). *J. Chem. Phys.* **47**, 343.
Winograd, N. (1984). *In* "Resonance Ionization Spectroscopy, 1984," (G. S. Hurst and M. G. Payne, eds.), Phys. Conf. Ser. *No.* 71. p. 161. The Institute of Physics, Bristol and Boston.
Winograd, N., Baxter, J. P., and Kimock, F. M. (1982). *Chem. Phys. Lett.* **88**, 581.
Woodworth, J. R. and Rice, J. K. (1978). *J. Chem. Phys.* **69**, 2500.
Worden, E. F., Paisner, J. A., and Conway, J. G. (1978a). *Opt. Lett.* **3**, 156.
Worden, E. F., Solarz, R. W., Paisner, J. A., and Conway, J. G. (1978b). *J. Opt. Soc. Am.* **68**, 52.
Worlock, J. M. (1972). *In* "Laser Handbook," (F. T. Arecchi and E. O. Shulz-DuBois, eds.), Vol. 2, p. 1323. North-Holland Publ., Amsterdam.
Wunderlich R. and Kirsten, T. (1984). *Chem. Phys. Lett.* **106**, 142.
Wynne, J. J. and Armstrong, J. A. (1979). *IBM. J. Res. Dev.* **23**, 490.
Wynne, J. J. and Herman, J. P. (1979). *Opt. Lett.* **4**, 106.
Yeung, E. S. and Moore, C. B. (1973). *J. Chem. Phys.* **58**, 3988.
Young, J. P., Hurst, G. S., Kramer, S. D., and Payne, M. G. (1979). *Anal. Chem.* **51**, 1056A.
Young, W. A., Mirza, M. Y., and Dyley, W. W. (1980). *Opt. Comm.* **34**, 353.
Zacharias, H., Rottke, H., and Welge, K. H. (1981). *Appl. Phys.* **24**, 23.
Zandee, L. and Bernstein, R. B. (1979). *J. Chem. Phys.* **71**, 1359.
Zare, R. N. and Dagdigian, P. J. (1974). *Science*, **182**, 541.
Zatsepin, G. T. (1978). *In* "Proceedings of an Informal Conference on the Status and Future of Solar Neutrino Research," (G. Friedlander, ed.). BNL Report 50879. Brookhaven National Laboratory, Upton, New York.
Zel'dovich, Ya. B. (1973). *Usp. Fiz. Nauk* (Russ.), **110**, 139; *Sov. Phys. Usp.* **16**, 427.
Zeman, H. D. (1976). *In* "Proceedings of an International Symposium on Electron and Photon Interactions with Atoms, Stirling, Great Britain, 1974," (P. G. Burke, H. Kleinpoppen, and M. R. C. McDowell, eds.), p. 581. Plenum, New York.
Zherikhin, A. N., Letokhov, V. S., Mishin, V. I., Belayev, V. P., Evtunin, A. N., and Lesnoi, M. A. (1981). *Kvantovaya Elektron.* **8**, 1340 (in Russian).
Zherikhin, A. N., Letokhov, V. S., Mishin, V. I., Muchnik, M. E., and Fedoseyev, V. N. (1983). *Appl. Phys. B* **30**, 47.
Zherikhin, A. N., Kompanetz, O. N., Letokhov, V. S., Mishin, V. I., Fedoseyev, V. N., Alkhazov, G. D., Barzakh, A. E., Berlovich, E. E., Denisov, V. P., Dernjatin, A. G., and Ivanov, V. S. (1984a). *Zh. Eksp. Teor. Fiz.* **86**, 1249 (in Russian).
Zherikhin, A. N., Mishin, V. I., and Fedoseyev, V. N. (1984b). *Opt. Spektrosk.* **57**, 783 (in Russian).
Zimmerman, M. L., Ducas, I. W., Littman, M. G., and Kleppner, D. (1978). *J. Phys. B* **11**, L11.
Zmbov, K. F., Wu, C. H., and Ihle, H. R. (1977). *J. Chem. Phys.* **67**, 4603.
Zuyikov, A. D., Fedoseyev, V. N. and Mishin, V. I. (1986). Preprint of Institute of Spectroscopy, Troitzk, USSR.

Index